AF577009

Mutagen-induced

Chromosome Damage

in Man

*

Mutagen-induced Chromosome Damage in Man

edited by

H.J.EVANS AND D.C.LLOYD

New Haven

Yale University Press

1978

*

Published in the United Kingdom by
Edinburgh University Press
Published in the United States of America and Canada by
Yale University Press
Library of Congress catalog card number: 78-60354
International Standard Book Number: 0-300-02315-4
Printed in Great Britain

Contents

Contents

Contents

Preface

Eleven years have elapsed since the first international radiation cytogenetics symposium in Edinburgh, which proved to be a landmark in the study of the effects of ionising radiation on chromosomes in human lymphocytes. In the intervening years induced chromosome damage in lymphocytes has, under certain conditions, become established as a reliable indicator of exposure to ionising radiation, and has also been used to detect exposure to other mutagens. Considerable improvements have also been made in cytological techniques, and it therefore seemed an appropriate time to hold a second symposium, broadened to include the effects of both non-ionising radiation and chemical mutagens.

The meeting was jointly organised by the Medical Research Council and the National Radiological Protection Board and took place on 7 and 8 July 1977 at the MRC Clinical and Population Cytogenetics Unit in Edinburgh. Financial support was received from the Commission of the European Communities, Imperial Chemical Industries Ltd and the World Health Organisation. The meeting involved 60 invited participants from eighteen countries, and the 39 papers presented included both reviews and current research. Together they illustrate the vigorous development of this branch of cytogenetics in the past decade and demonstrate the considerable diversity of current activities in the field.

The first day of the meeting was mainly devoted to the effects of radiation, with a short session on the development of methods to automate the scoring of chromosome damage. The basic dose-response relationships of radiation-induced chromosome damage in human lymphocytes are now sufficiently established and cytogenetic dosimetry has taken its place in radiological protection. Outstanding problems discussed at the meeting included: the interpretation of the dose-response relationship after partial-body exposures; the need for a common denominator that would enable extrapolation between species used in experimental studies and man; and the continuing development of the technique as a fundamental radiobiological tool. It has long been evident that the circulating lymphocytes in the body form a heterogeneous cell population and the advent of methods enabling the separate study of T and B lymphocytes has led to a remarkable increase in research on the kinetics of these cells, and also on their radiation responses, both *in vivo* and *in vitro.*

The effects of chemical mutagens on human chromosomes, which occupied the second day of the meeting, is a rapidly increasing field of research. It has attracted a number of former radiation cytogeneticists as well as a new generation of researchers from the realms of cancer studies, environmental mutagenesis and toxicology. In some respects the research is at the stage reached in human radiation cytogenetics at the time of the first symposium, but many of the problems are different.

With a uniform whole-body exposure to ionising radiation approximately equivalent amounts of energy are absorbed in all parts of the body, so that the damaging events at the molecular level are, generally speaking, randomly distributed. Chemical mutagens, by contrast, comprise a wide variety of different substances, which vary in their degree of absorption, metabolism, location in the body and mode of action at the molecular level. Indeed, many chemicals may not be mutagenic *per se*, but when taken up by the body become transformed to yield active mutagenic/carcinogenic metabolites.

Ionising radiation may induce chromosome damage at all stages of the cell cycle and it is quickly transformed into aberrations, so that when cells are stimulated to enter into a mitosis, perhaps many years after irradiation, they reveal a history of their previous resting-stage exposure in the form of chromosome aberrations. This is not the case with most chemical mutagens, where DNA damage is only transformed into aberrations when the cell proceeds through a DNA replication phase. *In vivo* exposure of human peripheral blood lymphocytes to such mutagens will result in DNA damage, which may undergo repair before the cells are stimulated to enter mitosis and hence never give rise to aberrations; nevertheless, aberrations would have been developed in other cells in the body that were proliferating at the time of exposure. Chromosome aberrations in lymphocytes may therefore provide at best an inadequate, and often a misleading, measure of acute exposure of an individual to certain chemical mutagens. On the other hand, and as discussed in a number of papers at the symposium, chromosome damage in lymphocytes may provide a useful indicator of chronic exposure to chemical mutagens and a powerful method for *in vitro* studies on the actions of such substances. Studies of this kind have received a considerable impetus with the introduction of simple cytological methods to detect sister chromatid exchanges, and these techniques are being applied in efforts to monitor mutagen exposure and for *in vitro* testing of possible mutagens/carcinogens. Here, cytogenetics is clearly making a major contribution in the general upsurge of interest in the development of rapid methods for examining the mutagenic/carcinogenic potential of drugs and environmental pollutants.

An important consideration for cytogenetic monitoring is the cost, in terms of time and skill, of scoring for damaged chromosomes. Significant progress was reported at the meeting in the application of automated techniques to aberration scoring and karyotyping, although the advances have not been as rapid as many people would have wished. However, there have been dramatic developments in the use of flow systems for chromosome sorting and detecting aberrations, and the future here is most encouraging.

Finally, although the contributions to the meeting raised some new problems and answered some old ones, a number of basic questions that have concerned researchers over many years still remain unanswered. For example, despite our considerable knowledge of the effects of radiation on human lymphocytes we still do not know how an exposure distorts the normal pooling and recycling of lymphocytes in the body, and just how representative are the few hundred cells analysed of the million or more that may be present in a small blood sample? We are naturally interested in the use of chromosome damage as a measure of levels of exposure, but, as biologists, we are really more interested in the biological effects of exposure and its consequences to the individual and to exposed populations and their descendants. A proportion of the chromosome damage that we observe is cell lethal, but a great deal is not, so that clones of karyotypically abnormal cells are detectable in individuals many years after exposure to a mutagen. This prompts the question whether the level of chromosome damage in peripheral blood lymphocytes can be equated with an individual's risk of developing a malignancy within his life expectancy. More specifically, the question has been asked whether any particular mutagen-induced chromosome changes increase the chance of occurrence of a malignant transformation in the cells that carry them. The beginnings of answers to these and other questions emerged during the meeting, and it is to be hoped that they will form the basis of an equally successful symposium to be held in the not-too-distant feature.

H. J. Evans
D. C. Lloyd

G. W. DOLPHIN

A Review of in vitro Dose-Effect Relationships

There are two principal reasons for investigating the relationship between absorbed dose and the number of chromosome aberrations per cell in lymphocytes taken from samples of human peripheral blood. The first is to obtain information about the interaction of tracks of ionising particles and biological molecules, which will incidentally throw light on chromosome structure. The other is to obtain a calibration curve for biological dosimetry, and this paper will concentrate on those aspects of the dose-effect relationship that are of importance in a dosimetry system.

As all the work described involves the stimulation of lymphocytes in culture by means of phytohaemagglutinin (PHA) it is presumed that T-type lymphocytes are the cells under consideration. Factors affecting the radiation-induced aberration yield *in vitro* of T lymphocytes are discussed briefly below.

Temperature

From the early work carried out using plant material, comprehensively reviewed by Giles (1954), it was known that temperature during irradiation affected aberration yield. Recently Bajerska and Liniecki (1969) demonstrated an effect of temperature on aberration yield in lymphocytes. Work in progress at the cytogenetic laboratory of the National Radiological Protection Board (NRPB) shows that the dicentric aberration yield increases with dose, and at 300 rad of 250 kV X-rays delivered at the rate of 100 rad min^{-1} it changes from 0.63 to 0.80 per cell as the temperature is raised from 16° to 40.5°C. Results from this type of work will be of interest to radiotherapists using a combination of local heating and radiation, which has some advantage in the treatment of cancer.

Oxygen effect

The variation of the survival of reproductive capacity in cells after irradiation at different levels of oxygen tension has been studied for many years (Okada 1970). It is now recognised that loss of reproductive capacity, at least at doses below 1000 rad, is strongly correlated with the chromosome damage that can be observed at first post-irradiation metaphase. Hence an oxygen enhancement ratio (OER) of between 2 and 3 might be expected for chromosome aberrations. Freshly-taken venous blood for *in vitro* irradiation normally has sufficient oxygen for the aberration yield to be unaffected. Preliminary results from experiments at NRPB suggest OER values for 250 kV X-rays delivered at 100 rad min^{-1} vary from 4.7 to 2.0 as the aberration yield changes from 0.01 to 1.0 dicentrics per cell. This preliminary finding suggests that the oxygen effect is greater at low doses of low LET radiation where the dicentrics are mainly produced by single electron tracks, but more work must be done to firmly establish this observation.

Inter-Mitotic Death

It is important to know whether selective death of lymphocytes occurs in culture before they reach the first mitosis, because if this selective death occurred in cells with the higher amounts of chromosome damage it could affect the shape of the dose-response curve. However, it has always been observed that the chromosome aberrations are randomly distributed among the cells and that they follow a Poisson distribution. This observation is reassuring because it indicates that heavily damaged cells are not selectively dying in interphase. The amount of interphase death was estimated in experiments by Lloyd, Purrott and Dolphin (1973) who used a mixed culture technique. In this technique the number of irradiated cells is compared with the number of unirradiated cells reaching first metaphase in the same culture. It was found that inter-mitotic death varied exponentially with dose and could be represented by the equation $S/N = e^{-0.693D/270}$, where N is the fraction of unirradiated cells, and S is the fraction of irradiated cells reaching first metaphase, after a dose of D rad. From this equation it can be seen that, for doses of about 40 rad, 90% of the cells are capable of reaching first metaphase relative to unirradiated cells.

The cause of inter-mitotic death is not known, but it could be due to damage to the membrane or some other vital cell structure. Mixed culture experiments have not been carried out with high LET radiation and it should be interesting to observe the amount of interphase death with these radiations.

Mitotic Delay

Mitotic delay has been observed in irradiated cell cultures, and if there were selective delay of the cells containing the highest amount of chromosome damage this could affect the observed aberration yields at various culture times. As the chromosome aberrations are distributed at random among the irradiated cells cultured for 48 h this suggests that the heavily damaged cells are not at a selective disadvantage in passing through the cell cycle.

Lloyd *et al.* (1977) have investigated the effects of mitotic delay in some detail for two doses of X-rays, 150 and 400 rad, and for culture times varying from 36 to 120 h. At both the doses used the number of complete dicentrics per cell, i.e. dicentrics with the appropriate fragment visible in the metaphase, remained constant from 36 to 52 h in culture but fell off at longer times. The number of incomplete dicentrics, i.e. dicentrics without the appropriate fragment present at metaphase, began to rise after 44 h in culture, indicating that some cells had already passed through into second mitosis by this time. It was also observed that complete dicentrics occurred even after 120 h at a level of about 20% of the yield at 48 h for both doses. This observation suggests that some cells take over four days to pass through to the first metaphase.

By analysing the change of the distribution of dicentrics among the cells with increasing time in culture, Lloyd *et al.* (1977) showed that cells containing dicentrics come through to first division more slowly than cells without aberrations. However the delay in reaching mitosis of cells with more dicentrics does not affect the measured aberrations per cell between 36 and 52 h in culture.

Dose Rate

A decrease of aberration yield is expected with a decrease of dose rate from the early results obtained with *Tradescantia*. Purrott and Reeder (1976a) have investigated the effects of dose rate by means of split dose experiments. In these experiments two equal doses were given separated by a few hours. When two doses of 100 rad were separated by more than 6 h there was no interaction between them. However, with two doses of 250 rad each there appeared to be interaction between the doses, as measured by dicentric damage, even when they were separated by 48 h. A possible interpretation of this latter finding is that some breaks remain open for a long time, as discussed by Lea (1946), and that the effects of these long-term breaks are more readily seen at higher total doses.

The effect of changes in dose rate on the yield of dicentrics has been investigated over a range from 1.9 to 2.8×10^{12} rad h^{-1}. Purrott and Reeder (1976b) found that the dicentric aberration yield at 100 rad of low LET radiation remains constant at rates above 25 rad h^{-1} but decreases below this rate. At 500 rad the yield was constant at rates above 200 rad h^{-1} but fell off at lower dose rates. In experiments in which the aberration yield in *Tradescantia* microspores was measured, Kirby-Smith and Dolphin (1958) found that the yield fell off at very high dose rates. They attributed this finding to the instantaneous depletion of oxygen near the chromosomes, which reduced the amount of dicentric damage. However Purrott, Reeder and Lovell (1977) did not observe this effect in chromosomes of human lymphocytes with similarly high dose rates.

Background of Aberrations in Normal Humans

When investigating the aberration yield at low doses of radiation it is essential to know the naturally occurring aberration frequency in unirradiated blood; this is particularly important in biological dosimetry where low exposure doses are most frequently encountered. The normally accepted background of dicentrics is about 1 in 3000 cells but, in a continuing study at NRPB, 11 dicentrics and 40 acentrics have been found in 6600 cells from 173 blood samples taken from new recruits. A similar study with new recruits to the Windscale site has yielded 8 dicentrics in 8900 cells from 89 men. Hence values for the background aberrations may vary from country to country depending on the frequency of diagnostic radiological procedures, prevalence of virus disease and other agents producing aberrations in chromosomes. However the levels are low and would not significantly affect biological dosimetry except at the lowest doses.

Mathematical Representation

In the late 1960s it was customary to express the relationship between yield and dose as $Y=aD^n$, where n varied between 1 and 2. For fission neutrons $n=1$ but for low LET radiation many workers found a value of n close to 2, the exact value depending on the range of dose used in the experiment. More recently the quadratic equation $Y=\alpha D+\beta D^2$, with α and β as constants, has been used in the analysis of data on aberration yields. This equation is preferred because it has some biological significance, in that it represents aberrations formed by single tracks and by two separate tracks.

An important feature of this equation is represented by the quotient α/β; this is the dose at which equal numbers of aberrations are produced by one and two tracks. Below this dose the majority of the aberrations are produced by single tracks. These low doses are of considerable importance in biological dosimetry, and if this equation truly represents the aberration data then, at these doses, a dose-rate effect would not be expected; the aberration yield would be proportional to the total energy deposited in the body if the lymphocytes are in a reasonably uniform distribution.

Another aspect of representing aberration yields by the quadratic equation is that the RBE for high LET radiation, particularly fission neutrons, may be obtained at low doses. The aberration yield for fission neutrons is given by the equation $Y=\alpha' D$, so that at low doses the RBE for these neutrons is α'/α, which has values of 47 and 23 for fission neutrons and cyclotron-produced neutrons with a mean energy of 7.6 MeV, respectively. It also follows that there will be some high doses at which the aberration yield from neutrons will be the same as that from γ-rays. The dose at which this equality occurs is given by the quotient $(\alpha+\alpha')/\beta$, which equals 1700 rad for fission neutrons.

As the chromosome aberrations are distributed at random, Poisson statistics apply and the fraction of cells without aberrations is given by e^{-Y}, where Y is the mean number of aberrations per cell. The curve obtained when this expression is plotted against doses up to 500 rad lies close to those determined in survival experiments for human cell types. The value of the initial slope of the curve represented by e^{-Y} is given by $1/\alpha$, which has the value of about 1000 rad for 250 kV X-rays and about twice this value for ^{60}Co γ-rays (Lloyd *et al.* 1975). The slopes of the curves for X-rays and ^{60}Co γ-rays at 400 rad are 110 and 130 rad from the work of Lloyd *et al.* (1975). These values of slope are close to the D_0 values obtained in survival experiments with human cells, and lie in the range 50 to 200 rad (Whitmore and Till 1964).

Review of the Published Data on Dose-Effect Relationships

Values for α and β obtained by various authors are summarised in tables 1 and 2. The data in both tables are for dicentric or dicentric plus ring yields obtained from whole blood irradiated at 37°C. In table 1 the values of α for low LET radiation tend to be larger for the lower voltage X-rays, and most of the values of β are between 5 and 8×10^{-6} rad^{-2} for all radiations. The values for α and β depend on the dose range used in the experiments, and several dose points between 5 and 50 rad are required to establish the value of α with certainty from the measured yields. From table 1 it can be seen that there is an RBE of 3 for ^{60}Co γ-rays relative to 250 kV X-rays in the work of Lloyd *et al.* (1975), but Sasaki (1971) found a higher value for 200 kV X-rays.

The relationship between yield and dose for fission neutrons is linear and, apart from the data of Todorov *et al.* (1973), there is considerable agreement for all laboratories on the value of α, about 80×10^{-4} rad^{-1}. At higher neutron energies the value of β becomes more important and there is less agreement among the workers.

Summary

In the ten years since the previous conference at Edinburgh there has been considerable progress in harmonising dose-effect relationships, but still more is needed before this biological dosimetry system can be accepted with-

Table 1. Data on aberration yields of dicentrics, or dicentrics plus centric rings, published by various workers in which unstimulated whole blood was irradiated at 37°C with X or γ-rays at high dose rates. The equivalent doses are those corresponding to a yield of 0.1 aberrations per cell.

uthor	$\alpha \times 10^{-4}$	$\beta \times 10^{-6}$	Energy/ source	Dose range (rad)	Including centric rings	Equivalent dose (rad)
-Rays						
iniecki *et al.* (1973)	4.03	2.62	180 kV	49–449	No	133
asaki (1971)	7.51	7.11	200 kV	20–400	Yes	77
chmidt *et al.* (1972)	7.8	4.15	220 kV	25–400	No	87
chmidt *et al.* (1976)	7.9	5.36	220 kV	25–400	No	82
ender & Brewen (1969)	5.64	5.52	250 kV	100–300	Yes	93
vans (1967)	29.0	0.66	250 kV	121–400	No	34
loyd *et al.* (1975)	4.76	6.19	250 kV	5–800	No	94
ulpis *et al.* (1976)	4.83	—	250 kV	5–60	No	—
eonard *et al.* (1976)	—	6.97	270 kV	100–400	No	120
orman & Sasaki (1966)	0.85	4.28	1.9 MeV	15–800	No	143
-Rays						
asaki (1971)	0.91	6.82	^{60}Co	20–400	Yes	115
rewen *et al.* (1972)	3.93	8.16	^{60}Co	50–400	Yes	89
loyd *et al.* (1975)	1.57	5.0	^{60}Co	25–800	No	127

Table 2. Values of the coefficients α and β for the equation $Y=\alpha D+\beta D^2$ given by various authors for dicentric aberrations in whole blood irradiated at 37°C with neutrons.

Author	Radiation type	$\alpha \times 10^{-4}$	$\beta \times 10^{-6}$	Dose range (rad)	Dose for $Y=0.1$ (rad)
Fission Spectra					
Biola *et al.* (1974)	Crac	90.1	—	68–317	11
	Nereide	87.4	—	100–300	11
	Harmonie 1.5 MeV max.	64.8	—	22–142	15
Scott *et al.* (1969)	BEPO $\bar{E}=0.7$ MeV	84.9	—	25–150	11
Carrano (1975)	Janus $\bar{E}=0.85$ MeV	78.4	—	25–150	13
Todorov *et al.* (1973)	IRT-2000	26.6	—	25–200	38
Lloyd *et al.* (1976)	BEPO $\bar{E}=0.7$ MeV	83.5	—	50–300	11
	AWRE $\bar{E}=0.9$ MeV	72.8	—	6–265	14
Accelerated Deuterons on Beryllium					
Sasaki (1971)	$\bar{E}=2.03$ MeV	74.5	—	14–250	13
Biola *et al.* (1974)	Louvaine $\bar{E}=6.2$ MeV	33.8	—	22–172	30
Lloyd *et al.* (1976)	Hammersmith $\bar{E}=7.6$ MeV	47.8	6.4	27–324	20
d-t Reaction					
Sasaki (1971)	$\bar{E}=14.1$ MeV	25.0	3.71	12–450	37
Bauchinger *et al.* (1975)	$\bar{E}=15.0$ Mev	14.1	3.77	31–375	60
Lloyd *et al.* (1976)	$\bar{E}=14.7$ MeV	26.2	8.8	5–303	32

out question. In particular, more work is needed to establish the factors that affect the stimulation of lymphocytes into the cell cycle and the time taken for them to reach first metaphase. These and other problems concerning lymphocytes in culture could be the subject of papers at a symposium in ten years' time.

Acknowledgements

The author wishes to thank his colleagues David Lloyd, Roy Purrott and Stuart Prosser for access to their unpublished data, and Mrs Elaine Henry for help with the manuscript.

References

Bajerska, A. & J.Liniecki (1969) The influence of X-ray dose and time of its delivery *in vitro* on the yield of chromosome aberrations in the peripheral blood lymphocytes. *Int. J. Radiat. Biol. 16*, 467–81.

Bauchinger, M., E.Schmid, G.Rimpl & H.Kühn (1975) Chromosome aberrations in human lymphocytes after irradiation with 15.0 MeV neutrons *in vitro*. I. Dose-response relation and RBE. *Mutat. Res. 27*, 103–9.

Bender, M.A. & J.G.Brewen (1969) Factors influencing chromosome aberration yield in the human peripheral leukocyte system. *Mutat. Res. 8*, 383–99.

Biola, M.T., R.LeGo, G.Vacca, G.Ducatex, J.Dacher & M.Bourguignon (1974) Efficacité relative de divers rayonnements mixtes gamma, neutrons pour l'induction *in vitro* d'anomalies chromosomiques dans les lymphocytes humains, in *Proc. Conf. on Biological Effects of Neutron Irradiation*, pp.221–36. Vienna: International Atomic Energy Agency.

Brewen, J.G., R.J.Preston & L.G.Littlefield (1972) Radiation-induced human chromosome aberration yields following an accidental whole-body exposure to Co-60 γ-rays. *Radiat. Res. 49*, 647–56.

Carrano, A.V. (1975) Induction of chromosomal aberrations in human lymphocytes by X-rays and fission neutrons: dependence on cell cycle stage. *Radiat. Res. 63*, 403–21.

Evans, H.J. (1967) Dose-response relations from *in vitro* studies, in *Human Radiation Cytogenetics* (ed. H.J.Evans, W.M.Court Brown & A.S.McLean) pp.20–36. Amsterdam: North Holland.

Giles, N.H. (1954) Radiation-induced chromosome aberrations in *Tradescantia*, in *Radiation Biology* (ed. A.Hollaender) vol. 1, pt.2. London: McGraw-Hill.

Kirby-Smith, J.S. & G.W.Dolphin (1958) Chromosome breakage at high radiation dose rates. *Nature 182*, 270–1.

Lea, D.E. (1946) *Actions of Radiations on Living Cells*. Cambridge: University Press.

Leonard, A., G.B.Gerber, D.G.Papworth, G.Decat, E.D.Leonard & Gh.Deknudt (1976) The radiosensitivities of lymphocytes from pig, sheep, goat and cow. *Mutat. Res. 36*, 319–32.

Liniecki, J., A.Bajerska & W.Karniewicz (1973) The influence of blood oxygenation during *in vitro* irradiation upon the yield of dicentric chromosomal aberrations in lymphocytes. *Bull. Acad. Pol. Sci. Cl. VI, 21*, 69–76.

Lloyd, D.C., R.J.Purrott & G.W.Dolphin (1973) Chromosome aberration dosimetry using human lymphocytes in simulated partial body irradiation. *Phys. Med. Biol. 18*, 421–31.

Lloyd, D.C., R.J.Purrott, G.W.Dolphin, D.Bolton, A.A.Edwards & M.J.Corp (1975) The relationship between chromosome aberrations and low LET radiation dose to human lymphocytes. *Int. J. Radiat. Biol. 28*, 75–90.

Lloyd, D.C., R.J.Purrott, G.W.Dolphin & A.A.Edwards (1976) Chromosome aberrations induced in human lymphocytes by neutron irradiation. ***Int. J. Radiat. Biol. 29***, 169–82.

Lloyd, D.C., G.W.Dolphin, R.J.Purrott & P.A.Tipper (1977) The effect of X-ray induced mitotic delay on chromosome aberration yields in human lymphocytes. ***Mutat. Res. 42***, 401–12.

Norman, A. & M.S.Sasaki (1966) Chromosome-exchange aberrations in human lymphocytes. ***Int. J. Radiat. Biol. 11***, 321–8.

Okada, S. (1970) Radiation-induced death, in ***Radiation Biochemistry. Volume I: Cells***, p.278. N.Y.: Academic Press.

Purrott, R.J. & E.Reeder (1976a) Chromosome aberration yields in human lymphocytes induced by fractionated doses of X-radiation. ***Mutat. Res. 34***, 437–46.

—— (1976b) The effect of changes in dose rate on the yield of chromosome aberrations in human lymphocytes exposed to gamma radiation. ***Mutat. Res. 35***, 437–44.

Purrott, R.J., E.J.Reeder & S.Lovell (1977) Chromosome aberration yields induced in human lymphocytes by 15 MeV electrons given at a conventional dose-rate and in microsecond pulses. ***Int. J. Radiat. Biol. 31***, 251–6.

Sasaki, M.S. (1971) Radiation-induced chromosome aberrations in lymphocytes: possible biological dosemeter in man, in ***Biological Aspects of Radiation Protection*** (ed. T.Sugaharo & O.Hug) pp.81–91. Berlin: Springer.

Schmid, E., M.Bauchinger & O.Hug (1972) Chromosome aberrations of human lymphocytes after X-irradiation ***in vitro***. I. Qualitative and quantitative aspects of dose-effect relationships. ***Mutat. Res. 16***, 307–17.

Schmid, E., M.Bauchinger & W.Mergenthaler (1976) Analysis of the time relationship for the interaction of X-ray-induced primary breaks in the formation of dicentric chromosomes. ***Int. J. Radiat. Biol. 30***, 339–46.

Scott, D., H.Sharpe, A.L.Batchelor, H.J.Evans and D.G.Papworth (1969) Radiation-induced chromosome damage in human peripheral blood lymphocytes ***in vitro***. I. RBE and dose-rate studies with fast neutrons. ***Mutation Res. 8***, 367–81.

Todorov, S., M.Bulanova, M.Mileva & B.Ivanov (1973) Aberrations induced by fission neutrons in human peripheral lymphocytes. ***Mutat. Res. 17***, 377–83.

Vulpis, N., G.Panetta & L.Tognacci (1976) Radiation-induced chromosome aberrations in radiological protection. Dose-response curves at low dose-levels. ***Int. J. Radiat. Biol. 29***, 595–600.

Whitmore, G.F. & J.E.Till (1964) Quantitation of cellular radiobiological responses, *Ann. Rev. Nucl. Sci. 14*, 347–74.

M. BAUCHINGER

Chromosome Aberrations in Human Lymphocytes as a Quantitative Indicator of Radiation Exposure

In 1962 Bender and Gooch suggested the use of chromosome aberration yields in human lymphocytes as a quantitative measure of radiation exposure of an individual. For radio-protection monitoring of exposed individuals it is essential to establish whether estimates of the absorbed dose can be based on dose-effect curves established from *in vitro* experiments. Studies from different laboratories have shown that the aberration yield in human peripheral lymphocytes can serve as a quantitative biological indicator of a radiation exposure. For this, two conditions are necessary: (a) calibration curves, which have been established under standardised and reproducible *in vitro* irradiation and culture conditions, must be available; (b) irradiation of lymphocytes *in vitro* and *in vivo* should result in essentially the same dose-response relationship.

If one compares published calibration curves, which generally are based on dicentric chromosomes, clear correlations between the induced yield and the absorbed dose are indeed obvious. However, interlaboratory discrepancies are found that could result in considerably different dose estimations of an *in vivo* exposure. An example of the great variability in the dose-dicentric relationship from thirteen laboratories is demonstrated in a WHO study (Abbatt *et al.* 1974). Consequently it is urgently recommended that the procedures necessary for the establishment of calibration curves should be standardised.

Materials and Methods

All of our experiments with 220 kV X-rays (Schmid, Bauchinger and Hug 1972; Schmid, Bauchinger and Mergenthaler 1976), 3 MeV electrons (Schmid, Rimpl and Bauchinger 1974), 15 MeV neutrons (Bauchinger *et al.* 1975) and ^{60}Co γ-rays (preliminary) are carried out with the following standardised schedule. 1 ml of whole blood is injected into a flat nylon chamber, which is closed by two Hostaphan foils 10 μm thick and is irradiated. During irradiation the chamber is kept in a temperature-controlled box at 37°C. From one chamber two duplicate cultures are established (0.5 ml whole blood, 4 ml F-10 medium, 0.5 ml FCS and 0.12 ml PHA) and incubated for 48 h. The lymphocytes are therefore kept at a constant temperature of 37°C throughout the whole experiment. After chromosome preparation and scoring a least-squares regression analysis is carried out on the data by fitting aberration yields to the common models of dose-effect relationships. A calibration curve comprises about 2–3000 cells.

G-bands were obtained by direct treatment of freshly prepared air dried slides with a Giemsa solution (Merck) diluted in Sörensen buffer (pH 6.7–6.8) with an ionic strength of 0.05–0.1.

Results and Discussion

For the four radiation qualities the data for dicentrics could be best fitted to a linear-quadratic model of dose-effect relationship:

$$Y_{Co}=(0.27\pm0.06)\times10^{-3}D+(4.78\pm0.18)\times10^{-6}D^2$$
$$Y_e=(0.26\pm0.03)\times10^{-3}D+(3.47\pm0.16)\times10^{-6}D^2$$
$$Y_X=(0.79\pm0.04)\times10^{-3}D+(5.36\pm0.22)\times10^{-6}D^2$$
$$Y_n=(1.41\pm0.07)\times10^{-3}D+(3.77\pm0.12)\times10^{-6}D^2$$

These curves now form the basis for assessing human exposure with different types of ionising radiation.

The second requisite for meaningful biological dosimetry is a similar dose-response to *in vitro* and *in vivo* irradiation. In an earlier study (Schmid *et al.* 1974) we made comparative chromosome analyses of blood samples of fifteen therapeutically whole-body irradiated cancer patients and of samples from the same subjects after identical irradiation conditions *in vitro*. The dose range was between 10 and 29 rad of ^{60}Co γ-rays. The dicentric plus ring data could be fitted to linear models

$$Y(\textit{in vitro})=2.98\times10^{-4}D$$
$$Y(\textit{in vivo})=3.03\times10^{-4}D$$

Using an analysis of variance (Schneeberger 1960) it could be demonstrated that the linear regression on dose was significant at 0.05% for both the *in vitro* ($F_{2,11}=1.24$) and the *in vivo* ($F_{1,12}=0.65$) data. A comparison of these linear regressions indicated no significant differences ($F_{2,26}=0.003$). Consequently, a dose estimation based on the frequency of chromosome aberrations is feasible at least in cases of acute whole-body irradiation or after exposure of larger regions of the body. This confirms the results of Buckton *et al.* (1971) from a study of six cancer patients after therapeutic whole-body exposure to 2 MeV X-rays and of several mammalian experiments (Brewen and Gengozian 1971; Clemenger and Scott 1973; Preston, Brewen and Jones 1972). Whether biological dosimetry can also give reliable results after partial body exposure still has to be established.

An essential problem for all irradiation conditions, *in vitro* as well as *in vivo*, is the low dose ranges that are particularly encountered in occupational exposures. Generally the yield of dicentrics or dicentrics plus rings are used for dose estimation. However, dicentrics are infrequently induced at low doses so that a significant increase in yield may not be detected. We have repeatedly suggested that other aberration types be recorded and that total aberrations be used for dose estimations.

Table 1. Dose-effect curves for dicentric chromosomes and acentric fragments after X-irradiation; established with conventional staining (c) and G-banding (b).

k	n	Goodness-of-fit χ^2	d.f.	P	$Y = kD^n$
$(1.0 \pm 0.6) \times 10^{-5}$	1.95 ± 0.21	9.07	5	0.09	dic-b
$(1.07 \pm 0.73) \times 10^{-5}$	1.87 ± 0.23	11.41	5	0.05	ace-b
$(1.71 \pm 0.62) \times 10^{-5}$	1.87 ± 0.20	8.32	5	0.15	dic-c
$(4.45 \pm 1.07) \times 10^{-5}$	1.63 ± 0.21	5.23	5	0.41	ace-c
$(1.76 \pm 0.93) \times 10^{-4}$	1.86 ± 0.10	10.8	6	0.10	dic-c (8)
$(12.70 \pm 3.40) \times 10^{-5}$	1.43 ± 0.05	2.6	6	0.85	ace-c (8)

The newly developed staining techniques assist in the detection of more aberrations but it needs to be determined whether these techniques make a practical improvement to biological dosimetry. In our laboratory Wehner (1977) carried out an *in vitro* experiment with X-rays in the dose range 40–280 R. 830 cells were analysed after conventional staining followed by Giemsa banding. Clearly the goodness of fit of a dose-effect curve depends upon the number of analysed cells and 830 is rather low compared to about 2–3000 cells for our standard dose-effect curves. However, it is a high number when considering the effort needed for banding analysis. It is not surprising, therefore, that the data for dicentrics and acentric fragments could be fitted only to a power law model (table 1). A comparison of the data for dicentrics after conventional staining and G-banding, according to the identity test of Schneeberger (1960), indicated that the corresponding dose effect curves do not differ significantly at the 5% level. This is also true for a comparison with the dicentric data of one of our earlier X-ray experiments (Schmid, Bauchinger and Hug 1972), which is based on 2500 cells. By contrast the dose-effect curves for acentric fragments were different in both studies (table 2).

What conclusions can be drawn from this experiment? In principle the banding technique is an excellent method for establishing improved calibration curves, provided that a sufficient number of cells is analysed. In addition, with

Table 2. Comparison of the dose-effect curves (Schneeberger 1960) for dicentric chromosomes and acentric fragments after X-irradiation; established with conventional staining (c) and G-banding (b).

Aberration type	d.f.	F	$P_{0.05}$	
dic-b/dic-c	2; 10	0.141	4.10	identical
dic-b/dic-c (8)	2; 11	3.080	3.98	identical
ace-b/ace-c (8)	2; 11	7.69	3.98	different

acentric fragments one gets information on whether they originated as deletions or whether they belong to an exchange. Unfortunately this extra information entails a considerable increase in time spent in analysis. This is due to the more stringent scoring criteria: spreading and banding must be excellent in nearly all cells in order to warrant a random selection. In addition, a photographic karyotype analysis is more time-consuming. The dicentric data show that even with banding analysis no improvement of dose estimation is obtained. Consequently, for routine cytogenetic analyses in radioprotection monitoring, the application of banding cannot be recommended without suitable automated techniques. This is especially so for low dose estimations where many cells are required. Further studies are necessary to establish the principle of whether an improvement of biological dosimetry could be obtained with conventional staining by including total aberrations, i.e. total breakage yield. For this definitive criteria must be determined based on extensive banding analyses to allow an exact evaluation of different aberration types after conventional staining.

Summary

Calibration curves that have been established under standardised and reproducible irradiation and culture conditions are essential for biological dosimetry based on chromosome aberration yields in human lymphocytes. The dose-dicentric relationships for 220 kV X-rays, 3 MeV electrons, ^{60}Co γ-rays and 15 MeV neutrons from our laboratory are presented. Generally only dose-effect curves of dicentrics or dicentrics plus rings are established. However, in cases of radiation exposures in low dose ranges the dicentric yield is very low and rarely distinguishable from control levels. It is therefore suggested that other aberration types such as deletions should be recorded or total chromosome damage used, e.g. as a breakage yield, for dose estimation. The application of banding techniques allows an exact analysis and classification of all aberration types, but the question whether G-banding is feasible for biological dosimetry, and whether it provides an improvement as compared with conventional methods based on dicentrics, is discussed.

References

Abbatt, J.D., K.C.Bora, M.R.Quastel & L.P.Lefkovitch (1974) International reference study on the identification and scoring of human chromosome aberrations. *Bull. World Health Organ. 50*, 373.

Bauchinger, M., E.Schmid, G.Rimpl & H.Kühn (1975) Chromosome aberrations in human lymphocytes after irradiation with 15 MeV neutrons *in vitro*. I. Dose-response relation and RBE. *Mutat. Res. 27*, 103–09.

Bender, M.A. & P.C.Gooch (1962) Types and rates of X-ray induced chromosome aberrations in human blood irradiated *in vitro*. *Proc. Nat. Acad. Sci.* (*U.S.A.*) *48*, 522–32.

Brewen, J.G. & N.Gengozian (1971) Radiation-induced human chromosome aberrations. II. Human *in vitro* irradiation compared to *in vitro* and *in vivo* irradiation of marmoset leukocytes. *Mutat. Res. 13*, 383–91.

Buckton, K.E., A.O.Langlands, P.G.Smith, G.E.Woodcock & P.C.Looby (1971)

Further studies on chromosome aberration production after whole body irradiation in man. *Int. J. Rad. Biol. 19*, 369–78.

Clemenger, J.F. & D.Scott (1973) A comparison of chromosome aberration yields in rabbit blood lymphocytes irradiated *in vitro* and *in vivo. Int. J. Rad. Biol. 24*, 487–96.

Preston, R.J., J.G.Brewen & K.P.Jones (1972) Radiation-induced chromosome aberrations in Chinese hamster leukocytes. A comparison of *in vivo* and *in vitro* exposures. *Int. J. Rad. Biol. 21*, 397–400.

Schmid, E., M.Bauchinger & O.Hug (1973) Chromosomenaberrationen menschlicher Lymphocyten nach Röntgenbestrahlung *in vitro.* I. Qualitative und quantitative Aspekte der Dosis-Wirkungs-Beziehung. *Mutat. Res. 16*, 307–17.

Schmid, E., M.Bauchinger & W.Mergenthaler (1976) Analysis of the time relationship for the interaction of X-ray-induced primary breaks in the formation of dicentric chromosomes. *Int. J. Radiat. Biol. 30*, 339–46.

Schmid, E., G.Rimpl & M.Bauchinger (1974) Dose response relation of chromosome aberrations in human lymphocytes after *in vitro* irradiation with 3 MeV electrons. *Rad. Res. 57*, 228–38.

Schmid, E., M.Bauchinger, E.Bunde, H.F.Ferbert & H.V.Lieven (1974) Comparison of the chromosome damage and its dose response after medical whole-body exposure to ^{60}Co γ-rays and irradiation of blood *in vitro. Int. J. Radiat. Biol. 26*, 31–7.

Schneeberger, H. (1960) Linearitätsteste der k-dimensionalen Regressionsgleichung und Anwendung. *Monatshefte für Mathematik 64*, 361.

Wehner, G. (1977) Die Bedeutung der Bandenfärbung für die Analyse strahleninduzierter Chromosomenaberrationen. Thesis of the ‘Fachbereich Biologie’, University of Munich, 1977.

The Effects of Recoiling Oxygen Nuclei on the Frequency of Chromosome Breakage in Human Lymphocytes after Fast Neutron Irradiation

The absorbed dose after neutron irradiation of mammalian cells is due to charged secondary particles of two different types. Elastically scattered hydrogen nuclei contribute 80–90% of the total dose at neutron energies around 1 MeV, while 10–20% is due to the recoils of heavy nuclei (mainly oxygen nuclei, but also carbon and nitrogen nuclei). The oxygen nuclei have maximal energies of 220 keV after being elastically scattered by 1 MeV neutrons and they have a linear energy transfer of 3–400 keV μm^{-1} (Northcliffe 1963), i.e. the particle tracks are short ($< 1\ \mu m$). Thus it is of interest to know the efficiency of these recoiling heavy nuclei in producing chromosome breakage as compared to the efficiency of the recoiling protons. The use of low energy particles with a high linear transfer and with short ranges within the cell nucleus might also give information of the spatial distribution of the lesions that could interact to give a chromosome exchange aberration.

Low energy oxygen ions produced in accelerators do not penetrate the cell nucleus, and therefore indirect methods have to be used if the effects of the heavy recoils on the chromosomes are to be studied. The cross section for neutron elastic scattering against oxygen has a peak in the cross section at 1.0 MeV of neutron energy and a marked trough at 2.35 MeV. Thus, the relative contribution to the total dose from these oxygen recoils can be changed by using monoenergetic neutrons and by varying the incident neutron energy. In the present experiment human lymphocytes were irradiated at 0.8, 1.0, 1.2 and 2.35 MeV of incident neutron energy. Data are given on the frequencies of chromosome exchange aberrations and the distributions of these aberrations between cells.

Materials and Methods

Cell Culture Technique

Heparinised peripheral blood from two apparently healthy donors was used. The leucocytes were separated (without the use of PHA) as buffy coats after centrifugation at 800 rev min^{-1} for 10 min and then transferred into sterile plastic tubes, 0.5 ml to each tube. A tube was placed 40 mm from the target beam spot of the accelerator and irradiated at 37°C. The contents from two tubes (irradiated at the same dose and at the same energy) were pooled and transferred into a culture bottle containing 6 ml Parker 199. PHA was added and the cells were cultured for 48 h. A more detailed account of the experimental method and the cytogenetic analysis has been published elsewhere (Holmberg and Jonasson 1973).

Neutron Irradiation Facility

The neutron irradiations were performed at the 5.5 MeV Van de Graaff accelerator at Studsvik. Neutrons were produced using the ^{7}Li(p,n)^{7}Be reaction. Li-metal targets were made by evaporation of Li onto Ag-backings. The thickness of the targets used was determined at the ^{7}Li(p,n) threshold by measuring the neutron yield in the forward direction. A long-counter and a ^{235}U fission chamber were used as neutron detectors. The total energy spread of the neutrons was $\pm$ 65 keV.

Neutron Dose Measurement

The neutron dose was determined in two ways for each energy of the incident neutrons: (a) The total neutron dose was measured with a homogenous ionisation chamber with its gas and walls of composition CH. The data were recalculated, and given as rad in water, using the transformation tables of Bach and Caswell (1968). (b) Neutron fluxes were determined by means of a ^{235}U fission chamber placed 8 cm from the target in the zero direction. The neutron flux was calculated from the known amount of fissile material and the fission cross-section of ^{235}U. The first collision dose was then taken from the transformation tables. To avoid extrapolation from these tables separate calculations of the dose component from oxygen recoils were made using the angular distributions of neutron elastic scattering against oxygen (BNL 400, 1962).

Results

The observed yields of dicentric chromosomes at the neutron energies used are given in table 1 and the distributions of these aberrations between cells are given in table 2. At the lowest dose (19.4 rad) at 1.0 MeV of incident

Table 1. Number of dicentrics observed after fast neutron irradiation of human lymphocytes.

Neutron energy (MeV)	Proton recoil dose (rad)	Oxygen recoil dose (rad)	Number of cells analysed	Dicentrics per cell	Dicentrics per cell normalised[1]
0.80	18.6	1.0	486	0.224 ± 0.021	0.345 ± 0.033
1.00	16.1	3.3	486	0.251 ± 0.023	0.446 ± 0.040
1.00	28.6	5.8	1000	0.431 ± 0.021	0.431 ± 0.021
1.20	28.65	2.4	1000	0.365 ± 0.019	0.364 ± 0.019
2.35	31.5	0.9	1000	0.311 ± 0.018	0.282 ± 0.018

[1] Normalised to a proton recoil dose of 28.6 rad, assuming a linear relationship between yield and dose (see text).

neutron energy all cells were analysed by the fluorescent technique. Thus the distribution of the interchange aberrations (dicentrics and translocations) between cells is also given in table 2. In order to compare the yields of dicentrics at the same dose for the different neutron energies, the yield at each energy was normalised to the same proton recoil dose, 28.6 rad, assuming a linear relationship between yield and dose. This seems justified, as neutron irradiation of human lymphocytes gives a linear dicentric yield as a function of dose (e.g. Lloyd *et al.* 1976). The present values for the two different doses at 1.0 MeV are in good agreement with each other after this normalisation.

As can be seen from table 1 and figure 1, the highest dicentric yield is observed at 1.0 MeV of incident neutron energy and the lowest at 2.35 MeV. Thus, an increase in the dose contribution from the oxygen recoils gives an increase in the yield of dicentric chromosomes, as can be seen from figure 1, where the cross section for neutron elastic scattering against oxygen also is displayed. For a fixed proton recoil dose the dicentric yield increases from 0.28 dicentrics per cell at 2.35 MeV to 0.43 dicentrics per cell at 1.0 MeV of incident neutron energy. The increase of 0.15 dicentrics per cell must largely

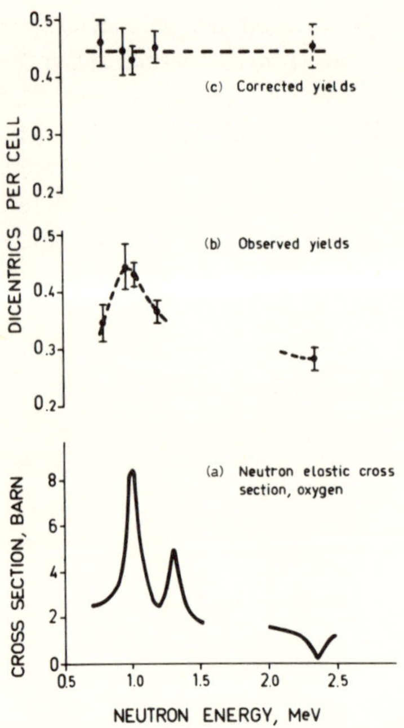

Figure 1. The number of dicentrics per cell as a function of incident neutron energy. (a) The neutron elastic scattering cross section against oxygen as a function of neutron energy. (b) Points normalised to the same proton recoil dose (data from table 1). (c) Points normalised to the same neutron dose (proton recoils + oxygen recoils) assuming that the oxygen recoils are 2.5 times more efficient than the proton recoils in producing chromosome breakage. The point at 2.35 MeV has been corrected upwards by 15% in accordance with Kellerer & Rossi (1972).

Table 2. Distribution of dicentrics among cell irradiated with neutrons.

Neutron energy (MeV)	Neutron dose (rad)	Number of cells analysed		Dicentric distribution in cells							Goodness-of-fit		
				0	1	2	3	4	5	>5	χ^2	d.f.	P
0.80	19.6	486	Observed	394	77	13	2						
			Exp. Poisson	388.4	87.0	9.7	0.7				4.8	2	<0.05
1.00	19.4	486	Obs.	391	70	23	2						
			Exp.	378.1	94.9	11.9	1.0				18.3	2	<0.001
			Obs.[1]	346	86	37	14	2	1				
			Exp.	312.3	138.1	30.5	4.5	0.5			44.8	2	<0.001
1.00	34.4	1000	Obs.	685	219	82	12	0	1	1			
			Exp.	649.9	280.1	60.4	8.7		0.9		24.2	2	<0.001
1.20	31.0	1000	Obs.	718	214	56	10	1	1				
			Exp.	694.2	253.3	46.2	5.6		0.8		12.5	2	<0.01
2.35	32.4	1000	Obs.	749	201	42	6	2					
			Exp.	732.7	227.9	35.4	3.7	0.3			6.20	2	<0.05

[1] Dicentrics and translocations per cell. The ratio between translocations and dicentrics was 0.76 ± 0.11.

be ascribed to the oxygen recoils since the yields of dicentric chromosomes observed both below (0.8 MeV) and above (1.2 MeV) the resonance in the neutron cross section against oxygen are lower than the peak values at 1.0 MeV. The efficiency with which oxygen recoils induce chromosome breakage was estimated to be 2.5 times the efficiency of the proton recoils, i.e. if the dose due to the oxygen recoils is multiplied by this figure, the dicentric yield as a function of the total dose is rather flat within the actual neutron energy range (figure 1c). It has been assumed that the dicentric yield is only slightly affected by the changes in the proton recoil spectra as neutron energy is varied. According to microdosimetric concepts the biological effect should be proportional to an effective LET, which is a smooth function of neutron energy and increases from 50 keV μm^{-1} at 2.5 MeV to about 60 keV μm^{-1} at 0.8 MeV (Kellerer and Rossi 1972). This agrees with the present observation that the dicentric yield at 2.35 MeV is 15–20% lower than the yields around 1 MeV.

As can be seen from table 2, the exchange aberrations induced by fast neutrons are not distributed between cells according to the Poisson law, since cells with multiple aberrations are over-represented in the observed distributions. The deviation from a Poisson distribution is most significant ($P < 0.001$, χ^2-test) at the peak at 1.0 MeV in the neutron cross section for the elastic scattering against oxygen, whereas the observed distribution at the trough at 2.35 MeV is more close to a Poisson distribution ($P < 0.05$). The deviations from the Poisson distribution are intermediate for the distributions observed below (0.8 MeV) and above (1.2 MeV) the peak at 1.0 MeV of incident neutron energy. Thus it seems reasonable to assume that the exchange aberrations due to the oxygen recoils do not follow a Poisson distribution between cells, whereas the proton recoils give a distribution rather close to a Poisson distribution.

Discussion

The present data show that after fast neutron irradiation the recoils from heavy nuclei (oxygen, nitrogen and carbon nuclei) are about 2.5 times more effective than proton recoils in producing chromosome breakage. These findings are in reasonable agreement with the data of Broerse, Barendsen and van Kersen (1968) on cell inactivation after irradiation with 15 MeV neutrons and also with the calculations by Bewley (1968). In an experiment by Moutschen *et al.* (1969), on chromosome aberrations in seeds of *Nigella damascena* irradiated with relatively monoenergetic fast neutrons, a higher frequency of two-break aberrations was observed at 0.44 and 1.0 MeV, at which energies there are peaks in the neutron elastic scattering cross section against oxygen.

The present data show that the distribution of aberrations between cells is rather close to a Poisson distribution for the proton recoils but not for the recoils of the heavy nuclei, which give a distribution where cells with multiple aberrations are over-represented. It has previously been reported (Schmid and

Bauchinger 1975, Lloyd *et al.* 1976) that the distribution of chromosome exchange aberrations in human lymphocytes irradiated with fast neutrons deviates from a Poisson distribution, and according to the present results this could be explained as an effect mainly due to the recoils of the heavy nuclei.

The spatial distribution of ionisation events after neutron irradiation (high-LET radiation) is different from the distribution after X-irradiation (low-LET radiation). In contrast to X-ray-induced lesions, the neutron-induced lesions are not homogenously distributed in the cell nucleus, but are clustered along the recoiling particle tracks with their high ion densities. According to the theory of Lea (1955), interactions between neutron-induced lesions occur within the same track and consequently the yield of aberrations follows a linear dose dependence after neutron irradiation. Independent of the radiation quality, however, the number of chromosome breaks per track unit is expected to follow a Poisson distribution as well as the number of tracks per cell nucleus. Thus, as pointed out by Read (1966), the number of chromosome breaks per cell might be non-Poisson distributed, if the number of tracks per nucleus is low (< 5). At the neutron energies used and the actual doses used in the present work the number of proton tracks was calculated to be 2–3 per cell nucleus (for a nuclear volume of 145 μm^3), whereas the number of oxygen recoils that originate within the nucleus was estimated to be less than 0.5 tracks per cell nucleus at 1.0 MeV of incident neutron energy. Thus, the observed distribution of aberrations per cell might deviate from a Poisson distribution even for the proton recoils, if the mean number of chromosome breaks per track exceeds some critical value. The mean number of exchange aberrations (dicentrics and translocations) induced by a single oxygen recoil was estimated from table 2 to be about 2–3, but the distribution of aberrations per track might well be a Poisson distribution. Thus, a low energy oxygen recoil is very efficient in producing chromosome breakage within a small track volume. It should be noted, however, that a comparison with accelerator produced high-LET particles is difficult to make, as these particles in general have much higher energies and the tracks have a larger radial extension, which in some cases could exceed the nuclear dimension.

To estimate the density of chromosome interphase fibres within the track volume of a low energy oxygen recoil, a crude calculation was made. Assuming a packing ratio of 30:1 for the DNA thread in the chromosome fibre and a fibre diameter of 100 Å (Schwarzacher 1976), the total length of the chromosome fibre is 6.2×10^4 μm and it occupies about 3.5% of the nuclear volume. A thin layer will be crossed by approximately 5×10^2 fibres per μm^2, and for a track with a length of 1 μm and a radial extension of the order of 50 Å, the number of fibres within the track volume should be of the order of 5. As two to three exchange aberrations might be induced by a single oxygen recoil, eventually all fibres within the track volume are involved in exchange aberrations. If so, this means that two broken chromosomes very close to each other preferentially misrepair to form an exchange aberration. Alternatively, the number of chromosome fibres within the actual track volume may be much higher than the calculated value, i.e. the chromosome threads are clustered to certain regions in the nucleus, and then a proportion of the chromosome breaks might restitute. There are indications that the interphase chromatin in human lymphocytes is preferentially located at the nuclear membrane (Berliner *et al.* 1975). In any case, several ionisation events might occur within the DNA molecule in the chromosome fibre, as the local energy density is very high due to the small radial extension of the track. If the repair of strand breakage involves heteroduplex formation, misrepair might be favoured by the fact that several lesions are induced in close spatial proximity in the DNA molecule.

Summary

Human lymphocytes were irradiated with monoenergetic neutrons at 0.8, 1.0, 1.2, and 2.35 MeV of incident neutron energy. The cross section for neutron elastic scattering against oxygen has a peak at 1.0 MeV of neutron energy and a marked trough at 2.35 MeV, and the relative contribution to the total dose from the oxygen recoils is different for the actual neutron energies. The heavy oxygen recoils have short ranges ($< 1\ \mu m$) within the cell nucleus and LET values of the order of 300 keV μm^{-1}. The oxygen recoils were found to be about 2.5 times more efficient than proton recoils in producing chromosome breakage. The distribution of dicentrics between cells was rather close to a Poisson distribution for the proton recoils, whereas multiple aberrations were over-represented in the observed distribution for the oxygen recoils. This implies that chromosome threads located within the volume limited by the ranges of an oxygen recoil are broken with a high efficiency, and the number of chromosome threads within such a volume was estimated. The data indicate that the interphase chromosome threads are clustered within the cell nucleus.

References

Bach, R.L. & R.S.Caswell (1968) Energy transfer to matter by neutrons. *Radiat. Res. 35*, 1–25.

Berliner, J., S.W.Himes, C.T.Aoiki & A.Norman (1975) The sites of unscheduled DNA synthesis within irradiated human lymphocytes. *Radiat. Res. 63*, 544–52.

Bewley, D.K. (1968) A comparison of the response of mammalian cells to fast neutrons and charged particle beams. *Radiat. Res. 34*, 446–58.

BNL 400 (1962). *Brookhaven National Laboratory Report*, 2nd ed., vol.1.

Broerse, J.J., G.W.Barendsen & G.R. van Kersen (1968) Survival of cultured human cells after irradiation with fast neutrons of different energies in hypoxic and oxygenated conditions. *Int. J. Radiat. Biol. 13*, 559–72.

Holmberg, M. & J.Jonasson (1973) Preferential location of X-ray induced chromosome breakage in the R-bands of human chromosomes. *Hereditas 74*, 57–68.

Kellerer, A.M. & H.H.Rossi (1972) The theory of dual radiation action. *Curr. Top. Radiat. Res. Q. 8*, 85–158.

Lea, D.E. (1955) *Actions of Radiations on Living Cells*. Cambridge: University Press.

Lloyd, D.C., R.J.Purrott, G.W.Dolphin & A.A.Edwards (1976) Chromosome aberrations induced in human lymphocytes by neutron irradiation. *Int. J. Radiat. Biol. 29*, 169–82.

Moutschen, J., M.Moutschen-Dahmen, R.Woodley & J.Gilot (1969) The relative biological effectiveness of different kinds of radiations on chromosome aberrations in *Nigella damascena* seed. *Int. J. Radiat. Biol. 15*, 525–40.

Northcliffe, L.C. (1963) Passage of heavy ions through matter. *Ann. Rev. Nucl. Sci. 13*, 67–102.

Read, J. (1966) Dependence of the numbers of 2-break chromosome aberrations on the size of the dose of ionizing radiation. *Radiat. Bot. 6*, 489–98.

Schmid, E. & M. Bauchinger (1975) Chromosome aberrations in human lymphocytes after irradiation with 15.0 MeV neutrons *in vitro*. II. Analysis of the number of absorption events and the interaction distance in the formation of dicentric chromosomes. *Mutat. Res. 27*, 111–17.

Schwarzacher, H.G. (1976) *Chromosomes in Mitosis and Interphase, Handbuch der mikroskopischen Anatomie des Menschen 1/3*, pp.56–86. Berlin: Springer-Verlag.

The Assessment of the Therapeutic Potential of High LET Beams by means of Chromosome Aberrations induced in Human Lymphocytes

A variety of radiobiological techniques are used in the pre-clinical assessment of new radiation modalities or treatment regimes. Many of the test systems, however, employ non-human material and problems of extrapolating the findings to man therefore arise. Furthermore the precision of some techniques, such as cell colony counting, is not sufficient to allow detailed work at low doses, e.g. to establish the initial slopes of survival curves. Cytogenetic data may be of value to the therapist because cells carrying unstable chromosome aberrations are usually unable to divide freely, owing to the formation of anaphase bridges and the loss of acentric chromosomes from the genome. Nevertheless, cytogenetic techniques have been largely overlooked by radiotherapists despite the ability to detect a wide range of doses from 1 or 2 krad down to 25 rad or less.

An early attempt to construct a radiotherapy treatment regime on the basis of optimising the frequency of dicentric aberrations in malignant and normal tissues was made by Wolff in 1972. The present paper reviews cytogenetic studies carried out in our laboratory on the therapeutic potential of negative π-mesons (pions) and of 7.6 and 14 MeV neutrons. The methods used to prepare and fix 48-h cultures of human venous blood lymphocytes and the scoring criteria are described by Purrott and Lloyd (1972). Because all types of unstable chromosome aberrations contribute to the loss of a cell's ability to divide, the data presented here are expressed in terms of total aberration yields.

Studies with Negative π-Mesons

In 1961 Fowler and Perkins drew attention to the potential advantages of using negative π-mesons in the treatment of deep tumours. Firstly, the advantageous depth-dose characteristics mean that by selecting particles of an appropriate momentum the dose to the normal tissues can be minimised whilst the Bragg ionisation peak is made to coincide with the tumour. Secondly, a large fraction of the tumour dose would consist of high LET events whilst the normal tissues in the plateau region of the ionisation curve would receive a spectrum of lower LET components. Using the calculations of Alsmiller *et al.* (1974) 25% of the peak dose results from particles with an LET > 10 keV μm^{-1} and 10% from > 100 keV μm^{-1}, whilst in the plateau the high LET contributions are reduced by a factor of 5.

Measurement of the Beam Profile

The pion beam used in these experiments was produced with the 7 GeV proton synchrotron, Nimrod, at the Rutherford Laboratory, Oxfordshire, using a tungsten target. It had a circular 80% dose contour in the transverse plane of 2.5 cm.

Aliquots (0.8 ml) of heparinised venous blood were placed in 2 ml sterile ampoules (Sterilin). These were positioned in a plastic phantom at depths shown in figure 1 and were maintained at 37°C by immersion in a water bath. Irradiations were performed at approximately 25 rad h^{-1} as measured in the peak, with doses in this position being 72, 129, 216 and 360 rad. The chosen momentum of the pions was 160 ± 7.5 MeV/*c*. Results of the aberration analysis are given by Lloyd *et al.* (1975b).

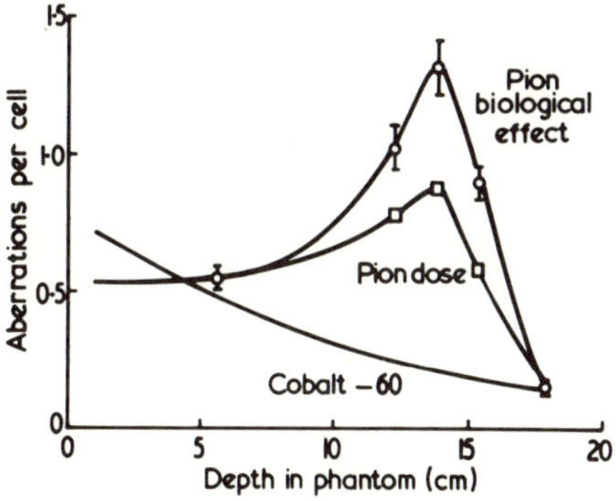

Figure 1. Aberration and ionisation profiles: (circles) aberration profile for a peak dose of 360 rad; (squares) ionisation profile in arbitrary units for 360 rad normalised at the plateau with the biological profile; (^{60}Co curve) aberration profile calculated from $Y = 4.8 \times 10^{-4} D + 4.4 \times 10^{-6} D^2$, assuming a 360 rad skin dose at 100 cm SSD and a 20 × 20 cm field.

Figure 1 illustrates how the total aberration yield for a dose of 360 rad varies with the depth in the phantom. Comparable profiles were obtained for the three other doses. The biological effect curve shows an enhancement of 1.5 over the ionisation curve in the peak position (14 cm depth) and gives a peak to plateau ratio of 2.5. Beyond the peak, which has a width at half height of 3 cm, the profile falls off sharply, and at a depth of 17.8 cm is 3.5 times lower than the plateau. For comparison with conventional depth-dose curves the profile of biological damage in terms of the total aberration yield is shown for ^{60}Co γ-radiation in a unit density material. In normal radiotherapy practice the shortcomings of the depth-dose curve associated with low LET radiation are minimised by using opposing fields. However, by carefully selecting the particle momentum such a procedure could also be adopted during treatment with pions, thereby further enhancing the peak to plateau ratio.

Table 1. Values of the coefficients α and β obtained by fitting the total aberration yield to dose in the quadratic expression $Y = \alpha D + \beta D^2$.

Radiation	$\alpha \pm$ S.D. $\times 10^{-4}$	$\beta \pm$ S.D. $\times 10^{-6}$	α/β (rad)
Negative π-mesons			
Peak	23.4 ± 1.9	4.8 ± 0.9	488
Plateau	13.4 ± 1.9	2.8 ± 1.2	479
Neutrons			
$\bar{E}$ = 7.6 MeV	87.8 ± 7.7	9.5 ± 4.3	922
$\bar{E}$ = 14.7 MeV	49.3 ± 6.2	14.7 ± 4.3	335
$\bar{E}$ = 0.7 MeV	178.9 ± 4.6	—	—
^{60}Co gamma			
50 rad min^{-1}	3.8 ± 0.9	9.4 ± 0.5	40
18 rad h^{-1}	4.8 ± 1.6	4.4 ± 0.8	109

Variation of Aberration Yield with Dose

Additional exposures in the plateau and peak positions alone were made at dose rates of 18, 25 and 70 rad h^{-1} so that dose-response relationships could be analysed over a wide range of doses (Lloyd *et al.* 1975b). The total aberration data were fitted to the quadratic equation $Y = \alpha D + \beta D^2$ by means

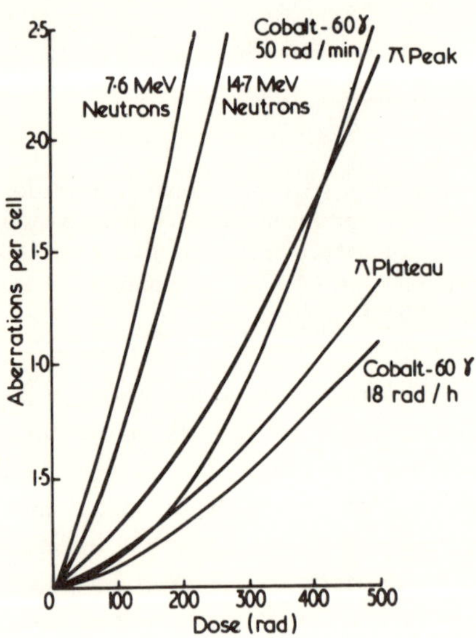

Figure 2. Variation of aberration yield with radiation dose for 7.6 and 14.7 MeV neutrons, low and high dose rate ^{60}Co γ-radiation, and the peak and plateau of the pion beam. The curves were obtained from the functions given in table 1.

of the least-squares analysis described by Edwards and Dennis (1973). Values of the coefficients α and β are given in table 1 and, for comparison, similar values obtained by Lloyd *et al.* (1975a) are given for ^{60}Co γ-radiation at dose rates of 18 rad h^{-1} and 50 rad min^{-1}. In the final column a value in rad is given for the quotient α/β, and this is the dose at which the contribution to the aberration yield from the linear (αD) term equals that from the square term (βD^2). At doses below this value the relationship between Y and D approaches linearity.

Ideally the same dose rate should have been used for the different dose exposures but this was not possible with the Nimrod facility. However, no systematic effect on the yield could be attributed to the different dose rates used. The data, together with those obtained for 7.6 and 14.7 MeV neutrons by Lloyd *et al.* (1975a, 1976) for two dose rates of ^{60}Co γ-radiation, are shown in figure 2. The two curves for the negative pions obtained in the peak and plateau positions are almost linear and lie between the low dose rate ^{60}Co and 14.7 MeV neutron curves. With respect to 150 rad of 18 rad h^{-1} γ-radiation, the RBE values for total aberrations were 2.3 in the peak and 1.5 in the plateau. The value of α/β for the peak (490) and plateau (480) regions of the pion beam are high compared with low LET ^{60}Co γ-radiation, and indicate that the production of aberrations by single track events predominates at doses that might be used as daily fractions in radiotherapy. It also suggests that little cellular recovery may occur between fractions of about 100 rad, and this was demonstrated subsequently by Purrott (1975) in a dose fractionation experiment.

Dose Fractionation

Blood samples were given a total dose of 200 rad either in a single exposure or in two equal fractions, 0.5, 4.25, 6.3 or 24 h apart. A slightly higher dose rate (120 rad h^{-1}) was used for this experiment. The blood was kept at 37°C during the interval between fractions.

The results given in table 2 show that the aberration yields in the peak

Table 2. Chromosome aberration yields induced by a dose of 200 peak rad of negative π-mesons delivered in a single exposure or in two equal fractions separated by various intervals of time.

Time intervals between fractions (h)	Total aberrations per cell ± S.E.	
	Peak	Plateau
0	0.79 ± 0.06	0.19 ± 0.01
0.5	0.79 ± 0.06	0.19 ± 0.02
4.25	0.64 ± 0.05	0.20 ± 0.02
6.3	0.81 ± 0.06	0.18 ± 0.02
24.0	0.73 ± 0.05	0.16 ± 0.02

and plateau positions remained approximately constant irrespective of the time interval between fractions. In contrast, when Purrott and Reeder (1976) exposed blood to 250 kV X-radiation the yield for a total dose of 200 rad was 35% higher than when two fractions of 100 rad were delivered 5 or more hours apart. The different response can be explained in terms of aberration induction. Most aberrations require two damaged sites and these may result from one or two ionising tracks. In the case of low LET X-radiation, where the two track mechanism is important, damaged sites induced by the first dose fraction gradually lose the ability to interact with sites induced by a second fraction. The longer the interval the more complete is the repair, until eventually no interaction can occur between the two increments of damage. In the case of pions, however, where single track mechanisms appear to predominate, the lesions involved in aberration formation must be produced virtually instantaneously.

Production of a Pion Beam with a Broad Ionisation Peak

Because of their narrow Bragg peak, monoenergetic beams of pions are unsuited to most therapeutic applications. One way of overcoming this disadvantage is to produce an expanded peak by the superimposition of several pion momenta. This has recently been attempted with the Nimrod pion facility using beams of 128, 146 and 166 MeV/c, and the resultant physical and biological profiles compare reasonably well.

Thirty-five polypropylene ampoules each containing 0.7 ml of venous blood were irradiated along the axis of the pion beam in a water phantom. By varying the exposure at each of the three momenta an expanded ionisation peak of about 10 cm was predicted over depths ranging from 8 to 17 cm. The model chosen to merge the three momenta and produce an approximately flat-topped biological response assumed a biological enhancement factor of 1.7 over the ionisation peak to plateau ratio. This value was obtained from the dose response experiment described above and by Lloyd *et al.* (1975b).

A total of 150 rad was given to an ampoule in the middle of the projected peak region at 12.4 cm depth. This likely therapeutic dose was given in three cycles of the three momenta to provide more even irradiation conditions. The results are shown in figure 3, with up to 800 cells being scored for each experimental point (Lloyd *et al.* 1978). Figure 3 also shows the predicted response (A), the ionisation profile (C) and ^{60}Co γ-aberration curve, which is for a dose of 150 rad at 12.4 cm. The total aberration data show fair agreement with the prediction except for a fall off at the position corresponding to the peak of the highest momentum, a rather low response in the plateau region and an indication of an enhanced response to the middle momentum. By using a 14% lower yield at the plateau, and enhancements of 2.4, 2.8 and 1.5 for the three momenta, the dashed line (B) is obtained, which shows a better fit to the observed data. In fact the biological enhancement of 1.5 for the highest momentum (166 MeV/c) would be predicted from the single momentum (160 MeV/c) profile described earlier. However, the value of 1.7 was used because

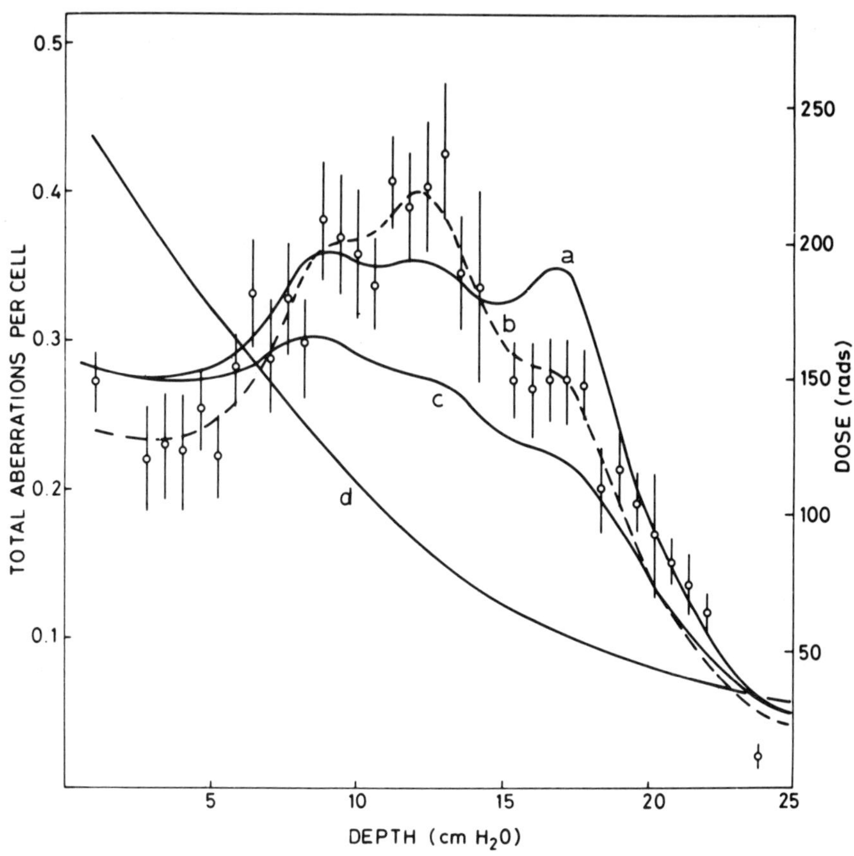

Figure 3. Observed yield of total aberrations: (a) prediction using an enhancement of 1.7 for each momentum; (b) taking a 14% lower yield at the plateau and enhancements of 2.4, 2.8 and 1.5 for the three momenta; (c) the pion dose profile; (d) the ^{60}Co γ aberration profile.

the original objective was to obtain a broad flat-topped peak for dicentric aberration yield where an enhancement factor of 1.7 was indicated. It is clear from these results that a single enhancement value may not be generally applicable, and it will be necessary to determine the values for individual momenta before further attempts are made to widen the peak. The ratio of observed damage between the highest point in the peak (12 cm) and the lowest in the plateau (3 cm) is approximately 1.8. The average values across the peak (7–14 cm) and plateau (1–5 cm) give the lower ratio of 1.6 compared with 2.5 for the single momentum at 160 MeV/c. This reduction in peak to plateau ratio is an inevitable consequence of superimposing momenta, but the resulting depth-dose distribution still compares favourably with γ-radiation.

Two further points of interest are the relatively high yield observed at a depth of 1 cm compared with the subsequent five samples, and the low yield with an RBE of 1 at the final tail position. A high entrance dose has been described with two other biological systems (Winston *et al.* 1974, Nias 1976) and may indicate the need for an absorber for skin sparing.

7.6 and 14.7 MeV Neutron Beams

In common with pions, fast neutrons should have a much greater effect on tumours with hypoxic foci than conventional therapy with low LET radiation for the same degree of damage to normal tissues. Radiotherapists have considerably more experience in the use of fast neutrons than pions, which have so far only been used to treat selected superficial tumours in terminal cancer patients (Kligerman 1975). Interest in the therapeutic application of fast neutrons led to the first clinical trials being started in California in 1938 (Stone and Larkin 1942) but severe late skin reactions occurred (Stone 1948). The clinical use of fast neutrons is currently being reappraised at several treatment centres (Catterall, Sutherland and Bewley 1975) but there is still a need for parallel radiobiological studies with these particles.

Dose-Response Relationships

Cytogenetic investigations have been carried out at NRPB with neutrons having a mean energy of 7.6 MeV and 14.7 MeV (Lloyd *et al.* 1976). The total aberration data for both energies were fitted to a quadratic expression (figure 2) and the α and β coefficients are shown in table 1. The responses are approximately linear, but with a greater β term for the 14.7 MeV particles. Data for fission spectrum neutrons, by comparison, show no dose-squared component. With the strongly linear tendency, the α/β values are high at 922 rad for 7.6 MeV and 335 rad for 14.7 MeV neutrons, this latter value being similar to that for the pion plateau. The implication of the high α/β values for neutron therapy is that less inter-fraction repair of chromosome damage is expected and hence less recovery of the reproductive capacity of the cancer cells. It is interesting, therefore, that fractionation effects have been found with cyclotron neutrons in Ehrlich ascites tumour cells (Hornsey and Silini 1962) and P-388 murine leukaemic cells (Winston *et al.* 1974). The RBE of neutrons with respect to ^{60}Co γ-radiation varies with the dose and is greater at low doses (23 and 13) than at high doses (4.1 and 2.7) for 7.6 and 14.7 MeV neutrons respectively.

In vitro Survival Curves

In vitro cell survival curves were first reported in 1956 and appeared to provide a quantitative system for understanding radiotherapy practice. Survival in its radiobiological context may be defined as the retention by cells of their reproductive integrity. Non-survivors are not necessarily dead but lack the capacity to pass through more than a few mitoses as a result, for example, of the presence of unstable chromosome aberrations.

Many studies have been carried out of the effect of various treatments on the shapes of survival curves for various cell lines. Curves similar in shape to those obtained by colony counting may be obtained from cytogenetic dose-response data such as those described in this paper, and provide relatively accurate information at the low doses involved in the initial slope region of survival curves. The data are assessed in the following way. In an homogeneous exposure the aberrations are distributed more or less randomly among the scored cells. Cells that do not contain unstable aberrations are regarded as being survivors, and the surviving fraction is given by e^{-Y}, where Y is the aberration yield per cell. This may be expanded to $S=e^{-(\alpha D+\beta D^2)}$. A feature of curves plotted from this expression is that there is normally no straight portion to the curve. Therefore, at higher doses greater cell survival is indicated by conventional cell survival curves for which the expression $S=[1-(1-e^{-D/D_0})^n]$ is used. Survival curves for peak and plateau pion radiation, for the two neutron energies and for ^{60}Co γ-radiation at 50 rad min^{-1} are shown in figure 4. It can be seen that much less survival is observed for a given dose of neutrons and that for the 7.6 MeV neutrons the relationship between log S and dose is almost a straight line. In the case of ^{60}Co γ-radiation more survival occurs following doses below 150 rad than for neutrons and pions, but above 400 rad the surviving fraction is smaller than for peak pions.

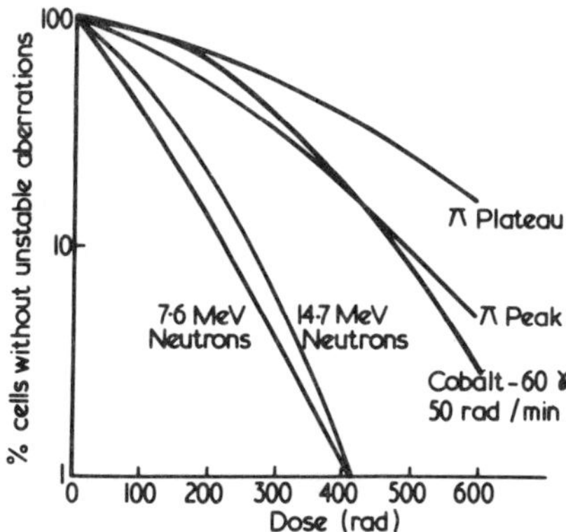

Figure 4. The percentage of cells without unstable chromosome aberrations plotted against dose for 7.6 and 14.7 MeV neutrons, peak and plateau pions, and ^{60}Co γ-rays at 50 rad min^{-1}.

Values for initial slopes are calculated as $1/\alpha$, and are 114 and 203 rad for 7.6 and 14.7 MeV neutrons and 473 and 750 rad for peak and plateau pions. There are no published estimates of the initial slopes of shouldered cell survival curves after neutron irradiation for comparison, as cell survival experiments cannot be performed with precision at low doses. Values for the D_0 at 200 rad are 75 and 83 rad for neutrons of 7.6 and 14.7 MeV. These values are in good agreement with other published values (see Lloyd *et al.* 1976). For pions the D_0 values are higher, at 195 in the peak and 295 in the plateau, but no other estimates have been published.

Future Work

Interphase Death

It is important to stress that besides loss of reproductive capacity owing to the presence of chromosome aberrations the radiotherapist is also interested in cell death, which occurs following irradiation and before the first mitosis. Indeed, in tumours that are not characterised by a high rate of cell division, interphase death may be primarily responsible for the success of treatment. Interphase death can be investigated cytogenetically by mixing equal quantities of irradiated and unirradiated blood. The resultant aberration yield should be 50% of that for the dose normally given if no cell death occurs. This procedure has so far been tested with low LET X-radiation and high LET α-particles. In the former case the amount of cell death occurring was dependent on dose with 94% of irradiated cells surviving at 50 rad, 86% at 200 rad, but only 12% at 700 rad (Lloyd *et al.* 1978). In the case of the α-irradiations preliminary unpublished data (Prosser, personal communication) indicate, as might be expected, much higher levels of cell death.

Experiments are under way to determine the amount of cell death associated with pion and neutron irradiations. These data can be combined with those for chromosome damage to give a more complete picture of the likely therapeutic efficiency.

Oxygen Enhancement Studies

For low LET X-radiation chromosome aberration levels are known to be dependent on the degree of oxygenation of the irradiated cells (Watson and Gillies 1970). For neutrons, where most of the damage at 200 rad results from single track mechanisms, it is likely that any effect of oxygen levels on aberration yields will be small. In the case of pions, however, where a significant number of low LET events occur, even in the peak, oxygenation levels may be more important. In view of the hypoxic condition of many tumour cells, cytogenetic experiments are under way to examine the OER for neutrons and pions.

Conclusions

Chromosome aberration analysis is a useful technique, which warrants further development as a source of human radiobiological data. Being a direct measurement of the effect of radiation on the cell nucleus, the system is particularly relevant to considerations of mitotic cell death. It can provide radiotherapists with a new biological end-point for the characterisation of radiation beams and for assessing treatment procedures. Cytogenetic techniques are adaptable and can also be used to investigate interphase cell death and oxygen enhancement effects.

Summary

Chromosome aberration analysis is a useful method of assessing the cell sterilisation potential of new techniques of radiotherapy. Depth-dose profiles have been constructed for 160 MeV/c pions, which show an ionisation ratio in the Bragg peak of 1.5 times the plateau value, but for chromosome aberrations the ratio was more advantageous at 2.5. The RBE with respect to 150 rad ^{60}Co γ-radiation was 2.3 in the peak and 1.5 in the plateau region. Dose-response data show a predominantly linear relationship in both positions. This is consistent with the absence of a fractionation effect on the total aberration yield with two 100 rad doses given up to 24 h apart.

A broad peak, such as would be needed for radiotherapy, was obtained by irradiating with pions of three different momenta. The increase in peak width to about 10 cm was accompanied by a reduced peak to plateau ratio for chromosome damage of about 1.6.

Cytogenetic data may be used to construct a form of cell survival curve based on the frequency of cells without unstable aberrations. As doses of a few rad may be detected, accurate data on 'initial slopes' can be obtained. Initial slopes for 7.6 and 14.7 MeV neutrons, and peak and plateau pions, were 114, 203, 473 and 750 rad respectively. D_0 values at 200 rad were 75, 83, 195 and 295 rad.

References

Alsmiller, R.G., R.T.Santoro, W.T.Armstrong, J.Barish, K.C.Chandler & G.T. Chapman (1974) Calculations related to the use of photons, neutrons, negatively charged pions, protons and alpha particles in cancer radiotherapy. *Oak Ridge National Laboratory Report*, ORNL-TM-4369.

Catterall, M., I.Sutherland & D.K.Bewley (1975) First results of a randomised clinical trial of fast neutrons compared with X or gamma rays in the treatment of advanced tumours of the head and neck. *Br. Med. J. 2*, 653–6.

Edwards, A.A. & J.A.Dennis (1973) Polyfit—a computer program for fitting a specified degree of polynomial to data points. National Radiological Protection Board, Harwell, NRPB-M11.

Fowler, P.H. & D.H.Perkins (1961) The possibility of therapeutic applications of beams of negative π mesons. *Nature 189*, 524–8.

Hornsey, S. & G.Silini (1962) Recovery of tumour cells cultured *in vivo* after X-ray and neutron irradiations. *Radiat. Res. 16*, 712–22.

Kligerman, M.M. (1975) Meson radiobiology and therapy, in *Proc. 7th Int. Conf. on Cyclotrons and their Applications*, pp.419–26. Basel: Birkaüser.

Lloyd, D.C., G.W.Dolphin, R.J.Purrott & P.A.Tipper (1977) The effect of X-ray induced mitotic delay on chromosome aberration yields in human lymphocytes. *Mutat. Res. 42*, 401–12.

Lloyd, D.C., R.J.Purrott, G.W.Dolphin, D.Bolton, A.A.Edwards & M.J.Corp (1975a) The relationship between chromosome aberrations and low LET radiation dose to human lymphocytes. *Int. J. Radiat. Biol. 28*, 75–90.

Lloyd, D.C., R.J.Purrott, G.W.Dolphin & D.H.Reading (1975b) An investigation of the characteristics of a negative pion beam by means of induced chromosome aberrations in human peripheral blood lymphocytes. *Int. J. Radiat. Biol. 27*, 223–36.

Lloyd, D.C., R.J.Purrott, G.W.Dolphin & A.A.Edwards (1976) Chromosome aberrations induced in human lymphocytes by neutron irradiation. *Int. J. Radiat. Biol. 29*, 169–82.

Lloyd, D.C., D.H.Reading, R.J.Purrott, M.A.Hynes, W.S.Spinks & B.D.Stephenson (1978) Expansion of a negative pi-meson peak to cover a range of depths suitable for radiotherapy. *Br. J. Radiol. 51*, 41–5.

Nias, A.H.W. (1976) Determination of the negative pion dose response curves for frozen Hela cells at various positions along the depth dose profile. *Rutherford Laboratory Report* RL–76–092, pp.43–4.

Purrott, R.J. (1975) Chromosome aberration yields in human lymphocytes exposed to fractionated doses of negative π mesons. *Int. J. Radiat. Biol. 28*, 599–602.

Purrott, R.J. & D.C.Lloyd (1972) The study of chromosome aberration yield in human lymphocytes as an indicator of radiation dose. I. Techniques. National Radiological Protection Board, Harwell, NRPB—R2.

Purrott, R.J. & E.Reeder (1976) Chromosome aberration yields in human lymphocytes induced by fractionated doses of X-radiation. *Mutat. Res. 34*, 437–46.

Stone, R.S. (1948) Neutron therapy and specific ionisation. *Am. J. Roentgenol. 59*, 771–85.

Stone, R.S. & J.C.Larkin (1942) Treatment of cancer with fast neutrons. *Radiology 39*, 608–20.

Watson, G.E. & N.E.Gillies (1970) The oxygen enhancement ratio for X-ray induced chromosomal aberrations in cultured human lymphocytes. *Int. J. Radiat. Biol. 17*, 279–83.

Winston, B.M., R.J.Berry, D.R.Perry & D.H.Reading (1974) Effects of the radiation at the entrance point of a negative π-meson beam. *Br. J. Radiol. 47*, 201–2.

Wolff, S. (1972) Genetic effects and radiation-induced cell death, in *Frontiers of Radiation Therapy and Oncology: Radiation Effect and Tolerance, Normal Tissue* (ed. J.M.Vaeth) pp.459–69. Basel: Karger and Baltimore: University Park Press.

R. J. PRESTON and J. G. BREWEN

X-Ray-Induced Chromosome Aberrations in the Leucocytes of Mouse and Man

In the past few years there has been considerable discussion of the sensitivity of the leucocytes of different species to the induction of chromosome aberrations by X-rays (see Sankaranarayanan 1976 for a summary). This, in part, stemmed from our report (Brewen *et al.* 1973) that there was a considerable variation in dicentric yield in the species that we studied, and also that for these particular species there appeared to be a linear relationship between dicentric yield and effective chromosome arm number at any particular dose. As stated in the paper, it was the relationship between aberration yields in man and mouse that was of particular interest; no explanation of the relationship between the dicentric yield and the effective chromosome arm number could be offered. Subsequently, Savage and Papworth (1973) and Clifford (1976) have offered explanations. However, many other species have been studied, and it has become fairly clear that in some the yield is approximately that which would be predicted from the effective chromosome arm number, whereas in others it is not. We feel that our original observation could in part be accounted for by a fortuitous choice of the species that were studied, as regards their DNA content, interphase chromosome volume, and perhaps repair characteristics. However, the facts that the dicentric yield was twice as high in human as in mouse leucocytes, and that acentric fragment yields were similar in the two species, appeared to be worth further study, particularly since the fixation times used for the mouse were rather late and could result in a reduced yield due to the presence of second-division cells. This became more important as a result of the study by de Boer *et al.* (1977) in which they concluded that dicentric yields were the same in mouse and human leucocytes and that the acentric yield was higher in the mouse at 200 rad. The present paper presents a comparison made at early fixation times for both human and mouse.

Materials and Methods

Buffy coat cultures were established from three human donors for 42-hour fixations, from one human donor for 48-h fixation, and from several mice for 36- and 48-h fixation. Human leucocytes were cultured according to our standard procedure as described elsewhere (Preston, Brewen and Gengozian 1974). The technique used for culturing the mouse leucocytes was that of Triman, Davisson and Roderick (1975).

The cultures were irradiated prior to the addition of phytohaemagglutinin. All irradiations were with a 250 kVp General Electric X-ray machine operated at 250 kV and 30 mA with 3 mm of additional Al fixation (h.v.l. 0.45 mm Cu). The dose rate was 100 R min^{-1}.

A minimum of 300 cells were analysed at any one dose point except for the mouse 36-h fixations. All aberrations were recorded, and a note was made of any exchange aberration that was not accompanied by an acentric fragment, so that some correction of the aberration yield could be made as a result of the presence of cells in their second mitosis after treatment.

Results

The yields of dicentrics and acentric fragments (terminal deletions, acentric rings, and small interstitial deletions) are given in table 1 for human 42-h and 48-h fixations, and the data for dicentrics are shown graphically in figure 1. The data were fitted to three models, $Y=bD$, $Y=cD^2$ and $Y=bD+cD^2$, by least-squares regression using Poisson variances and weights.

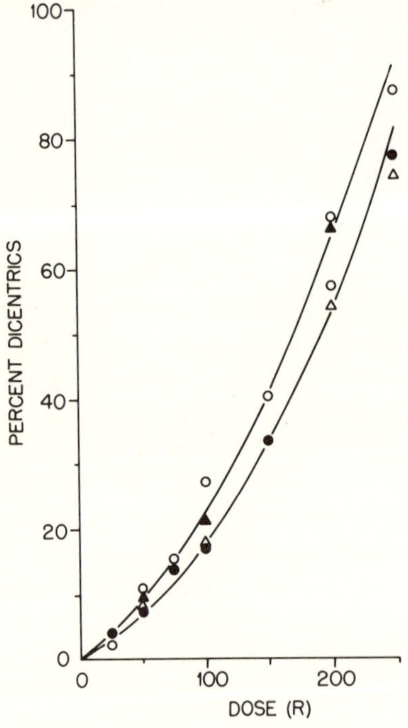

Figure 1. Dose-response curves for dicentric aberrations following exposures of human leucocytes to X-rays: (solid circles) 48-h fixation, donor J.P.; (open circles) 42-h fixation, donor J.P.; (solid triangles) 42-h fixation, donor T.H.; (open triangles) 42-h fixation, donor J.A. The curves are the fit of $Y=bD+cD^2$ to the 42-h and 48-h fixation data for donor J.P.

Table 1. Chromosome aberration frequencies following exposure of human leucocytes to X-rays.

Dose (R)	No. of cells scored	% Dicentrics (± S.E.)	% Acentric fragments (± S.E.)	Corrected % dicentrics[1]
48-h fixation, donor J.P.				
0	500	0	0.8 ± 0.4	0
25	500	3.8 ± 0.8	1.6 ± 0.6	4.0
50	500	7.2 ± 1.2	5.8 ± 1.1	7.6
75	500	13.6 ± 1.6	6.8 ± 1.2	14.4
100	500	16.8 ± 1.8	11.8 ± 1.5	17.8
150	500	33.4 ± 2.6	19.2 ± 2.0	35.3
200	500	57.4 ± 3.4	26.6 ± 2.3	62.2
250	500	77.4 ± 3.9	46.4 ± 3.0	81.8
42-h fixation, donor J.P.				
0	500	0	0.6 ± 0.3	
25	300	2.3 ± 0.9	1.7 ± 0.7	
50	300	11.0 ± 1.9	5.3 ± 1.3	
75	300	15.3 ± 2.3	7.0 ± 1.5	
100	300	27.3 ± 3.0	14.3 ± 2.2	
150	300	40.3 ± 3.7	11.7 ± 2.0	
200	300	68.0 ± 4.8	25.0 ± 2.9	
250	250	87.6 ± 5.9	26.4 ± 3.2	
42-h fixation, donor J.A.				
0	300	0	0	
50	300	8.0 ± 1.6	4.0 ± 1.1	
100	300	17.7 ± 2.4	11.0 ± 1.9	
200	300	54.3 ± 4.3	22.3 ± 2.7	
250	300	74.3 ± 5.0	40.3 ± 3.6	
42-h fixation, donor T.H.				
0	300	0	0	
50	350	9.4 ± 1.6	6.0 ± 1.2	
100	300	21.0 ± 2.6	11.3 ± 1.9	
200	300	66.3 ± 4.7	37.7 ± 3.5	

[1] The corrected frequency is calculated by multiplying the observed frequency by the total proportion of dicentrics without fragments (49/864 = 0.057) and adding this value to the observed frequency.

The best fit was to the model $Y = bD + cD^2$. The spontaneous rate was in most cases zero, and the curve was constrained to pass through the origin.

There is a significant difference in the dicentric yields at the two fixation times, with the yield at 42 h being higher than that at 48 h for two donors (J.P. and T.H.). Yields from donor J.A. fixed at 42 h were essentially the same as those from a different donor fixed at 48 h; no direct comparison could be made, however, as no data were available for the same donor at the later fixation time.

Dicentric yields for the 48-h fixation were corrected on the assumption that a cell containing a dicentric without an accompanying acentric fragment was at its second mitosis after treatment. For this correction, it is assumed that a dicentric aberration has a 50% probability of surviving a mitotic division to be analysed at the next division (Carrano and Heddle 1973). Thus for every dicentric without an accompanying fragment scored, one will be lost at the first division after treatment. The proportion of dicentrics lost at division will be equal to the product of the frequency of dicentrics analysed at any dose and the frequency of cells containing dicentrics without fragments. This product must then be added to the observed dicentric frequency to give the frequency expected if all cells were at their first mitosis after treatment. This correction also assumes that the frequency of second-division cells is independent of dose, and, from the rather low numbers obtained at each dose, this appears to be the case.

The data are shown in table 1, and it can be seen that the corrected yield was still significantly different from the yield in cells fixed at 42 h. This suggests that the increased yield at the earlier fixation time was not entirely due to the fact that only first-division cells were being analysed. Some other pattern of differential sensitivity of leucocytes to radiation is indicated.

Data on acentric fragments are somewhat more confusing, although it appears that there is essentially no difference in yield between cells fixed at 48 h and those fixed at 42 h.

Data for dicentrics and acentric fragments induced in mouse leucocytes by X-rays are shown in table 2 and figure 2, for both 48- and 36-h fixations. These data were also fitted to the three models, with the best fit at the 48-h fixation being to the model $Y = bD + cD^2$.

Table 2. Chromosome aberration frequencies following exposure of mouse leucocytes to X-rays.

Dose (R)	No. of cells scored	% Dicentrics (± S.E.)	% Acentric fragments (± S.E.)	Corrected % dicentrics[1]
48-h fixation				
0	500	0	0	0
25	500	2.2 ± 0.7	2.2 ± 0.7	2.6
50	400	3.0 ± 0.9	5.5 ± 1.2	3.5
75	500	5.8 ± 1.1	4.6 ± 1.0	6.8
100	500	7.8 ± 1.3	10.4 ± 1.4	9.1
150	500	14.4 ± 1.7	11.0 ± 1.5	16.8
200	400	22.5 ± 2.4	22.2 ± 2.4	26.3
250	500	38.8 ± 2.8	29.0 ± 2.4	45.3
36-h fixation				
0	300	0	0	
50	100	9.0 ± 3.0	5.0 ± 2.2	
100	300	12.7 ± 2.0	12.7 ± 2.0	
200	250	48.0 ± 4.4	27.2 ± 3.3	

[1] The corrected frequency is calculated by multiplying the observed frequency by the total proportion of dicentrics without fragments (75/447 = 0.168) and adding this value to the observed frequency.

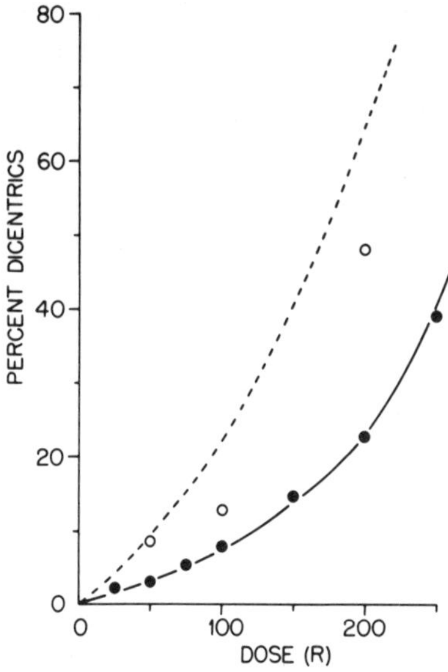

Figure 2. Dose-response curves for dicentric aberrations following exposures of mouse leucocytes to X-rays: (solid circles) 48-h fixation; (open circles) 36-h fixation. The dotted line is for the human 42-h fixation data for donor J.P.

When the dicentric yield at the 48-h fixation was corrected for loss at the first division, as described above, the yields obtained were still lower than those observed at 36 h (at which time all cells with a dicentric also had an accompanying acentric fragment). The proportion of cells containing dicentrics without an accompanying acentric fragment in the mouse cells fixed at 48 h was higher than that obtained for human cells fixed at the same time (0.16 vs. 0.06), suggesting a more rapid first-division cycle in mouse leucocytes. This was borne out by the fact that sufficient metaphases were obtained in mouse cultures fixed at 36 h, whereas it was necessary to fix human cultures at 42 h to obtain sufficient metaphases.

A comparison of dicentric and acentric fragment yields in the human and the mouse showed that for dicentrics the yield at 42 h in the human was between 1.5 and 2.1 times as high as in the mouse fixed at 36 h; the acentric fragment yields were comparable in the two species. If the corrected dicentric yields at 48-h fixation were compared, the human yield was about twice as high as that of the mouse; the acentric fragment yields were similar in the two, with perhaps a slightly higher value for the human.

Discussion

It is clear that the original comparison of aberration frequencies induced by X-rays in the leucocytes of mouse and man (Brewen *et al.* 1973) was based on cells that were in both their first and second mitoses after irradiation. It is also probable that more second-division cells were present in the mouse cultures than in the human, since it is shown here that a higher proportion of cells containing dicentrics without fragments was present in mouse cultures than in human cultures when both were fixed at 48 h. This factor alone could influence the aberration frequencies differently in mouse and man, making the ratio of dicentrics and acentric fragments in the two different from those originally reported. However, when the influence of second-division cells was accounted for, as in the 48-h fixations, or when only first-division cells were compared at the earliest fixations, a similar result to that already reported was obtained, i.e. human leucocytes in G_0 were twice as sensitive to dicentric production as mouse leucocytes, and the two had similar sensitivities to acentric fragment induction.

There are obviously some differences between these observations and those of de Boer *et al.* (1977), who from a large study concluded that mouse and human leucocytes were equally sensitive to dicentric production and that the mouse was more sensitive to deletion production than man. How can we reconcile these results with the ones presented here? The data obtained for the mouse at 36 h in the present study show very good agreement at 100 and 200 R for dicentrics and deletions with those reported by de Boer *et al.* (1977). The difference is in the results for human leucocytes. If comparisons are made at the same fixation time (48 h), it is clear that the yields reported here are considerably higher than those of de Boer *et al.* (1977). For example, at 200 R the dicentric yields are 57.4 vs. 34.0%, and the deletion frequencies at the same dose are 26.6 vs. 18.5%. If comparisons are made between the yields at the earliest fixation used here (42 h) and the 48-h fixation of de Boer *et al.* (1977), the only one used, then the contrast is greater. The differences at the different fixation times and between the two sets of data need not represent only a difference in the proportions of first- and second-division cells, since when the data reported here at the 48-h fixation are corrected for second-division cells, the yields obtained are still lower than those actually observed at 42 h after treatment and stimulation. It appears that those cells reaching mitosis earliest are slightly more sensitive to aberration induction than those arriving later. This does not necessarily imply that there is a differential sensitivity of lymphocytes in G_0, but perhaps that there is a differential stimulation of cells by phytohaemagglutinin. It appears, therefore, that the differences between the present results and those of de Boer *et al.* (1977) lie in the frequencies of aberrations in the human leucocytes, either as a result of different proportions of first- and second-division cells, or as a result of a different segment of the irradiated population being analysed, or a combination of both.

It should be added that the yields reported here for the 42-h fixation were similar for two different donors, and only slightly reduced for the third. This suggests that a donor-to-donor variation is not the most likely explanation of the differences discussed above.

From the data presented here, we suggest that indeed the yields of dicentrics induced by X-rays in human leucocytes are at least 50% higher than in mouse leucocytes, and that acentric yields are similar in the two species.

These data are not presented as lending support to the so-called 'arm number' hypothesis. Too many exceptions have been reported to make such a conclusion feasible. To use only chromosome aberration data obtained in leucocytes to make estimates of genetic hazard of radiation or chemicals to man could also be very misleading. The data of Sasaki (1975) for synchronised skin cells showed that mouse embryos and newborns were about one-half as sensitive to aberration induction by X-rays as human skin cells, but that the cells from 11-day-old mice were equally sensitive to aberration induction as human cells. More will need to be known about the comparative induction of initial lesions, and the extent and rate of their repair in different cells from different species, before the full usefulness of data such as those presented here will be realised.

Summary

In our earlier studies we showed that the frequency of dicentrics induced by X-rays in human leucocytes was about twice that induced in mouse leucocytes. The frequencies of deletions were similar in both species. However, the mouse cultures were fixed at 60 h and the human cultures at 54 h. In both cases it was likely that some of the cells analysed were in their second post-treatment mitosis. Further studies were carried out using fixation times of 48 h for both mouse and human cultures (three different human donors were used). The same relationships held here, namely twice as many dicentrics in humans, and similar deletion frequencies in both. The aberration frequencies observed were corrected to take account of second-division cells by assuming that cells containing a dicentric without an accompanying fragment were in their second division. There were more such cells in mouse than in human cultures. To further increase reliance on the conclusions, we fixed cultures at the earliest times that 300 cells per dose could be obtained—36 h for the mouse, 42 h for the human. The frequencies of dicentrics were increased in both, and a relationship of about 2:1 for human to mouse was obtained. Deletion frequencies were similar in both. Since no dicentrics without fragments were obtained, it appeared that aberration frequencies in first-division cells only were being compared.

Acknowledgements

Research supported by the Energy Research and Development Administration under contract with the Union Carbide Corporation.

References

Brewen, J.G., R.J.Preston, K.P.Jones & D.G.Gosslee (1973) Genetic hazards of ionizing radiations: cytogenetic extrapolations from mouse to man. *Mutat. Res. 17*, 245–54.

Carrano, A.V. & J.A.Heddle (1973) The fate of chromosome aberrations, *J. Theor. Biol. 38*, 289–304.

Clifford, P. (1976) Chromosome arm number and radiation-induced dicentric yield. *Mutat. Res. 37*, 141–4.

de Boer, P., P.P.W. van Buul, R. van Beek, F.A. van der Hoeven & A.T.Natarajan (1977) Chromosomal radiosensitivity and karyotype in mice using cultured peripheral blood lymphocytes, and comparison with this system in man, *Mutat. Res. 42*, 379–94.

Preston, R.J., J.G.Brewen & N.Gengozian (1974) Persistence of radiation-induced chromosome aberrations in marmoset and man. *Radiat. Res. 60*, 516–24.

Sankaranarayanan, K. (1976) Evaluation and re-evaluation of genetic radiation hazards in man. II. The arm number hypothesis and the induction of reciprocal translocations in man. *Mutat. Res. 35*, 371–86.

Sasaki, M.S. (1975) A comparison of chromosomal radiosensitivities of somatic cells of mouse and man. *Mutat. Res. 29*, 433–48.

Savage, J.R.K. & D.G.Papworth (1973) The relationship of radiation-induced dicentric yield to chromosome arm number. *Mutat. Res. 19*, 139–43.

Triman, K.L., M.T.Davisson & T.H.Roderick (1975) A method for preparing chromosomes from peripheral blood in the mouse. *Cytogenet. Cell Genet. 15*, 166–76.

J. LINIECKI, A. BAJERSKA and K. WYSZYŃSKA

Animal Models for Studies of Chromosome Aberration Induction in PHA-Stimulated Lymphocytes

There is a need for animal models to study the effects of ionising radiation and other factors on lymphocyte chromosomes. These models should permit a systematic *in vivo* investigation of the influence of such variables as radiation dose, dose-rate and fractionation, and partial-body irradiation, which for obvious reasons cannot be assessed directly in man.

Studies comparing the response of the cells to *in vitro* vs. *in vivo* whole-body irradiation have already been performed on animals, e.g. rabbits (Clemenger and Scott 1973, Bajerska and Liniecki 1975), swine (McFee, Banner and Sherill 1972), marmosets (Brewen and Gengozian 1971) and hamsters (Preston, Brewen and Jones 1972). However, a sound knowledge of the biological system employed, mainly of the proliferative kinetics of lymphocytes, has not often been available. For instance, the optimum culture time for swine lymphocytes (mostly 48 h) has not been experimentally determined. In the rabbit, experimental data have provided a basis for the culture times used (Bajerska and Liniecki 1975, Clemenger and Scott 1973), but the information is scanty, and even the influence of such factors as the choice of culture medium remains unknown.

The purpose of this study has been to analyse the mitotic kinetics of rabbit and porcine lymphocyte cultures and to make a direct comparison with the more thoroughly investigated system involving the human lymphocyte. The study has been limited to the culture method currently employed in this laboratory (whole blood microculture with a particular medium). The method for determining the culture harvest times was that of Heddle, Evans and Scott (1967), with some support from differential sister chromatid staining (FPG). The study is still in progress, but a clear basic pattern of results has already emerged.

Materials and Methods

Blood donors. Human venous blood was obtained from two healthy adult males aged 34 and 25. Rabbit blood was obtained from the marginal ear vein of two young healthy Chinchilla males. Porcine blood was sampled in a slaughterhouse by incision of the aorta immediately after sacrificing males of unknown age.

Culture procedure. Whole blood microcultures were established by adding 0.3–0.4 ml blood to a culture fluid consisting of 4 ml Ham's F-10 medium and 1 ml of heat-inactivated calf serum containing 100 i.u. penicillin and 0.1 mg streptomycin per ml. Throughout the study Wellcome phytohaemagglutinin (0.2 ml per culture) was used. From 10 to 15 h after culture initiation the medium was removed by gentle suction and replaced by fresh medium free from PHA. All the cells were cultured at 38 ± 0.5°C.

The cultures of human, rabbit and porcine lymphocytes were terminated at various times (see below); 3 h prior to this colcemid had been added to give a final concentration of 0.1–0.2 μg ml^{-1}. The remaining steps of hypotonic treatment, fixation, staining and scoring using coded slides have been described previously (Bajerska and Liniecki 1967, Liniecki *et al.* 1977).

For differential sister chromatid staining the FPG method of Perry and Wolff (1974) was used. 5′-bromodeoxyuridine was used at a final concentration of 5 μg ml^{-1} in the cultures. The latter were maintained for a given time in darkness in a thermostat and the metaphases arrested with colcemid. After treatment with hypotonic KCl the cultures were fixed, dried and stained with Hoechst 33258 for 12 min, exposed to daylight in a humid atmosphere for 24 h, warmed in SSC at 62°C for 2 h, dried and stained for 5 min with 3% Giemsa.

End-points studied. To establish for each species the culture time at which only cells in their first post-PHA stimulation mitosis would be harvested, the following end-points were studied as a function of culture duration: (a) mitotic index; (b) tetraploidy index; (c) yield of dicentrics; (d) frequency of cells with dicentrics and/or centric rings not accompanied by acentric fragments, CuX_2 cells (Buckton and Pike 1964); (e) frequency of mitotic figures with similarly and differentially stained sister chromatids. End-points (a) and (b) were studied both in irradiated (300 rad γ-rays) and non-irradiated cultures; points (c) and (d) exclusively in irradiated ones, and (e) so far in non-irradiated cells at selected times when, based on the other end-points, no cells in second mitosis were expected, and 12 h later.

Human whole-blood cultures were terminated at 40, 44, 48, 52, 56 and 60 h; porcine cultures at 16, 20, 24, 28, 32, 36, 40, 44, 48, 52, 56, 60, 64, 68, 72 and 96 h; and the rabbit cells were harvested at 32, 36, 40, 44, 48, 52, 56 and 60 h.

Dose-effect relations were also studied for the dicentric yields in a few individuals of the three species.

Irradiation and dosimetry. Prior to its injection into culture medium well aerated whole blood was irradiated at 38°C with ^{60}Co γ-rays at a dose-rate of 108 rad min^{-1}. Dosimetry has been based upon measurements of exposure (R) with a Siemens dosimeter and conversion to the absorbed dose (rad) by using the factor 0.94.

Results

Lymphocyte mitotic and tetraploidy indices vs. duration of culture in the irradiated and non-irradiated blood are presented in figures 1, 2 and 3 for

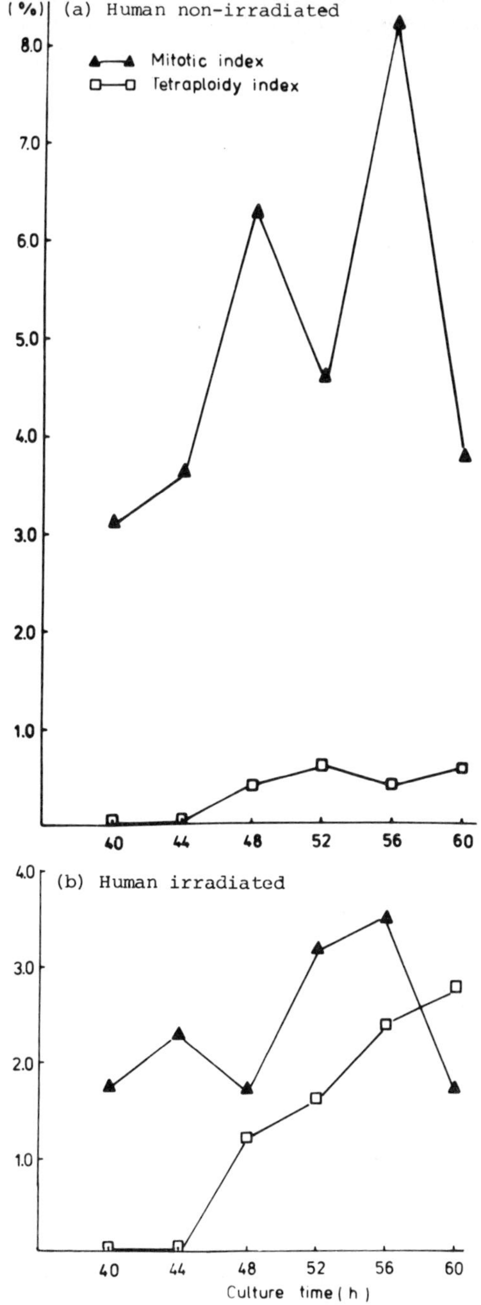

Figure 1. Mitotic and tetraploidy indices in cultures of human lymphocytes as a function of culture duration (one individual): (a) non-irradiated; (b) irradiated prior to culture with 300 rad γ-rays.

man, rabbit and pig respectively. The dicentric yields and frequency of CuX_2 cells after 300 rad gamma-rays are presented in figures 4, 5 and 6. From these data it follows that appreciable mitotic activity occurs in the porcine cultures from 20–24 h, somewhat later in rabbit cultures, from 32–36 h, and later still in human cultures. Tetraploids increase significantly after irradiation and start to appear unambiguously at 40 and 36 h in rabbits and swine respectively. In human cultures they were already present at 48 h.

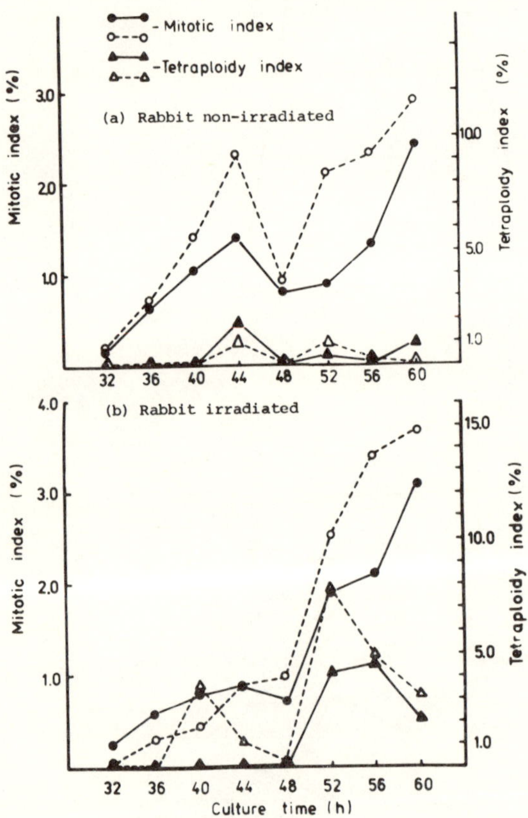

Figure 2. Mitotic and tetraploidy indices in cultures of rabbit lymphocytes as a function of culture duration: (a) non-irradiated; (b) irradiated prior to culture with 300 rad γ-rays. Open and closed symbols denote cultures of blood from two individual animals.

When dicentric yields and CuX_2 cells are considered, a common pattern is seen in all three species studied, in that the former start to decline from early high values when CuX_2 cells appear with a frequency above 1 or 2%. This happens at 36, 44 and 48–52 h in porcine, rabbit and human cultures respectively. It is suggested, therefore, that in rabbit and swine respectively 40 and 31 h should be the latest harvest times at which significant contamination with second mitosis cells would be avoided.

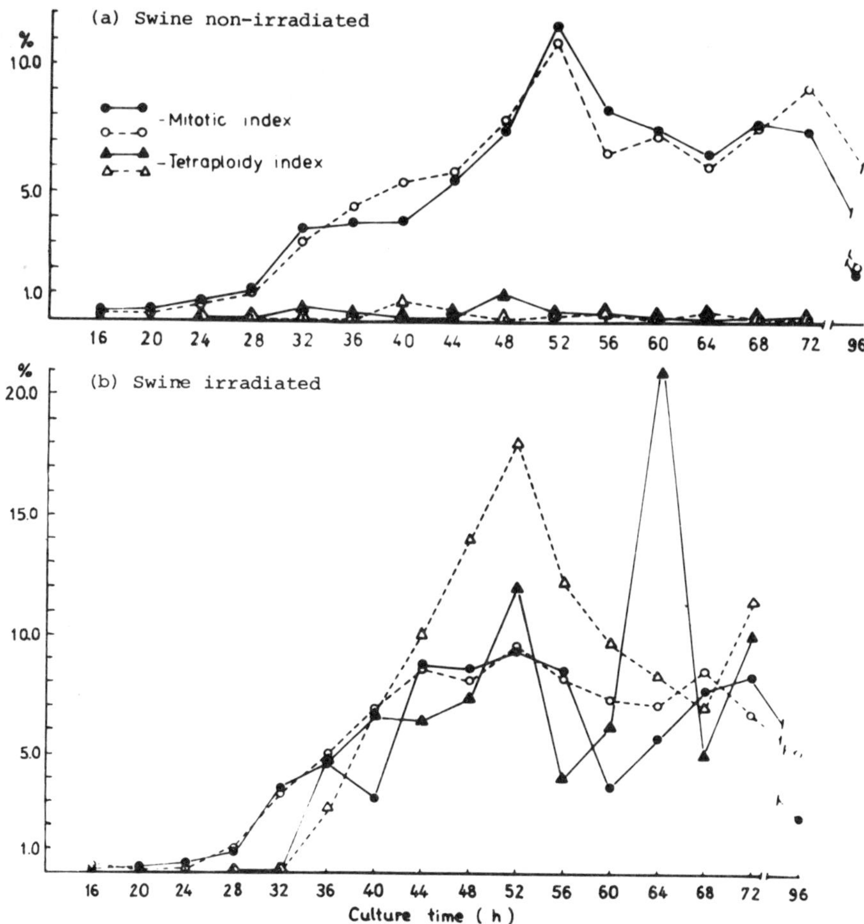

Figure 3. Mitotic and tetraploidy indices in cultures of porcine lymphocytes as a function of culture duration: (a) non-irradiated; (b) irradiated prior to culture with 300 rad γ-rays. Open and closed symbols denote cultures of blood from two individual animals.

Table 1. Yield of dicentrics in porcine and rabbit lymphocytes cultured for 31 and 40 h respectively. Number of cells scored in parentheses.

Dose (rad)	Rabbits 1	Rabbits 2	Pig
100	3.2 (280)	3.4 (500)	2.4 (210)
200	17.1 (140)	12.0 (500)	10.0 (100)
300	23.7 (160)	25.0 (300)	18.0 (400)
400	40.3 (114)	45.7 (300)	37.3 (300)

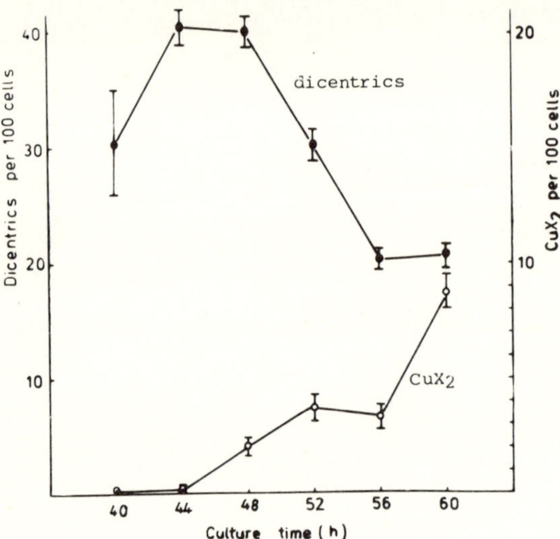

Figure 4. Dicentric yield and frequency of CuX_2 in cells in cultures of human lymphocytes as a function of culture time (one individual). 300 rad prior to culture initiation.

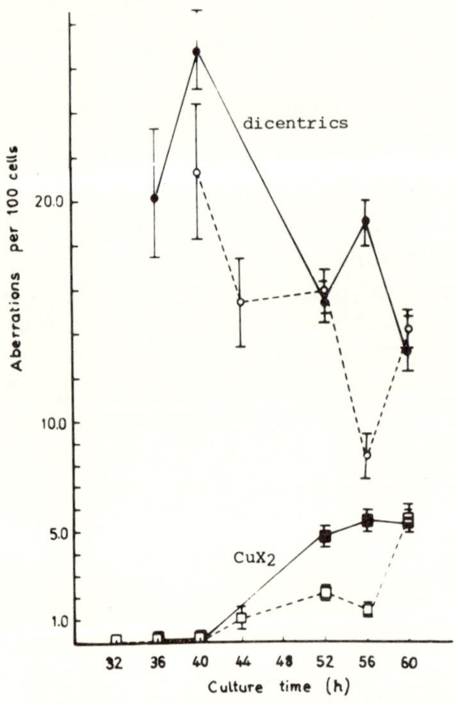

Figure 5. Dicentric yield and frequency of CuX_2 in cells in cultures of rabbit lymphocytes as a function of culture time. Open and closed symbols denote two individual animals. 300 rad prior to culture initiation.

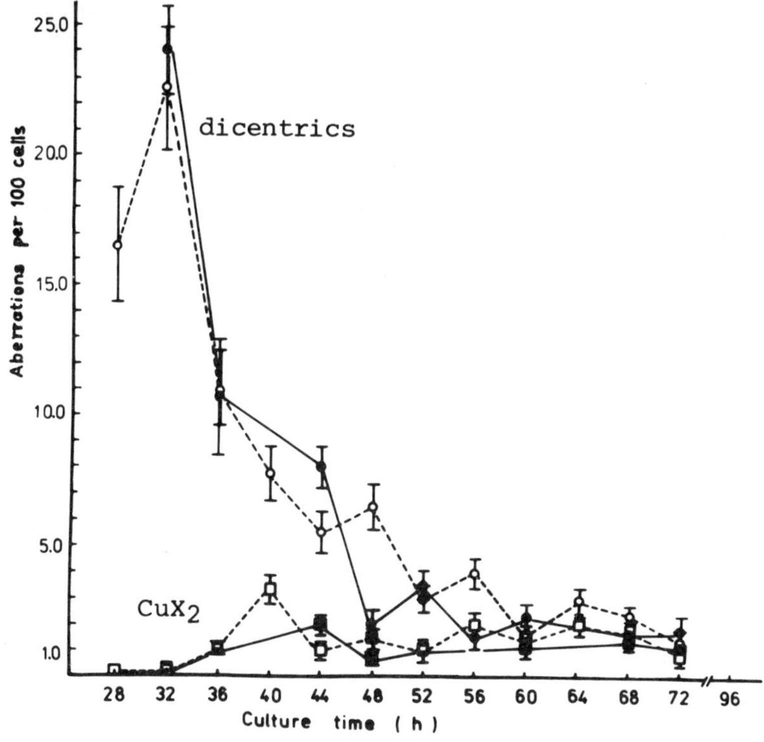

Figure 6. Dicentric yield and frequency of CuX_2 cells in cultures of porcine lymphocytes as a function of culture time. Open and closed symbols denote two individual animals. 300 rad prior to culture initiation.

Table 2. Yield of dicentrics in lymphocytes from two individuals cultured for varying periods prior to the harvest. Number of scored metaphases in parentheses.

Dose (rad)	Culture time (hours) 44	48	53
100	5.6 (500)	6.6 (320)	5.5 (500)
	8.0 (162)	8.2 (85)	—
200	20.1 (500)	12.2 (320)	10.8 (500)
	26.4 (110)	—	—
300	40.0 (300)	32.0 (300)	30.0 (500)
	41.2 (80)	46.5 (120)	—
400	70.7 (300)	61.5 (200)	55.8 (430)
	67.0 (106)	67.5 (83)	—

Sister chromatid differential staining of rabbit and porcine lymphocytes confirmed that at these times practically only first mitosis cells were present (97.5 and 100%); 12 h later 70 and 17% of the cells were demonstrated to be in their second mitosis.

Porcine and rabbit blood were irradiated with doses of 100, 200, 300 and 400 rad γ-rays. The cultures were terminated at 31 and 40 h, and the data are presented in table 1. Human cells were cultured similarly for 44, 48 and 53 h and the results are shown in table 2. There is very good agreement in the dicentric yields for human cells at 44 h, a slight tendency for the yields to decline at 48 h and significantly lower results when cultures were harvested only 5 h later (53 h).

Discussion

The results presented in this paper show that to obtain cells exclusively in their first mitosis the duration of culture for rabbit and porcine lymphocytes should be 38–40 and 28–31 h respectively. For the rabbit these culture times agree very well with that used by our group previously (41 h), when Parker TC-199 medium was employed (Bajerska and Liniecki 1975). Similarly the yields of dicentrics at comparable doses in both studies are practically the same. Clemenger and Scott (1973) have measured mitotic activity in individual cultures of blood taken from the rabbit; the harvest times varied from 40 to 48 h, being selected on the basis of time of appearance of mitoses. For swine, the culture duration postulated here is much shorter than that which has been used so far by McFee, Banner and Sherill (1972), McFee, Sherill and Banner (1973), McFee *et al.* (1972), and Leonard *et al.* (1976), where both groups have used a conventional time of 48 h. It is puzzling that they obtained comparable dicentric yields to that observed by us at 31 h despite the steep decline of the frequency of dicentrics with culture duration noted in this study. This discrepancy cannot be explained at present. However, to some extent this late harvest at 48 h might have been responsible for the rather unusual dose-effect relationship obtained by Leonard *et al.* (1976), of the form $Y = 0.057\,D^{2.74}$.

We have yet to study the uniformity of the response of rabbit and porcine cells within the earliest mitotic wave, i.e. when harvested at times less than 40 and 31 h respectively. A common feature of the observations presented above remains, however, that the observed yield of dicentrics is critically dependent upon culture time, declining sharply when cells in their second mitosis start to appear in significant numbers in the cultures. Therefore, if quantitative studies on the kinetics of aberration induction in peripheral blood lymphocytes are being undertaken, meticulous selection of an early culture harvest time must precede the proper study to avoid contamination of the scored metaphases with cells in second mitosis. The most unequivocal method for scoring aberrations only in first mitosis would appear to be the application of FPG sister chromatid staining. For routine purposes however, the experimental effort could appear prohibitive.

A rather enlightening lesson in this study seems to come from the most unexpected direction. We have been using a 51 h culture time over many years for human lymphocytes with the consistent result that our dicentric yields, which were otherwise reproducible, have been lower by 25–30% than those reported by others (summarised by Lloyd *et al.* 1975). When, in this study, 44 h cultures were applied, the dicentric frequency became higher and now agrees very well with that of most authors using similar culture and scoring procedures (Lloyd *et al.* 1975; Sasaki 1971; Brewen, Preston and Littlefield 1972). Of course the reproducibility of this finding remains to be extensively tested. However, taking the results at their face value it appears that under the experimental conditions employed in our laboratory the proper termination time of human cultures would be 44–46 h.

One can hardly believe that 3 or 7 h difference in culture time could make such a systematic shift in the results. If confirmed, this observation would point to the fact that biological dosimetry, based on the dicentric yield in lymphocytes, may yield comparable data in various laboratories provided the principle of meticulous selection of critical parameters is observed. It appears that to secure evaluation of truly first mitosis cells for the lymphocytes of all species studied, rigorous tests should be performed in each laboratory.

When dicentric yields after similar doses in rabbit, porcine and human lymphocytes, harvested respectively at 40, 31 and 44–48 h are compared, the data do not fit the predictions of the arm-number hypothesis (Brewen *et al.* 1973; table 3). These data therefore support the conclusions of Scott and Bigger (1974) that the induction of dicentrics in human and rabbit lymphocytes is not strictly proportional to the number of chromosome arms in the karyotypes of these species.

Table 3. Yield of dicentrics per 100 cells and per 100 chromosome arms in three species. Number of scored metaphases in brackets.

Dose (rad)	Dicentrics per 100 cells per 100 arms		
	Pig; 64 arms	Rabbit; 80 arms	Man; 81 arms
100	3.7 (210)	4.2 (780)	9.9 (162)
200	15.6 (100)	16.4 (640)	32.6 (110)
300	28.1 (400)	30.7 (460)	50.8 (80)
400	58.3 (300)	55.2 (414)	82.7 (106)

Summary

To assess the appropriate time for harvesting cultures of rabbit and swine lymphocytes, whole blood of these animals was irradiated with 300 rad γ-rays and microcultures were established using Ham's F-10 medium. Mitotic and tetraploidy indices, dicentrics per cell and the percentage of dicentric or ring-carrying cells, unaccompanied by acentrics, were determined as a function of culture duration. The same procedure was applied to human blood. The percentage of cells in first and second mitosis was determined in rabbit and swine lymphocyte cultures at selected times after stimulation using the FPG technique for differential staining of sister chromatids.

The first mitotic waves appear at 30 ± 1, 36 ± 2, and 45 ± 1 h after PHA stimulation for pig, rabbit and man respectively. Correspondingly, in the three species a significant percentage of cells in second mitosis is already present by 36, 44 and 48–52 h and is accompanied by a steep reduction in the dicentric yields.

These proposed culture times for rabbit and swine lymphocytes are shorter than those at which the majority of relevant studies reported in the literature have been performed.

Acknowledgements

The work has been supported by the International Atomic Energy Agency (Contract 1741/RB) by the World Health Organisation, Division of Environmental Health, and by the National Oncological Programme (PR-6/1316).

References

Bajerska, A. & J.Liniecki (1967) The influence of X-ray dose and time of its delivery *in vitro* on the yield of chromosomal aberrations in the peripheral blood lymphocytes. *Int. J. Radiat. Biol. 16*, 467–81.

——(1975) The yield of chromosomal aberrations in rabbit lymphocytes after irradiation *in vitro* and *in vivo*. *Mutat. Res. 27*, 271–84.

Brewen, J.G. & N.Gengozian (1971) Radiation-induced human chromosome aberrations. II. Human *in vitro* irradiation compared to *in vitro* and *in vivo* irradiation of marmoset leukocytes. *Mutat. Res. 13*, 383–91.

Brewen, J.G., R.J.Preston & L.G.Littlefield (1972) Radiation-induced human chromosome aberration yields following an accidental whole-body exposure to Co^{60} gamma-rays. *Radiat. Res. 49*, 647–56.

Brewen, J.G., R.J.Preston, K.P.Jones & D.G.Gosslee (1973) Genetic hazards of ionizing radiations: cytogenetic extrapolations from mouse to man. *Mutat. Res. 17*, 245–54.

Buckton, K.E. & M.C.Pike (1964) Time in culture. An important variable in studying *in vivo* radiation-induced chromosome damage in man. *Int. J. Radiat. Biol. 8*, 439–52.

Clemenger, J.F.P. & D.Scott (1973) A comparison of chromosome aberration yields in rabbit blood lymphocytes irradiated *in vitro* and *in vivo*. *Int. J. Radiat. Biol. 24*, 487–96.

Heddle, J.A., H.J.Evans & D.Scott (1967) Sampling time and the complexity of the human leukocyte culture system, in *Human Radiation Cytogenetics* (eds H.J. Evans, W.H.Court Brown & A.S.McLean) pp.6–19. Amsterdam: North-Holland.

Leonard, A., G.B.Gerber, D.G.Papworth, G.Decat, E.D.Leonard & Gh. Deknudt (1976) The radiosensitivities of lymphocytes from pig, sheep, goat and cow. *Mutat. Res., 36*, 319–32.

Liniecki, J., A.Bajerska, K.Wyszyńska & B.Cisowska (1977) Gamma-radiation-induced chromosomal aberrations in human lymphocytes: dose-rate effects in stimulated and non-stimulated cells. *Mutat. Res. 43*, 291–304.

Lloyd, D.C., R.J.Purrott, G.W.Dolphin, D.Bolton, A.A.Edwards & M.J.Corp (1975) The relationship between chromosome aberrations and low LET radiation dose to human lymphocytes. *Int. J. Radiat. Biol. 28*, 75–90.

McFee, A.F., M.W.Banner & M.N.Sherrill (1972) Induction of chromosome aberrations by *in vivo* and *in vitro* gamma-irradiation of swine leukocytes. *Int. J. Radiat. Biol. 21*, 513–20.

McFee, A.F., M.W.Banner, M.N.Sherrill & J.B.Mailhes (1972) Disappearance rates of radiation-induced chromosome aberrations from swine leukocytes. *Mutat. Res. 15*, 325–30.

McFee, A.F., M.N.Sherrill & M.W.Banner (1973) Radiation-induced aberrations in aneuploid vs. diploid swine leukocytes. *Mutat. Res. 18*, 311–14.

Perry, P. & S.Wolff (1974) New Giemsa method for the differential staining of sister chromatids. *Nature 251*, 156–8.

Preston, R.J., J.G.Brewen & K.P.Jones (1972) Radiation-induced chromosome aberrations in Chinese-hamster leukocytes, a comparison of *in vivo* and *in vitro* exposure. *Int. J. Radiat. Biol. 21*, 397–400.

Sasaki, M.S. (1971) Radiation-induced chromosome aberrations in lymphocytes: possible biological dosimeter in man, in *Biological Aspects of Radiation Protection* (eds T.Sugahara & O.Hug) pp.81–91. Berlin: Springer.

Scott, D. & T.R.L.Bigger (1974) The relative radiosensitivities of human, rabbit and rat-kangaroo chromosomes. *Chromosoma Berl. 49*, 185–203.

D.G.HARNDEN and A.M.R.TAYLOR

The Effects of Radiation on the Chromosomes of Patients Susceptible to Cancer

Radiosensitivity at the Clinical Level

At least three syndromes have been identified in which there is clinical evidence of an unusual response to ionising radiation. These are ataxia telangiectasia, basal cell naevus syndrome and retinoblastoma. All three are associated with an increased incidence of malignant disease, and it would be of great interest to know whether or not the observed radiosensitivity is directly linked to the primary susceptibility to cancer as well as to the secondary radiation-induced susceptibility observed in two of these syndromes (retinoblastoma and basal cell naevus syndrome). Further, if variation in response to ionising radiation can be under genetic control (since all three of these conditions are clearly inherited) there will certainly be a variation in radiosensitivity in the general population, and this may be of both practical significance and of theoretical importance in considering particularly the effects of low doses of radiation. There is some evidence of such variation, mostly anecdotal, but some attempts have been made to put this on a quantitative basis (Glicksman *et al.* 1960; Weichselbaum, Epstein and Little 1976).

Ataxia telangiectasia (AT) is an autosomal recessive disorder characterised by progressive cerebellar ataxia, oculocutaneous telangiectases and a susceptibility to develop malignant neoplasms (table 1) particularly of the lymphoid system (Harnden 1974). There are three reports of unusual radiosensitivity. Gotoff, Amirmokri and Liebner (1967) report a ten-year-old boy with AT and a malignant lymphoma who, after receiving a maximum dose of 3000 rad to the nasopharynx, out of a planned tumour dose of 4000 rad, was noted to show marked cutaneous erythema and clinical signs indicated deep tissue damage. Following his death eight months later, autopsy revealed deep tissue radiation necrosis and it is concluded that the radiation was directly responsible for his death. Morgan, Holcomb and Morrissey (1968) report a nine-year-old boy with AT and Hodgkin's disease, who again received a partial dose of 2843 rad out of a planned dose of 4000 rad to the mediastinum. The patient developed severe oesophagitis, skin damage and respiratory embarrassment and died four months later. Lastly, Cunliffe *et al.* (1975) report a

Table 1. Malignancies in patients with ataxia telangiectasia.

Type of neoplasm	No. of cases
Reticulum cell sarcoma	8
Undifferentiated round cell sarcoma	1
Lymphosarcoma	13
Hodgkins	7
Generalised reticuloendotheliosis	1
M Gammopathy	1
Lymphoma	12
Histiocytosarcoma	1
Leukaemias	
lymphoblastic	8
stem cell	1
myelogenous	1
unidentified	3
Carcinoma skin	2
Carcinoma stomach	2
Ovarian dysgerminoma	2
Ovarian dysgerminoma and gonadoblastoma	1
Mucoepidermoid carcinoma	1
Glioma	2
Medulloblastoma	1
Hepatoma	1
Colloid carcinoma pylorus	1
Unspecified	4
	74

These cases include all those reported by Kersey, Spector & Good (1973) and in addition individual reports by Al Saadi and Palutke (1976), Al Saadi *et al.* (1975), Bochkov *et al.* (1974), Cawley and Schenken (1970), Chuler *et al.* (1971), Cohen *et al.* (1975), Cunliffe *et al.* (1975), Datau *et al.* (1975), Goldsmith and Hart (1975), Harris and Seeler (1973), McCaw *et al.* (1975), Olive *et al.* (1975), Ozden *et al.* (1974), Rosenthal *et al.* (1965). Van Hemel (1976), Warot *et al.* (1963). There is one possible case of AT with carcinoma of the stomach reported also by Siegel *et al.* (1976).

seven-year-old boy with AT and a malignant lymphoma in the upper lobe of the right lung. After 2000 rad dysphagia and erythema were noted and, at 3000 rad, the treatment was stopped because of worsening symptoms. He slowly deteriorated, the skin sloughing off the irradiated areas of his chest, and died three weeks later. Again death appears directly attributable to the radiation treatment.

Basal cell naevus syndrome (BCNS) is an autosomal dominant condition characterised by multiple basal cell naevi, which frequently develop into carcinomas, and a variety of minor malformations such as odontogenic keratocysts, bifid ribs, pitting of the palms of the hands and the soles of the feet (Gorlin and Goltz 1960). There are several reports in the literature of unusually severe responses to radiation therapy in BCNS (e.g. Berendes 1971) and two reports of the production of multiple basal cell carcinomas within the irradiation field; one on a five-year-old boy with an enlarged thymus who was irradiated several years previously (Scharnagel and Pack 1949), and one on a patient with ankylosing spondylitis treated by irradiation to the spinal axis (Marsden, pers. comm.) The diagnosis of BCNS in the first patient is not clear cut, since this paper predates the description of the syndrome. There is also evidence to suggest that the latent period for the induction of basal cell carcinomas by ionising radiation is reduced in these patients (Strong 1977).

Retinoblastoma cases may be sporadic or occur in familial associations. In the latter cases the susceptibility appears to be inherited in an autosomal dominant manner (Smith and Sorsby 1958). There are no other clinical features regularly associated with the occurrence of retinoblastoma, but it has recently been shown that there is an increased incidence of osteosarcoma in patients who also have retinoblastoma (Kitchin and Ellsworth 1974, Miller 1977). This appears to be a real association, since some of these cases clearly do occur outside the irradiation field. However, most do occur in patients whose primary tumour has been irradiated (Sagerman *et al.* 1969) and this could suggest that the irradiation enhances a predisposition to develop osteosarcoma in these patients.

Knudson (1971, 1977) has suggested that in retinoblastoma (and other childhood neoplasms) two mutational steps are necessary for the development of neoplasm. In familial cases one of these mutational events is already in the genome and hence only a single mutation event is required post-zygotically. This hypothesis would be consistent with an unusual sensitivity to tumour induction by ionising radiation, and it is not incompatible with other hypotheses that stress the importance of regulatory events and clonal evolution.

Radiosensitivity at the Cellular Level

We have previously reported a reduced survival of fibroblasts cultured from patients with AT following exposure to ionising radiation when compared to fibroblasts derived from normal subjects (Taylor *et al.* 1975). This result has since been confirmed on further cases in this and other laboratories (Weichselbaum, Nove and Little 1977; R. Cox, pers. comm.; M. C. Paterson, pers. comm.). All cases tested where there is a firm diagnosis of AT appear to be sensitive to cell killing by ionising radiation. In the same report we show, however, that cells from three patients with basal cell naevus syndrome fall within the normal range in their sensitivity to cell killing after exposure to γ-rays. Recently there has been a report (Weichselbaum, Nove and Little, 1977) of an unusual radiosensitivity of cells from a single patient with retinoblastoma and an interstitial deletion of a chromosome 13. However, other patients who had retinoblastoma but did not have the D-deletion syndrome did not differ from controls in their radiation response. The single sensitive patient shows a response to ionising radiation intermediate between controls and a patient with ataxia telangiectasia. The interstitial deletion of the long arm of chromosome 13, including band q14, has been reported in a total of thirteen cases of retinoblastoma (Howard *et al.* 1974, Franke 1976), but it seems likely that the majority of retinoblastoma patients have normal chromosomes (Atkin 1976). If this difference, between these cases with the 13-deletion and those without, can be confirmed it could prove most valuable in studying the mechanisms both of radiosensitivity and of cancer induction. In figure 1 it can be seen that there is good concordance for the sensitivity of the

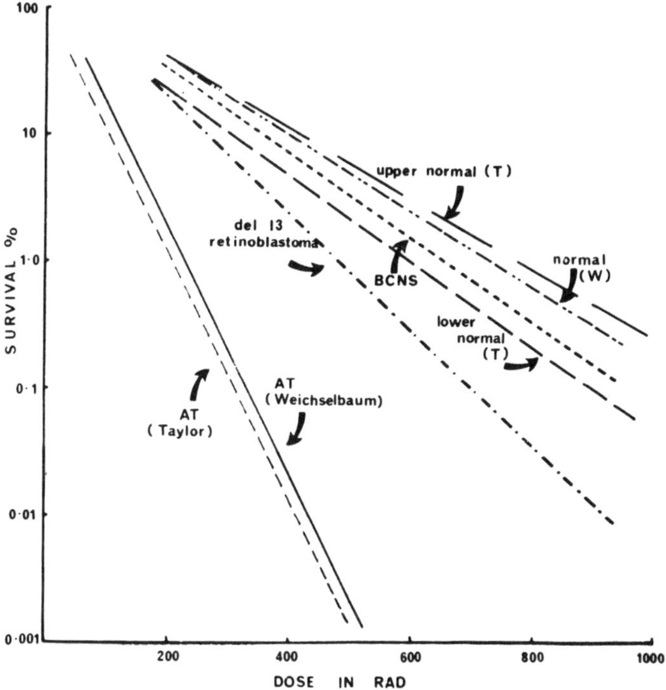

Figure 1. Sensitivity of cultured fibroblasts to ionising radiation. Composite figure showing relationship between the results of Taylor *et al.* (1975) on ataxia telangiectasia (AT) and those of Weichselbaum, Nove & Little (1977) on a case of 13-deletion retinoblastoma. Normal (W) is the control value given by Weichselbaum; upper normal (T) and lower normal (N) are the upper and lower limits of control values given by Taylor. BCNS is basal cell naevus syndrome.

AT cells and that Weichselbaum's controls fall within the control range given by Taylor *et al.* (1975), while the deleted-13 retinoblastoma case falls clearly outside the range of normal. However, in their earlier paper Weichselbaum, Epstein and Little (1976) found no consistent correlation between clinical radiosensitivity and cell survival. In that report, however, the range is much wider and two of their clinically sensitive patients have survival curves very close to that of the deleted-13 retinoblastoma.

Radiosensitivity at the Chromosome Level

So far we have evidence of radiosensitivity at the chromosome level only in ataxia telangiectasia. It is well established that in this disease there is a high spontaneous incidence of chromosome damage in cells cultured from the peripheral blood (Hecht *et al.* 1966) and that in these patients clones of cytogenetically marked lymphocytes are found (Hecht, McCaw and Koler 1973; Oxford *et al.* 1975; McCaw *et al.* 1975). These clones increase in size in the peripheral blood over a period of several years (Hecht, McCaw and Koler 1973; Harnden 1977). The origin of these 'spontaneous' chromosome abnormalities is not known. It is now clear that the chromosomes of patients with AT are more susceptible to damage by ionising radiation than are the chromosomes of normal subjects (Rary, Bender and Kelly 1974; Taylor *et al.* 1975). In the studies in this laboratory the lymphocytes of eight AT patients were studied together with eight controls following exposure to varying doses of X-rays. These results are reported in detail by Taylor (1977). X-irradiation was carried out both at G_0 and at the G_2 phase of the cell cycle. Following G_0 irradiation at 100, 200 and 400 rad, the main findings were that the number of rings and dicentrics induced was approximately the same in AT patients and controls. Fragment incidence, however, was doubled, while gaps and breaks were induced ten times more frequently and chromatid interchanges twenty times more frequently in the AT patients. These findings held good at all dose levels. It is important to note that it is not simply an overall increase in all types of aberrations in the AT patients but a clearly disproportionate increase in fragments and chromatid aberrations. Further, following G_2 irradiation there is a dramatic increase in chromatid aberrations in the AT patients over and above those induced in the controls.

Taylor *et al.* (1975) postulated that the pattern of aberrations observed in AT suggested a defect of DNA repair, but observations on the repair of single-strand and double-strand breaks using sucrose gradients failed to show any defect. However, Paterson *et al.* (1976) have shown that the cells from some cases of AT have a low capacity to excise γ-ray-induced base damage. Paterson (1977) has shown further that of those cases that do show this defect there may be complementation between different cases, suggesting that, as in xeroderma pigmentosum, there may be a number of different genetic complementation groups. Of four of our patients studied by Paterson only one case had a low repair capacity, the others responding as did the normal controls. In spite of Paterson's results, and the results on strand rejoining reported in our earlier paper (Taylor *et al.* 1975), there are a number of reasons for believing that the AT patients all have a defect in their ability to rejoin both single- and double-strand breaks. Taylor (1977) has suggested that the excess of fragments could be due to incomplete exchanges and terminal deletions that would result from the incomplete rejoining of a small fraction of the induced double-strand breaks. Likewise it is suggested that the excess of chromatid damage seen at mitosis following G_0 irradiation is derived from subchromatid

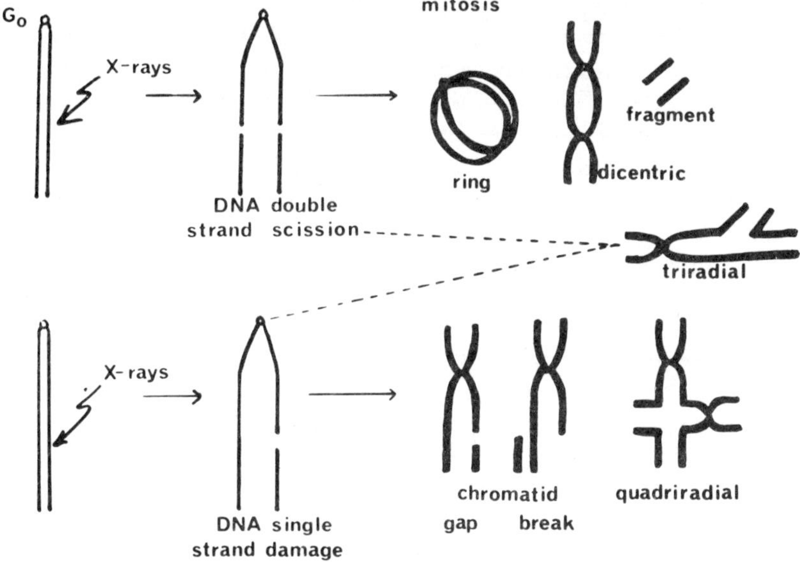

Figure 2. Possible aberrations following X-irradiation of the G_0 chromosome. Failure to repair double-strand DNA scissions could lead to an excess of fragments, while failure to repair single-strand DNA scissions could lead to chromatid gaps, chromatid breaks, and chromatid interchanges of both triradial and quadriradial types.

damage in the form of unrepaired single-strand breaks (figure 2). If these suppositions are correct, then the observation of chromosome damage in irradiated cells may be a more sensitive indicator of low numbers of unrepaired strand breaks than are the sucrose gradient techniques.

Relationship Between Induced and Spontaneous Chromosome Damage

It seems possible that the spontaneous chromosome damage observed in AT patients is a consequence of the exposure of these patients to background radiation or to radiomimetic substances, but it could also be due to abnormalities arising as a consequence of errors caused during cellular processes not induced by external chromosome-damaging agents. For example, errors could be introduced by the actions of abnormal enzymes rather than by the direct damaging action of the radiation on the DNA. There have been conflicting reports about the distribution of X-ray-induced lesions in human chromosomes, but most workers agree that, with the possible exception of some excess near the centromere, break-points are distributed randomly among and within the chromosomes.

Buckton (this volume) has, however, shown that some time after the initial damaging event the distribution of break-points is no longer random. Obviously unstable rearrangements will tend to be eliminated, but the preferential involvement of certain chromosome regions in the stable rearrangements that persist seems to imply that certain of the apparently stable rearrangements are also eliminated, presumably because they are incompatible with prolonged life of the cell. Another possible interpretation is that amongst the stable rearrangements there are some that confer on the cells some sort of proliferative advantage, so that these cells would be observed more frequently.

It has now been recognised for some time in patients with AT that the lymphocyte clones have specific chromosome rearrangements (McCaw *et al.* 1975). The clonal karyotype in these cases always appears to involve the chromosome band 14q12, with the segment of the long arm of chromosome 14 distal to that point being translocated to one of a small number of other sites, namely the end of the long arm of the other number 14, the end of the long arm of 7 or X, or the end of the short arm of 7. This situation bears certain resemblances to the translocation of the long arms of chromosome 22 to other sites, resulting in the Philadelphia chromosome in chronic myeloid leukaemia. The lymphoid clones in AT may be selected out because these particular rearrangements confer on lymphocytes some proliferative advantage. The finding of translocations between chromosomes 7 and 14 on rare but repeated occasions in the lymphocytes of normal individuals could support this notion (Welch *et al.* 1975).

The question must then arise whether these lymphoid clones are in any way associated with the neoplasms of the lymphoid system to which these patients are predisposed. In only one case has a cytogenetic study been carried out on an AT patient with a neoplasm (McCaw *et al.* 1975). In this case the clonal marker was found, together with further chromosome abnormalities in the lymphoid cells, after the patient had developed a leukaemia described as chronic lymphocytic. This at first seems anomalous, because chronic lymphocytic leukaemias are usually due to B-cell proliferation and the clones of lymphoid cells in AT patients are clearly T cells. However, in this case T-cell markers were found on the leukaemic cells (McCaw, pers. comm.) so that this appears to be one of the rare T-cell cases of chronic lymphocytic leukaemia. On the other hand, such clones of lymphoid cells have been observed in many AT patients over a prolonged period without any evidence of malignant disease (Hecht, McCaw and Koler 1973; Harnden 1977). Similarly, lymphoid clones have been recognised in patients who have received prolonged radiation from deposited thorotrast (Court Brown *et al.* 1967) and in patients with ankylosing spondylitis who have received therapeutic X-irradiation (Buckton, this volume), and these have so far not been shown directly to be involved in the lymphoid neoplasms or leukaemias to which these patients are prone. Further, in the case of cytogenetically marked murine lymphoid cells arising in radiation chimaeras, Ford has repeatedly passed such a clone through many generations of mice without a malignant transformation occurring.

Thus, while it is tempting to speculate that the radiosensitivity to chromosome damage in AT patients is in some way linked to the observed spontaneous chromosome damage and clone formation, and that these clones are in some way linked to the predisposition to lymphoid neoplasms, the evidence to make these links firm is as yet lacking.

Conclusion

The study of these three diseases, ataxia telangiectasia, basal cell naevus syndrome and retinoblastoma, has yielded very interesting information at the clinical level. In one, AT, there is some understanding of the possible mechanisms of susceptibility to malignant disease at the cellular, chromosomal and molecular level, but so far this is very incomplete. There is some suggestion that in retinoblastoma similar information may be beginning to emerge, but in the case of basal cell naevus syndrome there is not even a beginning of an understanding of the mechanisms.

Acknowledgement

We would like to thank the Cancer Research Campaign for their continued support.

References

Al Saadi, R. & M.Palutke (1976) Evolution of chromosomal aberrations in ataxia telangiectasia in *Proc. V Int. Congr. Hum. Genet.* Excerpta Medica.

Al Saadi, A., M.Palutke & K.Kumar (1975) Cytogenetic and immunologic studies in ataxia telangiectasia. *Am. J. Hum. Gen. 27*, 78A.

Atkin, N.B. (1976) Cytogenetic aspects of malignant transformation, in *Experimental Biology and Medicine Monographs on Interdisciplinary Topics 6*, pp.23–7. Basel: Karger.

Berendes, U. (1971) Die Klinische Bedeutung der onkotischen Phase des Basalzell-naevus-Syndromes. *Hautarzt 22*, 261–3.

Bochkov, N.P., Y.M.Lopukhin, N.P.Kuleshov & L.V.Kovalchuk (1974) Cytogenetic study of patients with ataxia-telangiectasia. *Humangenet. 24*, 115–28.

Cawley, L.P. & J.R.Schenken (1970) Monoclonal hypergamma-globulinaemia of the γM type in a nine-year-old girl with ataxia telangiectasis. *Am. J. Clin. Pathol. 54*, 790–801.

Chuler, D., L.Schongut & E.Cserhati (1971) Lymphoblastic transformation in ataxia telangiectasia. *Acta Paediatr. Scand. 60*, 66–72.

Cohen, M.M., M.Shaham, J.Dagan, E.Schmueli & G.Kohn (1975) Cytogenetic investigation in families with ataxia telangiectasia. *Cytogenet. Cell. Genet. 15*, 338–56.

Court Brown, W.M., K.E.Buckton, A.O.Langlands & G.E.Woodcock (1967) The identification of lymphocyte clones with chromosome structural aberrations in irradiated men and women. *Int. J. Radiat. Biol. 13*, 155–68.

Cunliffe, P.N., J.R.Mann, A.H.Cameron, K.D.Roberts & H.W.C.Ward (1975) Radiosensitivity in ataxia telangiectasia. *Br. J. Radiol. 48*, 374–6.

Dutau, G., J.Corberand, M.Abbal, M.Blanc, P.Claverie & P.Rochiccioli (1975) Ataxie télangiectasie avec déficit immunitaire mixte et formes apparentées. *J. Genet. Hum. 23*, 281–99.

Franke, U. (1976) Retinoblastoma and chromosome 13. *Cytogenet. Cell Genet. 16*, 131–4.

Glicksman, A.S., F.C.H.Chu, H.N.Bane & J.J.Nickson (1960) Quantitative and qualitative evaluation of skin erythema. II. Clinical study of patients on a standardised irradiation schedule. *Radiol. 75*, 411–15.

Goldsmith, C.I. & W.R.Hart (1975) Ataxia telangiectasia with ovarian gonadoblastoma and contralateral dysgerminoma. *Cancer 36*, 1838–42.

Gorlin, R.J. & R.W.Goltz (1960) Multiple nevoid basal-cell epithelioma, jaw cysts and bifid rib: a syndrome. *New Engl. J. Med. 262*, 908–12.

Gotoff, S.P., E.Amirmokri & E.J.Liebner (1967) Ataxia telangiectasia: neoplasia, untoward response to X-irradiation and tuberous sclerosis. *Am. J. Dis. Child. 114*, 617–25.

Harnden, D.G. (1974) Ataxia telangiectasia syndrome: cytogenetic and cancer aspects, in *Chromosomes and Cancer* (ed. J.German) pp.619–36. New York: Wiley.

—— (1977) The relationship between induced chromosome aberrations and chromosome abnormality in tumour cells, in *Proceedings V Int. Congr. Hum. Genet.* Excerpta Medica.

Harris, V.J. & R.A.Seeler (1973) Ataxia telangiectasia and Hodgkin's disease. *Cancer 32*, 1415–20.

Hecht, F., B.K.McCaw & R.D.Koler (1973) Ataxia telangiectasia–clonal growth of translocation lymphocytes. *New Engl. J. Med. 289*, 286–91.

Hecht, F., R.D.Koler, D.A.Rigas, G.S.Dahnke, M.P.Case, V.Tisdale & R.W.Miller (1966) Leukaemia and lymphocytes in ataxia telangiectasia. *Lancet 2*, 1193.

Howard, R.O., W.R.Breg, D.M.Albert & R.L.Lesser (1974) Retinoblastoma and chromosome abnormality. Partial deletion of the long arm chromosome 13. *Arch. Ophthalmol. 92*, 490–3.

Kersey, J.J., B.D.Spector & R.A.Good (1973) Primary immunodeficiency diseases and cancer: the immunodeficiency-cancer registry. *Int. J. Cancer 12*, 333–47.

Kitchin, F.D. & R.M.Ellsworth (1974) Pleiotropic effects of the gene for retinoblastoma. *J. Med. Genet. 11*, 244–6.

Knudson, A.G. (1971) Mutation and Cancer: statistical study of retinoblastoma. *Proc. Nat. Acad. Sci. USA 68*, 820–3.

—— (1977) Genetic and environmental interactions in the origin of human cancer, in *Genetics of Human Cancer* (eds J.J.Mulvihill, R.W.Miller and J.F.Fraumeni). New York: Raven Press.

McCaw, B.K., F.Hecht, D.G.Harnden & R.L.Teplitz (1975) Somatic rearrangement of chromosome 14 in human lymphocytes. *Proc. Nat. Acad. Sci. USA 72*, 2071–5.

Miller, R.W. (1977) Relationship between human teratogens and carcinogens. *J. Natl. Cancer Inst. 58*, 471–4.

Morgan, J.L., T.M.Holcomb & R.W.Morrissey (1968) Radiation reaction in ataxia telangiectasia. *Am. J. Dis. Child. 116*, 557–8.

Olive, D., A.M.Gehin & A.Boilletot (1975) Les complications néoplastiques de l'ataxie télangiectasie. *Arch. Fr. Pédiatr. 32*, 292.

Oxford, J.M., D.G.Harnden, J.M.Parrington & J.D.A.Delhanty (1975) Specific chromosome aberrations in ataxia telangiectasia. *J. Med. Genet. 12*, 251–62.

Ozden, K., K.Yalaz, K.Taysi & B.Say (1974) Immunological studies in ataxia telangiectasia. *Clin. Genet. 5*, 40–5.

Paterson, M.C., B.P.Smith, P.H.M.Lohman, A.K.Anderson & L.Fishman (1976) Defective excision repair of γ-ray-damaged DNA in human (ataxia telangiectasia) fibroblasts. *Nature 260*, 444–7.

Paterson, M.C. (1977) Environmentally induced DNA damage, its faulty repair and malignant genetic diseases, in *Dahlem Workshop on Neoplastic Transformation: Mechanisms and Consequences*. Berlin: Dahlem Konferenzen.

Rary, J.M., M.A.Bender & T.E.Kelly (1974) Cytogenetic studies of ataxia telangiectasia. *Am. J. Hum. Genet. 26*, 70A.

Rosenthal, I.R., A.D.Makowitz & R.Medenis (1965) Immunologic incompetence in ataxia telangiectasia. *Am. J. Dis. Child. 110*, 69–75.

Sagerman, R.H., J.R.Cassady, P.Tretter & R.M.Ellsworth (1969) Radiation induced neoplasia following external beam therapy for children with retinoblastoma. *Am. J. Roentgenol. 105*, 529–35.

Scharnagel, M. & G.T.Pack (1949) Multiple basal cell epitheliomas in a five-year-old child. *Am. J. Dis. Child. 77*, 647–51.

Siegel, S.E., D.M.Hays, S.Ramansky & H.Isaacs (1976) Carcinoma of the stomach in childhood. *Cancer 38*, 1781–4.

Smith, S.M. and A.Sorsby (1958) Retinoblastoma; some genetic aspects. *Ann. Hum. Genet. 23*, 50–8.

Strong, L.C. (1977) Theories of pathogenesis: mutation and cancer, in *The Genetics of Human Cancer* (eds J.J.Mulvihill, R.W.Miller & J.F.Fraumeni). New York: Raven Press.

Taylor, A.M.R., J.A.Metcalfe, J.M.Oxford & D.G.Harnden (1976) Is chromatid type damage in ataxia telangiectasia after irradiation at G_0 a consequence of defective repair? *Nature 260*, 441–3.

Taylor, A.M.R., D.G.Harnden, C.F.Arlett, S.A.Harcourt, A.R.Lehmann, S.Stevens & B.A.Bridges (1975) Ataxia telangiectasia: a human mutation with abnormal radiation sensitivity. *Nature 258*, 427–9.

Taylor, A.M.R. (1977) Unrepaired single and double strand breaks shown cytogenetically following X-irradiation of lymphocytes from patients with ataxia telangiectasia. *Mutat. Res.* (in press).

Van Hemel, J.O. (1976) Chromosomes in patients with ataxia telangiectasia, in *Proc. V Int. Congr. Hum. Genet.* Excerpta Medica.

Warot, P., C.Dupuis, R.Walbaum & L.Boniface (1963) Le syndrome d'ataxia télangiectasies de Louis-Bar. A propos de deux observations dans une fratrie. *Pédiatr. 18*, 340–5.

Weichselbaum, R.R., J.Epstein & J.B.Little (1976) *In vitro* radiosensitivity of human diploid fibroblasts derived from patients with unusual clinical responses to radiation. *Radiology 121*, 479–82.

Weichselbaum, R.R., J.Nove & J.B.Little (1977) Skin fibroblasts from a D-deletion type retinoblastoma patient are abnormally X-ray sensitive. *Nature 266*, 726–7.

Welch, J.P., C.L.Y.Lee, J.W.Beatty-De Sana, M.J.Hoggard, J.W.Cooledge, F.Hecht, B.K.McCaw, D.Peakman & A.Robinson (1975) Non-random occurrence of 7–14 translocations in human lymphocyte cultures. *Nature 255*, 241–4.

Radiation Damage and its Repair in the Formation of Chromosome Aberrations in Human Lymphocytes

According to the most widely accepted theory of chromosome aberration formation, radiation breaks chromosomes, and these 'breaks' either reconstitute the original configuration (restitutional repair), unite with illegitimate ends to form exchange aberrations (illegitimate repair) or remain open to give rise to terminal deletions. Although many plausible suggestions for the chromosome aberration formation have been proposed, based on the modern concept of radiation damage to chromosomal DNA (Evans 1966; Bender, Griggs and Bedford, 1974; Leenhouts and Chadwick 1974; Comings 1974), the nature of basic damage and the primary processes involved in the formation of chromosome aberrations are by no means understood.

In the present work an attempt has been made to analyse the effects of the quality of radiation, free radical scavengers, and fractionated dose delivery on radiation-induced chromosome aberrations in human lymphocytes, with attention directed to the possible nature of the basic damage and its relevance to the primary process in the formation of chromosome aberrations.

Materials and Methods

Throughout the experiment, heparinised venous blood obtained from healthy adult persons aged 22–39 was used. Detailed descriptions of methods of irradiation and establishment of cultures adopted in each experiment is given with the results. Unless specified, the lymphocyte cultures were established by the use of culture medium NCTC-109 supplemented with 20% fetal calf serum and 3% rehydrated phytohemagglutinin (PHA). Chromosomes were analysed in cells harvested 50 h after the commencement of PHA culture, with a 20 h prefixation treatment with colcemid. The chromosome preparations were scored for the types and frequencies of chromosome aberrations. Each dicentric and centric ring was assigned one acentric fragment. The acentric fragments not associated with dicentrics and centric rings were scored as terminal deletions. Abnormal monocentrics and minute 'dot' deletions were also scored, but they were not included in the present analysis.

Results

Physical Property of Basic Damage

Table 1 shows the dose-response relations of chromosome aberrations in lymphocytes irradiated, as whole blood, with radiations of different quality. The parameters giving the best fit to the dose exponential model, $Y=aD^n$, and the linear-quadratic model, $Y=\alpha D+\beta D^2$, are presented. As seen in the table, the frequency of dicentrics and rings was profoundly influenced by the radiation quality. The decrease in the dose exponent, n, with the increase in the linear energy transfer (LET) is consistent with the idea that an exchange aberration arises as a consequence of the interaction of two lesions, which are produced within a limited distance either by one track or two independent tracks, so that the yield is a sum of the one- and two-track events (Lea and Catcheside 1942). With the increase in LET, one-track events increase while two-track events decrease.

Based on the microdosimetric concept of energy-loss events in the production of primary damage leading to exchange aberrations, Neary (1965) developed a general model to account for the observed dependence of aberration yield on dose and LET. Analysis of dose-yield relations according to this model will offer an opportunity to assess a possible event size of primary damage responsible for the exchange aberration formation. According to his model, the yields of one-track events, Y_1, and two-track events, Y_2, are given as follows:

$$Y_1=NEg(pAtD)(pst)[(1-e^{-pst})/pst]^2$$
$$Y_2=NE(pAtD)^2[(1-e^{-pst})/pst]^2$$

The factors incorporated in this model are: t, diameter of interphase chromosomes in μm; h, radius of site in μm; A, mean projected area of chromosome segment within a site, $A\simeq ht$; L, LET in keV μm^{-1}; s, mean number of energy-loss events per μm, $s=L/\varepsilon$, where ε is mean size of event in eV; D, radiation dose in rad; p, probability that a single energy-loss event in a target will produce a primary damage; g, probability that a track that traversed one target chromosome will also traverse the other, $g\simeq t/\pi h$; E, probability that two lesions in a site will undergo exchange to form an exchange aberration; N, limitation in number of exchange aberrations per cell (Neary 1965).

The ratio of Y_1 to Y_2 is thus given by $Y_1/Y_2=s/\pi h^2D=L/\varepsilon\pi h^2D$. Similarly, from the general linear-quadratic dose-response relation, the Y_1/Y_2 ratio is given by $Y_1/Y_2=\alpha D/\beta D^2=\lambda/D$, where $\lambda=\alpha/\beta$ is the dose at which the one-track and two-track components are equal. Thus, the energy expended per primary damage is given by $\varepsilon=L/\lambda\pi h^2$. According to Kellerer and Rossi (1972), the interaction distance (site diameter, $d=2h$) is expressed as a function of λ and LET as $d=4.8(L/\lambda)^{\frac{1}{2}}$. The interaction distance, d, and event size, ε, for primary damage calculated from the λ value for each radiation are also given in table 1. They do not vary greatly with the radiation

Table 1. Parameters giving the best fit to the dose-response relations.

			Terminal deletions			Dicentric and rings						
			$Y=aD^n$		$Y=\alpha D$	$Y=aD^n$		$Y=\alpha D+\beta D^2$				
Radiation	LET (keV μm^{-1})	Dose range (rad)	$a(\times 10^{-4})$	n	$\alpha(\times 10^{-4})$	$a(\times 10^{-6})$	n	$\alpha(\times 10^{-4})$	$\beta(\times 10^{-6})$	λ (rad)	d (μm)	ε (eV)
^{60}Co γ-ray	0.28	30–500	1.55	1.19	4.12	25.50	1.78	0.91	6.82	13.3	0.80	41.8
200 kVp X-ray (1 mm Cu)	2.5	20–400	2.69	1.16	5.75	81.14	1.66	7.51	7.11	105.6	0.81	46.0
14.1 MeV T(d,n) fast neutron	11.0	12–450	11.33	0.98	10.66	1039.70	1.24	25.00	3.71	673.9	0.61	55.9
2.03 MeV Be(d,n) fast neutron	45.7	14.6–234.9	19.86	0.89	12.71	5125.36	1.09	74.55	—	∞	—	—

$\lambda=\alpha/\beta$, d = interaction distance, ε = energy expended per primary lesion leading to dicentrics and rings.

quality. The interaction distance obtained in this experiment is in good accord with Bauchinger and Schmid (1973) and Bauchinger, Schmid and Rimpl (1974) who obtained $d=0.8\ \mu m$ in their experiments with 220 kV X-rays and 3 MeV electrons. The estimated event size of about 50 eV is consistent with the idea that the primary damage responsible for the exchange aberration can be produced by a single energy-loss event (Neary, Savage and Evans 1965).

The frequency of terminal deletions was also influenced by radiation quality. However, the LET effect was less pronounced as compared with the exchange aberrations. They were directly proportional to the dose even in cases of sparsely ionising radiations, indicating that the majority of them were single track events less dependent on radiation LET.

Chemical Property of Basic Damage

Certain chemicals, including sulphydryl (SH) compounds, are known to protect against the formation of radiation-induced chromosome aberrations, possibly by interfering with the indirect action of radiation by scavenging highly reactive water radicals. Since changes in the aberration frequency afforded by scavengers are considered to result from protection against the formation of primary damage responsible for the chromosome aberration formation, the scavenger method offers an opportunity to probe further into the nature of the basic damage. The dose-response relation $Y=aD^n$ may be converted into $Y=a(N/\varphi)^n=\Phi N^n$ in terms of the number of primary lesions ($N=\varphi D$). When a scavenger is present at a concentration of $[S]$, protecting a fraction of damage p_i, the following relation will hold between aberration yield and p_i.

$$Y_i=\Phi[(1-p_i)N]^n$$
$$p_i=1-(Y_i/\Phi N^n)^{1/n}=1-(Y_i/Y_0)^{1/n}$$

where Y and Y_0 are aberration yields in the presence and absence of scavenger, respectively.

Figure 1 shows the changes in aberration frequency in lymphocytes irradiated with 300 rad of ^{60}Co γ-rays as a function of scavenger concentration. The cells were irradiated in Dulbecco's phosphate buffered saline (PBS) containing graded doses of scavengers. When the cells were irradiated in PBS alone, the frequency of dicentrics and rings (Y_0) was about 1.7 times higher than that in cells irradiated as whole blood, while the frequency of terminal deletions was not changed. As seen in figure 1, the scavengers afforded considerable protection against the formation of dicentrics and rings, whereas the terminal deletions were not significantly protected. The frequency of dicentrics and rings decreased with increase in scavenger concentration, but there seemed to be a maximum protection beyond which further increase in scavenger concentration showed no additional protection. The level of maximum protection differed between alcohols and SH-compounds. The frequencies of dicentrics and rings at maximum protection (Y_{max}) were about 0.25 for alcohols and about 0.07 for SH-compounds. Assuming that the dose-power law of $n=1.78$ also holds for the irradiation in PBS, p_{max} of 0.59 and

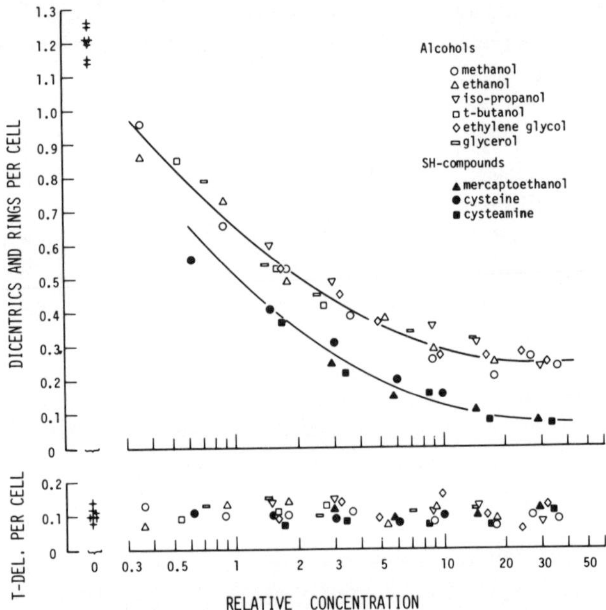

Figure 1. The change in the frequency of chromosome aberrations as a function of scavenger concentration. The scavenger concentration is given as concentration relative to that needed to afford half maximum protection ($Y_{\frac{1}{2}\max}$). The cells were suspended in Dulbecco's phosphate buffered saline (PBS) containing scavenger and exposed to 300 rad of ^{60}Co γ-rays. After irradiation, cells were washed and incubated in culture medium. The control values (Y_0) were 1.20 $\pm$0.04 and 0.11$\pm$0.02 per cell for dicentrics plus rings and terminal deletions (T-DEL), respectively. Each point is based on 100 cells.

Table 2. The reaction rates and effective concentrations of scavengers. (The reaction rates were taken from Johansen & Howard-Flanders 1965, Adams 1967, Roots and Okada 1972.)

Scavengers	MW	$[S]_{\frac{1}{2}\max}$ (mol)	k (10^8mol^{-1}sec^{-1}) OH	H	e^-_{aq}	$k[S]_{\frac{1}{2}\max}$ (10^8 sec^{-1}) OH	H	e^-_{aq}
Alcohols								
Methanol	32.04	0.175	4.8	0.016	0.0001	0.84	0.0028	0.000018
Ethanol	46.07	0.121	11	0.16	0.001	1.33	0.019	0.00012
iso-Propanol	60.10	0.056	39	0.4	—	2.18	0.022	—
t-Butanol	74.12	0.250	2.5	0.001	—	0.63	0.00025	—
Ethylene glycol	62.07	0.100	8.6	0.095	—	0.86	0.0095	—
Glycerol	92.10	0.077	9.5	0.2	—	0.73	0.015	—
SH-Compounds								
Mercaptoethanol	78.13	0.0044	120	—	—	0.53	—	—
L-Cysteine·HCl	175.63	0.0092	79	10	87	0.73	0.092	0.800
Cysteamine·HCl	113.61	0.0026	—	—	200	—	—	0.52

0.80 were obtained for alcohols and SH-compounds, respectively, i.e., about 60 and 80% of the primary damage leading to the exchange aberrations are prevented by these two types of scavengers.

If most of the protectable fraction is related to the indirect action of radiation (where the damage is produced in the target molecule, R, by reaction of free radicals with reaction rate k_r, and the protection by added chemical, S, is due to scavenging such free radicals with reaction rate k_s), the following relation will hold for all scavengers (Johansen and Howard-Flanders 1965, Sanner and Pihl 1969, Roots and Okada 1972):

$$k[R] = k_1[S_1]_{\frac{1}{2}\max} = k_2[S_2]_{\frac{1}{2}\max} = k_3[S_3]_{\frac{1}{2}\max} \ldots k_i[S_i]_{\frac{1}{2}\max} = \text{constant}$$

where $[S_i]_{\frac{1}{2}\max}$ is the scavenger concentration needed for half-maximum protection, $Y_{\frac{1}{2}\max} = Y_0(1-\frac{1}{2}p_{\max})^{1.78}$. In table 2 is listed the $k[S]_{\frac{1}{2}\max}$ value for each of three types of water radicals, i.e. the hydroxy radical (OH), hydrogen atom (H) and hydrated electron (e^-_{aq}). The constant value was obtained only for the OH radical, and no correlation was found between protective ability and the rate of reaction with reducing radicals, either H or e^-_{aq}. This indicates that the major radical species responsible for the primary damage leading to the exchange aberration is the OH radical.

Interaction of Two Doses

In order to study the post-irradiation sequence of events relating to chromosome aberration formation, changes in the chromosome aberration yield by dose fractionation were analysed in cells under various conditions, i.e. PHA stimulated or unstimulated lymphocytes, normal lymphocytes or lymphocytes from patients with Down's syndrome, and cells in culture with or without metabolic inhibitors. In these experiments, a total of 220 R ^{60}Co γ-rays was delivered in two equal fractions with varying rest intervals between doses. The results are shown in figures 2–4.

When cells were irradiated as whole blood, the decline in the frequency of dicentrics and rings was very small over periods of fractionation time intervals from 0 to 6 h (figure 2A). However, in cells suspended in Hanks' balanced salt solution (BSS) without PHA, significant changes in the frequency of dicentrics and rings were observed (figure 2B). With the increase in the rest time, the frequency first decreased, temporarily rose, and then decreased again to the nadir level. The difference in the response to the fractionated dose between cells in plasma and BSS suggests that lymphocytes can be activated to undergo repair by simply transferring the cells from plasma to artificial medium. The results presented in figure 2C are also consistent with the idea that the damage is not repaired until the cells are activated. In this experiment, whole blood was exposed to a first dose and kept for 12 h at 37°C before being exposed to the second dose either as whole blood, or mixed with culture medium containing PHA. The cells introduced into medium containing PHA were exposed to the second dose with varying time periods from the initiation of culture. The decline in the frequency of dicentrics and rings was not significant until the cells were mixed with culture medium; the decline occurred only

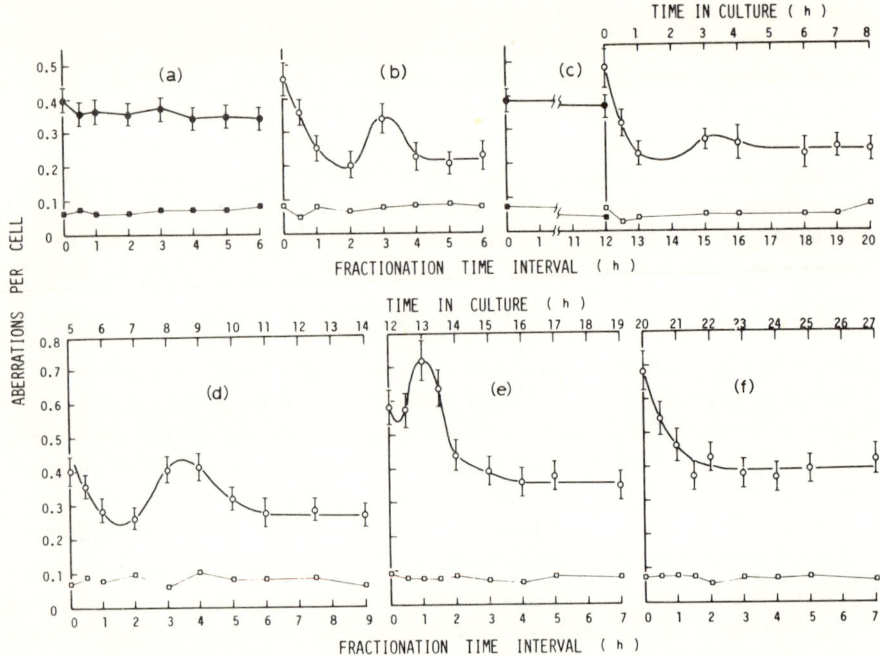

Figure 2. The changes in the frequency of chromosome aberrations as a function of fractionation time interval. A total of 220 R of ^{60}Co γ-rays was delivered in two equal fractions with indicated time intervals between doses. (a) Heparinised whole blood (1 ml in each plastic tube) was kept at 37°C and aerated with 5% CO_2 for 6 h with hourly gentle shaking. During this period, the dose was given in two fractions. (b) Whole blood was suspended in BSS with a ratio of 1:10, and then irradiated in the same way as in (a). After the experiment, cells were collected by centrifugation and incubated in culture medium containing PHA. (c) Whole blood (1 ml in each plastic tube) was irradiated with the first dose and kept for 12 h at 37°C. Then the blood was exposed to the second dose, either as whole blood (solid symbols) or at various times after mixing with PHA-containing medium (open symbols). (d,e,f) Dose-fractionation experiments in different stages in PHA culture. All cultures were set up in plastic tubes and the first dose was given either at 5 h (d), 12 h (e) or 20 h (f). Circles = dicentrics and rings; squares = terminal deletions; solid symbols = irradiation as whole blood. Vertical line indicates standard error of mean, based on Poisson distribution. Each point is based on 150–400 cells.

when the second dose was given after culture, but in this case the subsequent temporary rise was less pronounced.

In lymphocytes in PHA culture, the pattern of the response to the fractionated dose was dependent on the stage in culture. In the early stage, in which the first dose was given after 5 h in culture, the change in the frequency of dicentrics and rings followed essentially the same pattern as that observed in cells in BSS (figure 2D). In a more advanced stage, in which the first dose was given at 12 h, the initial fall was not apparent during the first 30 min, but it was soon followed by a temporary rise and a secondary fall (figure 2E). The pattern of the change in aberration yield seemed to be essentially the same as that found in BSS and the early stage of culture, except that the sequences occurred rather rapidly. When the dose fractionation effect was tested in further advanced cultures, in which the first dose was given at 20 h, the aberration yield simply declined with increasing interval up to 3 h (figure 2F), showing the response pattern commonly observed in other mammalian cell lines (Dewey and Humphrey 1964). However, if the assumption that the sequence occurs very rapidly in such fully transformed and metabolically very active cells is correct, the observed decline in aberration yield may be a manifestation of nothing more than a secondary fall in the sequence of changes, which can only be clearly resolved in the metabolically less active cells.

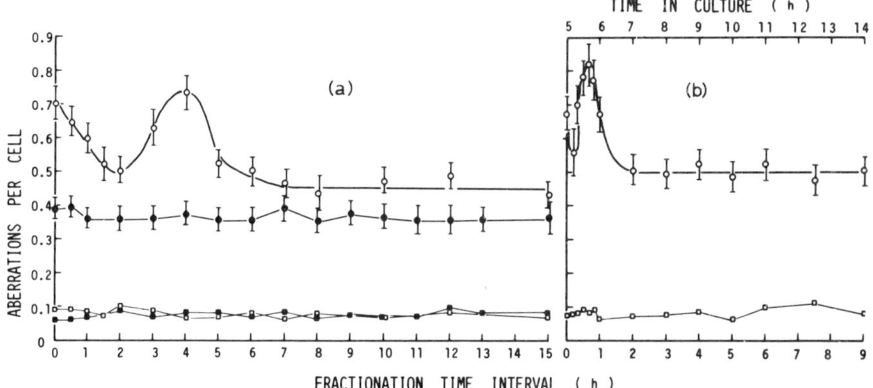

Figure 3. Comparison of the changes in the aberration frequency, after fractionated doses of γ-rays, between normal lymphocytes and lymphocytes from patients with Down's syndrome. A total of 220 R of ^{60}Co γ-rays was given in two equal fractions. (a) Dose fractionation in whole blood using the same procedure as in figure 2a. (b) Dose fractionation in lymphocytes from patients with Down's syndrome in their early stage in PHA culture. The first dose was given after 5 h in culture. Solid symbols: lymphocytes from normal person; open symbols: lymphocytes from patients with Down's syndrome. All patients were confirmed to have trisomy 21. Circles = dicentrics and rings; squares = terminal deletions. Each point is based on 150–400 cells.

Figure 3 shows the effect of dose fractionation in lymphocytes from patients with Down's syndrome. As has been noted previously (Sasaki and Tonomura 1969; Sasaki, Tonomura and Matsubara 1970), the lymphocytes from patients with Down's syndrome showed a high sensitivity to exchange aberration formation (figure 3A). A significant finding was that, even in the cells in plasma, the pattern of the response to fractionated dose was similar to that found in normal cells in BSS and in the early stage in culture. This was in marked contrast to the unstimulated normal control lymphocytes that were treated in parallel with the patients' cells. The normal cells in plasma again showed very little change in aberration frequency over a rest time up to 15 h.

When the lymphocytes from patients with Down's syndrome were tested for their response in the early stage of PHA culture, in which the first dose was given after 5 h in culture, the sequential response occurred very rapidly (figure 3B). In normal cells in a comparable stage in culture, the sequence (first fall/temporary rise/secondary fall) spanned a period of about 5 h as compared with a period of 1.5 h in the patients' cells.

These experimental data strongly suggest that the dual-step nature of the response pattern is the essential feature of the change in exchange aberration yield afforded by dose fractionation, and that the size and time course of the process may vary with the metabolic state of the cells. What is the nature of this sequential process? To answer this question, the effect of some metabolic inhibitors on the response pattern was studied in normal lymphocytes in the early stage of PHA culture. In this experiment, metabolic inhibitors were present throughout the period 5–13 h in culture, and 220 R ^{60}Co γ-rays were given in two equal fractions during this period, with the first dose given immediately after addition of inhibitors. The results are shown in figure 4. When cycloheximide was present, the first fall and the subsequent rise were not influenced, but the secondary fall was suppressed. In the presence of caffeine, the process seemed to proceed in a way comparable to that in its absence. When 5-fluorodeoxyuridine (FUdR) was present, the frequency of dicentrics and rings did not significantly change with the rest intervals, but stayed at a level comparable to that given by an unsplit single dose delivered at the beginning and at the end of FUdR treatment. The presence of any of these inhibitors did not in itself influence the aberration yield induced by a single dose.

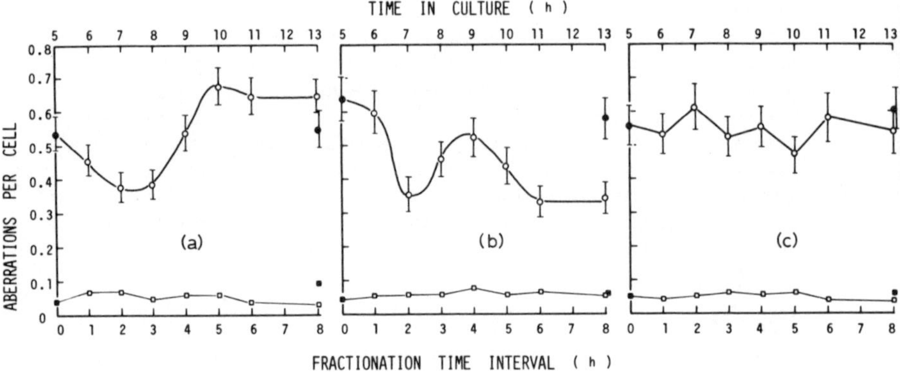

Figure 4. The effects of metabolic inhibitors on the response to fractionated dose in the early stage in PHA culture. After 5 h in culture, chemicals were added and cultures were incubated for 8 h in the presence of chemicals. During this period a total of 220 R of ^{60}Co γ-rays was given in two equal fractions, with the first dose given immediately after the addition of chemicals. After 13 h in culture, the cells were washed and reincubated. (a) cycloheximide, 10 μg ml^{-1}; (b) caffeine, 500 μg ml^{-1}; (c) FUdR, 10 μg ml^{-1}. Circles = dicentrics and rings; squares = terminal deletions; solid symbols = aberration frequencies by single unsplit dose. Each point is based on 150–200 cells.

Throughout the experiment, the yield of dicentrics and rings per unit dose was higher following irradiation of cells in BSS and culture medium, as compared with the irradiation of whole blood, and further increased with the blastic transformation. However, the frequency of terminal deletions was constant, irrespective of the change in the frequency of exchange aberrations.

Discussion

The event size of about 50 eV obtained in the present study for the energy required to produce primary damage responsible for exchange aberration formation is in good accord with Neary, Savage and Evans (1965). They studied chromatid aberrations in *Tradescantia* pollen tubes exposed to monochromatic soft X-rays, and concluded that the observed dependence of aberration yield on dose and LET was satisfactorily accounted for in terms of a primary lesion by a single energy-loss event in a continuous single macromolecule. Moreover, the results obtained by the scavenger method suggest that one can assign such damage to the oxidative radical intermediate introduced in the target molecule, possibly DNA (Sasaki and Matsubara, in preparation), the majority of the damage being due to the indirect action of radiation through the reaction with OH radicals. The experimental evidence is in line with the suggestion that DNA single-strand breaks may be plausible candidates for the primary damage responsible for the exchange aberration formation (Evans 1966), and meets some of the difficulties encountered in the postulation that DNA double-strand breaks initiate various types of aberrations (Bender, Griggs and Bedford 1974; Leenhouts and Chadwick 1974). The DNA double-strand breaks are produced as single-hit events for every 1000–3000 eV, while single-strand breaks are introduced for every 30–100 eV of energy deposition (see Ormerod 1976). Moreover, there is a dispute over the question about the rejoining of double-strand breaks in mammalian cells. The interaction of two single-strand breaks may be possible by a mechanism analogous to those of genetic recombination proposed by Holliday (1964) and by Whitehouse and Hastings (1965), and developed in more sophisticated form by Meselson and Radding (1975) and by Sobell (1972). The DNA double-strand breaks, if they are irreparable, would result in terminal deletions. However, simple comparison of the production rates of these two events gives roughly one terminal deletion for every 1000 double-strand breaks in the case of X-ray irradiation. The significance of this discrepancy might be more evident when sufficient resolution is possible between true double-strand breaks, with rupture of associated proteins, and two single-strand breaks on complementary strands several bases apart. Such true breaks of chromatin may be possible by local energy deposition from high LET particles, e.g. δ-rays from secondary electrons, and thus may not be simply proportional to the LET of radiation.

The radical nature of the transmutation of radiation energy to DNA, resulting in strand breaks (Roots and Okada 1972) and chromosome exchange aberrations, evokes the problem of the role of DNA-associated proteins. It is very reasonable to assume that any change in DNA micro-environment would result in a change in the sensitivity in a way that altered the amount of damage produced. It is well known that after brief exposure to PHA small lymphocytes, which are rather dormant cells, undergo a physical alteration in the DNA and associated proteins before the onset of RNA synthesis (Tanaka *et al.* 1963; Killander and Rigler 1965; Kleinsmith, Allfrey and Mirsky 1966; Pogo, Allfrey and Mirsky 1966). Such an initial step of nuclear activation also occurs when lymphocytes are transferred into artificial medium (Scornik, Conesa and Bachmann 1971; Sullivan and Garcia 1973). A change in the DNA micro-environment could provide the basis for increased sensitivity and initiation of the chromosome rejoining process in physiological saline and culture medium, even without PHA. The alteration in DNA and associated proteins could result in a decreased capacity to protect against radiation damage, whilst making ready access to repair enzymes. However, the lymphocytes from patients with Down's syndrome (trisomy 21) behaved differently. When they were irradiated as whole blood, chromosome rejoining processes occurred as for normal cells in the early stages in PHA culture. In addition to the altered enzyme level, the altered state of DNA micro-environment associated with delayed differentiation due to genetic imbalance, may be an attractive reason for the high sensitivity of chromosomes of trisomic cells to ionising radiation (Sasaki and Tonomura 1969), chemical mutagens (Schuler *et al.* 1972) and viral infection (Higurashi, Tamura and Nakatake 1973). In this connection, the recent finding of the synergistic effect of combined UV and X-ray irradiation in increasing chromosome aberration yield in unstimulated normal lymphocytes (Holmberg and Johansson 1974) and the absence of synergism in PHA stimulated normal (Holmberg 1976) and unstimulated trisomic lymphocytes (Holmberg 1974, Lambert *et al.* 1976) is interesting. In this context it may be attractive to consider not that there is a direct interaction between UV- and X-ray-induced damage, but rather that UV triggers nuclear activation in normal lymphocytes.

The repair of damage is a time-limited process. Thus, the results of dose fractionation and dose protraction experiments using sparsely ionising radiation would reflect the repair time of radiation-induced damage, if the amount of damage per unit dose is not modified by the post-irradiation conditions. Hitherto, several sets of experimental data have been published on the effect of dose fractionation and protraction in human lymphocytes. However, the reported non-uniform, and in some cases contradictory, results as to the repair time are rather puzzling (see Purrott and Reeder 1976, Liniecki *et al.* 1977). The present data strongly point to the change in the DNA micro-environment associated with the cell's metabolic state, as well as the radiation damage to non-DNA components, as being important. From the changes in aberration yields caused by dose fractionation, it seems probable that the damage remains unrepaired in lymphocytes in plasma until the cells are activated by artificial medium or PHA. When a dose was given in two fractions whilst the cells were in culture, the change in the aberration yield followed essentially the same pattern, fall/temporary rise/secondary fall, the process being comparable to the so-called Lane effect, which was first found by Lane (1951) in plant cells and noted more recently by Evans (1966) in human lymphocytes. The progression of this process is profoundly influenced by the stage of culture; it proceeds rather slowly in the early stage of culture so that each phase can easily be identified, then more rapidly in the later stages, finally becoming monophasic. The initial fall was suppressed by FUdR, and the secondary fall was specifically suppressed by cycloheximide. Thus, it is tempting to correlate the initial fall with the repair of damage to DNA, and the subsequent rise and fall with the radiosensitisation of chromosomal DNA due to the alteration of DNA micro-environment occurring in association with the repair of protein components. In the cells fully transformed, or in a metabolically active state, a monophasic fall would represent the repair process related either to DNA or its associated proteins. If this argument is correct, both the DNA synthesis inhibitors (Prempree and Merz 1969, Yamamoto and Yamaguchi 1969) and the protein synthesis inhibitors (Wolff 1959, 1972; Brewen 1963) would give an apparent lack of fall in aberration yield between two doses. However, it does not necessarily imply a requirement for protein synthesis for the DNA damage to be repaired. Elkind and Kamper (1970) reported that a DNA-protein complex dissociated by X-ray irradiation reassociated during the post-irradiation incubation time. The experimental evidence presented here thus strongly suggests that the interaction between doses is complex, and that the effects of dose rate on aberration yield cannot be explained by a simple hypothesis about the repair time necessary to rejoin broken DNA. The present results do, however, provide at least a partial explanation of the discrepancy in the reported data on the so-called repair time in human lymphocytes.

Summary

To obtain an insight into the nature of primary damage, and its repair, involved in the formation of radiation-induced chromosome aberrations in human lymphocytes, *in vitro* experiments were designed to study the effects of quality and quantity of radiation; modification by protective agents; and the kinetics of interaction between doses. The results indicated the following: (a) Exchange aberrations and terminal deletions come from different types of primary damage. (b) Primary damage involved in the formation of exchange aberrations is produced by a single energy-loss event of about 50 eV. Approximately 60% of this damage is due to indirect action and this is largely a consequence of the action of OH radicals. (c) In contrast, primary damage responsible for the formation of terminal deletions is produced by direct action as a single-hit event. These are irreparable and neither protected by radical scavengers nor influenced by the cell's metabolic state. (d) The repair of radiation damage to chromosomes is two-fold; one associated with the repair of DNA and another related to the repair or restoration of the DNA micro-environment, possibly associated proteins. The time sequence of these processes is influenced by the cell's metabolic state.

Acknowledgements

I acknowledge the continuous encouragement of Professor A. Tonomura and thank Dr S. Matsubara for kindly arranging irradiation. The work was supported by Scientific Research Grants from the Ministry of Education, Science and Culture, Japan.

References

Adams, G.E. (1967) The general application of pulse radiolysis to current problems in radiology, in *Current Topics in Radiation Research*, vol.3 (eds Ebert & Howard) pp.35–93. Amsterdam: North Holland.

Bauchinger, M. and E.Schmid (1973) Chromosome aberrations in human lymphocytes after X-irradiation *in vitro*. II. Analysis of primary processes in the formation of dicentric chromosomes. *Mutat. Res. 20*, 107–13.

Bauchinger, M., E.Schmid & G.Rimpl (1974) Interaction distance of primary lesions in the formation of dicentric chromosomes after irradiation of human lymphocytes with 3 MeV electrons *in vitro. Mutat. Res. 25*, 83–7.

Bender, M.A., H.G.Griggs and J.S.Bedford (1974) Mechanisms of chromosomal aberration production. III. Chemicals and ionizing radiation. *Mutat. Res. 23*, 197–212.

Brewen, J.G., (1963) Dependence of frequency of X-ray-induced chromosome aberrations on dose rate in the Chinese hamster. *Proc. Nat. Acad. Sci. USA 50*, 322–9.

Comings, D.E. (1974) What is a chromosome break? In *Chromosomes and Cancer* (ed. J.German) pp.95–133. New York: John Wiley & Sons.

Dewey, W.C. & R.M.Humphrey (1964) Restitution of radiation-induced chromosomal damage in Chinese hamster cells related to the cell's life cycle. *Exp. Cell Res. 35*, 262–76.

Elkind, M.M. & C.Kamper (1970) Two forms of repair of DNA in mammalian cells following irradiation. *Biophys. J. 10*, 237–45.

Evans, H.J. (1966) Repair and recovery from chromosome damage after fractionated X-ray dosage, in *Genetical Aspects of Radiosensitivity: Mechanism of Repair*, pp.31–48. Vienna: International Atomic Energy Agency.

Higurashi, M., T.Tamura & T.Nakatake (1973) Cytogenetic observations in cultured lymphocytes from patients with Down's syndrome and measles. *Pediatr. Res. 7*, 582–7.

Holliday, R. (1964) A mechanism for gene conversion in fungi. *Genet. Res. 5*, 282–304.

Holmberg, M. (1974) No interaction between ultraviolet and X-irradiation on chromosome aberrations in cells with trisomy 21. *Nature 249*, 448–9.

—— (1976) Lack of synergistic effect between X-ray and UV irradiation on the frequency of chromosome aberrations in PHA-stimulated human lymphocytes in G_1 stage. *Mutat. Res. 34*, 141–8.

Holmberg, M. & J.Jonasson (1974) Synergistic effect of X-ray and UV irradiation on the frequency of chromosome breakage in human lymphocytes. *Mutat. Res. 23*, 213–21.

Johansen, I. & P.Howard-Flanders (1965) Macromolecular repair and free radical scavenging in the protection of bacteria against X-rays. *Radiat. Res. 24*, 184–200.

Kellerer, A.M. & H.H.Rossi (1972) The theory of dual radiation action. *Curr. Top. Radiat. Res. Q. 8*, 85–158.

Killander, D. & R.Rigler (1965) Initial changes of deoxyribonucleoprotein and synthesis of nucleic acid in phytohemagglutinin-stimulated human leucocytes *in vitro*. *Exp. Cell Res. 39*, 701–4.

Kleinsmith, L.G., V.G.Allfrey & A.E.Mirsky (1966) Phosphorylation of nuclear protein early in the course of gene activation in lymphocytes. *Science 154*, 780–1.

Lambert, B., K.Hansson, T.H.Bui, F.Funes-Cravioto, J.Lindsten, M.Holmberg & R. Strausmanis (1976) DNA repair and frequency of X-ray and UV-light induced chromosome aberrations in leukocytes from patients with Down's syndrome. *Ann. Hum. Genet. 39*, 293–303.

Lane, G.R., (1951) X-ray fractionation and chromosome breakage. *Heredity 5*, 1–35.

Lea, D.E. & D.E.Catcheside (1942) The mechanism of the induction by radiation of chromosome aberrations in *Tradescantia*. *J. Genet. 44*, 216–45.

Leenhouts, H.P. & K.H.Chadwick (1974) Radiation-induced DNA double strand breaks and chromosome aberrations. *Theor. Appl. Genet. 44*, 167–72.

Liniecki, J., A.Bajerska, K.Wyszyńska & B.Cisowska (1977) Gamma-radiation-induced chromosomal aberrations in human lymphocytes: dose-rate effects in stimulated and non-stimulated cells. *Mutat. Res. 43*, 291–304.

Meselson, M.S. & C.M.Radding (1975) A general model for genetic recombination. *Proc. Nat. Acad. Sci. USA 72*, 358–61.

Neary, G.J. (1965) Chromosome aberrations and the theory of RBE. 1. General considerations. *Int. J. Radiat. Biol. 9*, 477–502.

Neary, G.J., J.R.K.Savage & H.J.Evans (1965) Chromatid aberrations in *Tradescantia* pollen tubes induced by monochromatic X-rays of quantum energy 3 and 1.5 keV. *Int. J. Radiat. Biol. 8*, 1–19.

Ormerod, M.G. (1976) Radiation-induced strand breaks in the DNA of mammalian cells, in *Biology of Radiation Carcinogenesis* (eds J.M.Yuhas, R.W.Tennant & J.D.Regan) pp.67–92. New York: Raven Press.

Pogo, B.G.T., V.G.Allfrey & A.E.Mirsky (1966) RNA synthesis and histone acetylation during the course of gene activation in lymphocytes. *Proc. Nat. Acad. Sci. USA 55*, 805–12.

Prempree, T. & T.Merz (1969) Does hydroxyurea inhibit chromosome repair in cultured human lymphocytes? *Nature 224*, 603–4.

Purrott, R.J. & E.J.Reeder (1976) Chromosome aberration yields in human lymphocytes induced by fractionated doses of X-radiation. *Mutat. Res. 34*, 437–46.

Roots, R. & S.Okada (1972) Protection of DNA molecules of cultured mammalian cells from radiation-induced single-strand scissions by various alcohols and SH compounds. *Int. J. Radiat. Biol. 21*, 329–42.

Sanner, T. & A.Pihl (1969) Significance and mechanism of the indirect effect in bacterial cells. The relative effect of added compounds in *Escherichia coli* B, irradiated in liquid and frozen suspension. *Radiat. Res. 37*, 216–27.

Sasaki, M.S. & S.Matsubara (1977) Free radical scavenging in protection of human lymphocytes against chromosome aberration formation by gamma-ray irradiation. *Int. J. Radiat. Biol. 32*, 439–45.

Sasaki, M.S. & A.Tonomura (1969) Chromosomal radiosensitivity in Down's syndrome. *Jpn. J. Hum. Genet. 14*, 81–92.

Sasaki, M.S., A.Tonomura & S.Matsubara (1970) Chromosome constitution and its bearing on the chromosomal radiosensitivity in man. *Mutat. Res. 10*, 617–33.

Schuler, D., M.Dobos, G.Fekete, T.Hachay & A.Nemeskéri (1972) Down's syndrome and malignancy. *Acta Pediatr. Acad. Sci. Hung. 13*, 245–52.

Scornik, J.C., L.C.Giraudo Conesa & A.E.Bachmann (1971) Increased RNA synthesis in unstimulated human leukocyte culture. *Exp. Cell Res. 69*, 444–7.

Sobell, H.M. (1972) Molecular mechanism for genetic recombination. *Proc. Nat. Acad. Sci. USA 69*, 2483–7.

Sullivan, P.A. & A.M.Garcia (1973) RNA synthesis in lymphocytes after hypotonic shock. *Acta Cytol. 17*, 502–6.

Tanaka, Y., L.B.Epstein, P.Brecher & F.Stohlman (1963) Transformation of lymphocytes in cultures of human peripheral blood. *Blood 22*, 614–29.

Whitehouse, H.L.K. & P.J.Hasting (1965) The analysis of genetic recombination on the polaron hybrid DNA model. *Genet. Res. 6*, 27–92.

Wolff, S. (1959) Interpretation of induced chromosome breakage and rejoining. *Radiat. Res., Suppl., 1*, 453–62.

—— (1972) The repair of X-ray-induced chromosome aberrations in stimulated and unstimulated human lymphocytes. *Mutat. Res. 15*, 435–44.

Yamamoto, K. & H.Yamaguchi (1969) Inhibitory effect of 5-fluorodeoxyuridine on the repair of γ-ray-induced lesions in barley chromosomes. *Mutat. Res. 8*, 424–7.

D.C.LLOYD

The Problems of interpreting Aberration Yields induced by in vivo Irradiation of Lymphocytes

During the ten years that have elapsed since the previous conference in Edinburgh (Evans, Court Brown and McLean 1967) most scientists concerned with radiological protection have accepted that the measurement of chromosome aberration yields in peripheral blood lymphocytes is the best available basis for biological dosimetry. It has often been stated that the technique is particularly applicable in cases of acute whole-body exposure to penetrating radiation (Bender 1969, Evans 1972) but most authors then add the rider that rarely, if ever, does an accident involve an homogeneous exposure to the whole body. Partial-body or non-uniform exposures are almost always experienced, and interpreting the aberration yields in such circumstances is one of the major outstanding problems for cytogenetic dosimetry.

Since our laboratory was established in 1967 we have examined chromosome preparations from about 300 people known or suspected of being accidentally exposed to radiation (Lloyd *et al.* 1977). Of the 110 people in whom evidence of exposure was found, none was thought to have received a uniform exposure to the whole body. For convenience we report our findings as equivalent whole-body dose estimates, which are defined as the doses to the whole body that would result in the observed aberration yield (Dolphin 1969). However this is an over-simplification, which ignores many of the *in vivo* and *in vitro* effects of radiation on lymphocyte kinetics. In the present paper an attempt is made to summarise some recent research on these aspects, which are relevant to cytogenetic dosimetry.

The Problem

Partial-body exposures result in the production of irradiated and unirradiated populations of lymphocytes, which are subsequently mixed by the circulation. Moreover, the irradiated cells may receive a wide spectrum of doses, depending on the volume of tissue irradiated and the uniformity of the exposure. For protracted irradiation the time over which the dose is spread is also important, as circulating cells will move within and through the exposed volume. Even in the unlikely event of a whole-body exposure in which the radiation dose is uniform at the skin, the monotonic reduction of dose with depth in tissue will result in a variety of doses being received by lymphocytes. *In vivo* the irradiated and unirradiated cell populations may behave differently in the recycling and pooling processes, so that in a blood sample taken some time later the ratio of damaged to undamaged cells no longer reflects the ratio of irradiated to unirradiated tissue volumes. Further distortion may arise *in vitro* due to the dose-dependent elimination of irradiated cells by interphase death, their failure to respond to phytohaemagglutinin, or by delay in their passage through the cell cycle.

In vitro Mixed Culture Experiments

The first *in vitro* cytogenetic experiments with human cells to examine the effect of selection in culture were made by Sharpe (1969). She mixed cultures of irradiated and unirradiated lymphocytes and showed that irradiation leads to a loss of lymphocytes at mitosis. The work was subsequently extended in our laboratory using Sharpe's original idea of mixed cultures. Essentially the mixed cultures simulate *in vitro* the somewhat artificial conditions of an acute 50%-body exposure in which the lymphocytes are static and the dose to the irradiated half is uniform. This is achieved by mixing equal volumes of irradiated and unirradiated blood taken from the same donor. The aberration yield is then compared with that from a control culture of wholly irradiated blood. If no selection against the irradiated cells occurred the aberration yield in the mixed culture would be simply 50% of the control. Any reduction from the expected 50% gives an indication of the amount of cell selection occurring. Mixed and control 48-h cultures were analysed for unstable chromosome aberrations after a series of seven doses ranging from 50

Table 1. The yields of dicentric aberrations and all cells with unstable chromosome damage in 100% and 50% irradiated blood.

Dose (rad)	Control or mixed 50%	No. of cells scored	Total no. of damaged cells	Dicentrics per cell	Damaged cells M/C %	Dicentrics per cell M/C %
50	C	2070	101	0.024	46.9	45
	M	4613	105	0.011		
100	C	872	111	0.084	41.7	34.5
	M	3990	211	0.029		
200	C	500	229	0.382	42.6	36.1
	M	1000	195	0.138		
300	C	305	238	0.928	27.7	26.1
	M	920	119	0.242		
400	C	200	163	1.26	28.6	26.4
	M	700	163	0.333		
500	C	60	58	2.38	16.7	14.8
	M	753	121	0.35		
700	C	42	42	4.24	6.2	5.3
	M	712	44	0.223		

to 700 rad (Lloyd, Purrott and Dolphin 1973). After 50 rad the dicentric yield in the mixed culture gave 45% of that from the control, indicating that 10% of the exposed cells had failed to reach metaphase due to the effect of radiation. At 700 rad this value rose to ~90% (table 1). It was suggested that the relatively small loss of cells at the lower dose meant that for the majority of radiation accidents, in which the dose to any part of the body does not exceed 50 rad, the selection of cells *in vitro* is a minor problem and in practice can be disregarded. However, for higher local doses some allowance for cell loss should be considered when it is known that a partial body exposure has occurred.

This experiment did not distinguish between the two principal causes of failure to reach metaphase at 48 h, namely interphase death and delayed entry into mitosis. In a subsequent mixed culture experiment (Lloyd *et al.* 1977) the effect of mitotic delay was considered. Here, using doses of 150 and 400 rad, mixed and control cultures were fixed at four-hourly intervals from 36 to 72 h. This experiment showed that mitotic delay did occur at these relatively high doses and that the delay was greater at the higher dose. However, the duration of the delay was of the order of only a few hours, suggesting that most of the reduction in the number of lymphocytes arriving at metaphase by 48 h may be ascribed to interphase death. Using the data at 400 rad the delay could be related to the presence of unstable aberrations. Cells containing these aberrations were delayed, relative to irradiated but aberration-free cells, in their rate of progress through the cell cycle. Furthermore, cells with two dicentrics were delayed for longer than those with just a single dicentric. This effect would lead to an under-estimate of dose in 48-h cultures following a very high localised exposure, but with partial-body doses of up to 400 rad mitotic delay probably does not seriously distort the aberration yield. Approximately constant dicentric yields were obtained at sampling times of 36 to 52 h at 150 rad and 44 to 52 h at 400 rad. After 52 h the yields fell, due to dilution by second division undamaged cells. Thus 48 to 52 h, which are the commonly used culture times for cytogenetic dosimetry, were shown to be still the optimum, and mitotic delay did not result in a late peak in aberration yield.

In vivo Animal Data

In vivo data relevant to the partial-body exposure problem have been obtained by McFee, Banner and Sherrill (1974) and McFee (1977) with pigs, and by Matsubara, Sasaki and Adachi (1974) with radiotherapy patients. McFee *et al.* (1974) showed that when pigs were given 400 rad at 58 R min^{-1}, to either the total body or to half the body, the aberration yield from the half-exposed animals was only one-third of that for the total. They were unable to show any significant difference when either the anterior or posterior half of the pig was irradiated. Furthermore, from unpublished data they were able to discount the effect of mitotic delay causing the lower than expected aberration yield in the half-exposed animals.

In a subsequent experiment, McFee (1977) examined whole- and half-body exposures to pigs at several doses. At the lower doses, 100, 150 and 200 rad, the half-exposures gave values that did not differ significantly from the expected 50%, but a marked decline was found at 300 and 400 rad. Thus a similar trend was observed as in the *in vitro* mixed culture experiments with human cells. The principal difference, however, is that Lloyd *et al.* obtained a significant reduction below the 50% level at 100 rad. Bearing in mind that species variability does occur, it is highly likely that the main factor accounting for the differences between the experimental results is a reduction in the effective dose to the pig lymphocytes due to body shielding. McFee calculated that this would result in the average lymphocyte dose in the exposed half of the pig being 45% of that measured in air at the mid-line. If one assumes that a similar effect will occur *in vivo* in a human accident it is suggested that the conclusion of Lloyd *et al.* (1973), that *in vitro* cell selection does not complicate dose estimates in accidents involving localised doses of 50 rad or less, is somewhat conservative and may well be applicable at slightly higher exposures.

In vivo Human Data

Matsubara, Sasaki and Adachi (1974) have studied a number of patients receiving partial body radiotherapy for various malignant conditions. They divided the patients into two groups; those with tumours of the abdomen and those treated for tumours in the head and neck or chest. Blood samples were taken 24 h after the first single exposure of the fractionated therapy regime. It was found that the observed yield of dicentrics and rings (Yp) was roughly proportional to the irradiated volume of tissue (v) and inversely proportional to the body weight (W). Thus $Yp=yv/W$, where y is the dose-response relationship of chromosome aberrations *in vivo*, which is probably identical to that *in vitro*. Their data also indicated that two further factors should be taken into account. Firstly, some consideration needs to be given to the non-uniform distribution of lymphocytes in the body. The data showed a consistent trend in which the aberration yield tended to be higher in patients irradiated on the head and neck or chest regions, even when corrections were made for the relative volumes of irradiated tissues. The second factor was the dose-dependent inactivation of lymphocytes, which Lloyd *et al.* (1973) primarily ascribed *in vitro* to interphase death. Thus Matsubara *et al.* expanded their model to $Yp=ykSv/W$, where k and S are the distribution and survival factors respectively for lymphocytes contained in the irradiated volume v.

Thus there is a broad agreement between the *in vivo* data of McFee *et al.* and Matsubara *et al.*, and the *in vitro* work of Lloyd *et al.* The mixed culture technique, which attempts to simulate partial body irradiation, has therefore been shown to reflect fairly accurately the general relationship for *in vivo* exposures. However, it must be acknowledged that the *in vivo* kinetics of lymphocytes are poorly understood, and even less is known about the perturbations that result from radiation.

In vivo Lymphocyte Kinetics

The rapid fall in blood lymphocyte count relative to other circulating cell types after exposure to radiation is well known (Vodopick and Andrews 1974). It is probably not due to the intrinsic radiosensitivity of lymphocytes, for Lloyd *et al.* (1975) have shown that the survival of T cells was similar to other types of irradiated human cells when cultured to test their colony-forming ability.

Stjernsward *et al.* (1972) showed a significant change in the ratio of T and B lymphocytes in the circulation following irradiation of patients for breast cancer. The T fraction was markedly reduced, and they pointed out that this might be related to the thymus gland being present in the treatment field. This work seems to be in conflict with later data from Blomgren *et al.* (1974a, b) who found a greater reduction of B cells in two groups of patients treated with X-rays for breast cancer or cancers of the prostate and bladder. The greater radiosensitivity of B cells was also observed in the *in vitro* experiments of Prosser (1977). Moreover, Chee, Ilbery and Rickinson (1974) showed that the lymphocyte-replicating ability of T cells, as determined by phytohaemagglutinin stimulation, was not especially related to thymic exposure as it occurred irrespective of the site irradiated within the trunk. It seems highly likely, therefore, that the *in vivo* fall in the lymphocyte count following irradiation may reflect an enhanced and differential migration of cells out of the vasculature and into other tissues.

It has been suggested that with protracted exposures the effect of partial-body irradiation is mitigated by the circulation of lymphocytes, because they receive a more uniform dose than the body itself (Lloyd, Purrott and Dolphin 1973). This may well be true for protracted exposures of a few hours, but with shorter times the effect is marginal. This is because in normal subjects less than one per cent of the lymphocytes in the body are in the circulation; the remainder are contained in the extra-vascular pool and are therefore relatively static. Sharpe *et al.* (1968), however, have demonstrated that there is a fairly rapid interchange of cells between the blood circulation and the other tissues. Lymphocytes labelled with chromosome aberrations in an extra-corporeal irradiation coil disappeared after a mean residence time in the circulation of only five minutes. Therefore, for relatively short exposures it is unlikely that the chromosome aberration yield is influenced by the blood volume of the irradiated site, but rather by the number of lymphocytes contained in other tissues. The distribution of cells in the extra-vascular lymphocyte pool is not homogeneous throughout the body (Osgood 1954) and this may well explain Matsubara, Sasaki and Adachi's observation of a significantly higher aberration yield in patients irradiated on the upper as opposed to the lower regions of the trunk.

An interesting observation was made by Tamura, Sugiyama and Sugahara (1974) on patients receiving pelvic irradiation. Repeated blood samples were taken, and the dicentric yields rose to a peak at 6 h post-irradiation and then declined, so that at 20 h the yield was the same as in samples taken immediately (20 min) after the exposures. They postulated that because the average radiation dose was larger to the static tissues than to the circulating blood the input of aberration-bearing cells to the blood from the extra-vascular pool, especially the lymphatic tissues, exceeded the reverse movement of damaged cells out of the circulation over the first six hours. By twenty hours all irradiated cells were mixed uniformly. These data indicate that some attention should be given to the correct blood sampling time following an accidental partial body irradiation, and ideally at least one day should elapse.

The Identification of Irradiated Tissues

In a recent case (January 1977) referred to our laboratory, an unclassified worker found a 6 Ci γ-radiography source and placed it in his shirt pocket. This resulted in an erythematous patch over the heart. However other parts of his body were also exposed as he had periodically handled the source when showing it to colleagues, tried to prise it apart with his teeth, and eventually taken it home over a weekend! The cytogenetic examination yielded 86 dicentrics in 1000 cells (table 2), which corresponded to an equivalent whole-body dose of 116 rad. However, the dose to his body was not uniform, as part of the chest wall probably received > 1000 rad.

Table 2. Cytogenetic data from an accidental non-uniform over-exposure to γ-radiation.

Cells scored	Dicentrics	Centric rings	Acentrics	Dose (rad)
1000	86	2	60	116

Dicentric Distribution in Cells

	0	1	2	3	4	5
Observed	932	56	9	1	1	1
Poisson	917	80	3	–	–	–

In cases of non-uniform exposure such as the above example, it is clearly desirable to identify which parts of the body in particular have been irradiated and to estimate the absorbed dose to those regions. Unless the localised skin dose is above the threshold for erythema (~600 rad acute) there is no biological technique that can readily give this information.

Fibroblasts from a series of skin biopsies traversing the probably irradiated sites may be cultured and stable chromosome aberrations identified (Bigger, Savage and Watson 1972). This is, however, unpleasant for the patient and requires longer culture periods than blood samples. The analysis would be very time-consuming as almost certainly banded preparations would be required. Whilst this method would serve to approximately delineate an exposed area, in the short term there seems little likelihood that the technique could provide an estimate of the local skin dose.

Dolphin (1969) has suggested that the equivalent whole-body dose data may be augmented by a consideration of the extent to which the aberrations deviate from the Poisson distribution that characterises a whole-body homogeneous exposure. This analysis would not enable one to determine which parts of the body were exposed, but may permit a rough estimate of the fraction of the body irradiated and its dose. This approach also suffers from a number of drawbacks. It takes no account of the uneven distribution of lymphocytes in the extra-vascular pool nor the non-linear dose-effect relationship for low LET radiation. In practice far more than the usual 500 cells would need to be examined in order to reduce the statistical errors on such an analysis. Thus this statistical approach, perhaps combined with skin fibroblast examinations, is at present impracticable but may become possible with the further development of computer-assisted cell finding and karyotyping.

Internal Emitters

Non-uniform irradiations also result from the internal deposition of radionuclides. This is currently one of the areas of greatest concern in the radiological protection of man and an area in which, to date, cytogenetic dosimetry has been of little value. Most dose estimates are based on metabolic models derived principally from animal experiments, and are thus subject to considerable uncertainty when applied to man. If cytogenetic estimates were able to supplement dose data from bioassay on, for example, urine or plasma, or from whole-body scanning, this would constitute a major advance.

Partial or non-uniform exposures occur because, with the possible exception of tritiated water, most incorporated radionuclides follow particular metabolic pathways that result in an uneven tissue distribution. The interpretation of aberration yields is particularly difficult in such circumstances because, as well as the partial body irradiation and the possible translocation of material around the body, one has to consider the protracted nature of the exposure at a dose rate that is decreasing due to radioactive decay and excretion.

Certain diseases are treated by the administration of various radionuclides, and surveys of groups of patients provide a valuable source of human data. Our laboratory has recently collaborated in two such studies. In the first, intra-articular injections of colloidal ^{198}Au (10 mCi) or ^{90}Y (5 mCi) were administered to synovial membranes especially of the knees, for the treatment of rheumatoid arthritis (Stevenson *et al.* 1973). In the majority of patients only a small aberration yield was found but in a few a large amount of chromosome damage was noted in the lymphocytes, indicating an equivalent whole-body dose of up to 250 rad. Such a dose is not possible; the amount of nuclide injected, if spread uniformly throughout the whole body, would give a dose of about 10 rad. These high aberration yields were found to correlate well with the amount of isotope leaking from the joint and passing along the lymphatic duct to the regional nodes, where it was measured by scintillation scanning. This correlation led Stevenson *et al.* to suggest that the high aberration yield was due to selective irradiation of lymphocytes as they passed through the nodes containing the activity.

The distribution of the aberrations scored in these patients did not follow Poisson statistics. A detailed analysis of the dicentric data from two patients (table 3) was consistent with 80% of the T cells being unexposed, so that the aberrations were contained in an exposed population of about 20% of the cells. These are the cells that would have circulated through the regional lymph nodes during the week or so required for most of the isotope to decay. The dose received by this population was approximately 400 rad, whereas the dose to the fixed cells of the nodes was much higher (> 6000 rad). This study is of special interest, because it serves to illustrate the problem common to most insoluble particulate radioactive materials that enter the body, as these tend to associate with the lymphatic system.

Table 3. Combined data on dicentric yield and distribution for two patients treated with ^{198}Au compared with the expected distribution from Poisson statistics and a hypothetical irradiation of 20% of the total mass of T cells in the body.

Data	No. of cells scored	Dicentric distribution 0	1	2	3	4	Dicentrics per cell	Dose (rad)
Score from patients	700	638	44	12	2	4	0.13	140
Poisson distribution	700	615	80	5	—	—	0.13	140
Hypothetical case of	562 (80%)	562	—	—	—	—	0	0
partial-body irradiation	138 (20%)	80	44	12	2	—	0.54	380
	700	642	44	12	2	—		

In the second study (Lloyd *et al.* 1976) ^{131}I in solution was given to eleven patients for the treatment of thyroid carcinoma. Each patient received two administrations of the isotope; firstly 80 mCi to ablate the cancerous thyroid followed some months later with 200 mCi to treat any remaining metastases. Thus at the time of the second iodine treatment no functional thyroid tissue capable of selectively taking up iodine remained. It was decided to make cytogenetic estimates of the equivalent whole-body dose, based on a chronic γ-dose-effect curve, for comparison with physical estimates of the dose obtained by standard techniques of measuring thyroid uptake, plasma activity variation and urine activity. Table 4 shows cytogenetic and physical dose data for a patient given 80 mCi and another given 200 mCi. After 80 mCi the physical estimates of whole-body dose ranged from 16 to 52 rad, whereas the cytogenetic estimates were considerably higher, 62 to 139 rad. In each patient the biological dose estimate exceeded twice the physical. At this stage of treatment the patients retained functional thyroid tissue, so that considerable uptake of iodine occurred, followed by the release of protein-bound

Table 4. Physical and cytogenetic dose data for two patients given 80 mCi of ^{131}I to ablate a thyroid carcinoma or 200 mCi ^{131}I to treat metastases.

Physical data									Cytogenetic data		
			Average whole-body radiation doses (rad)								
Patient	Dose (mCi)	Thyroid uptake % dose	PII[1] γ & β	Thyroid γ	Bladder γ	PBI[2] γ & β	Liver γ	Total	Pre or post ^{131}I	Dicentrics per cell	Dose (rad)
1	80	21	11.0	13.0	3.5	3.8	1.9	33	Pre	0	117
									Post	0.06	
2	200	0.3	33.5	0.6	12.9	—	—	47	Pre	0.08	45
									Post	0.094	

[1] Plasma inorganic iodide. [2] Protein-bound iodide.

iodide into the plasma, which then tended to accumulate in the liver. It was suggested that the cytogenetic estimates gave a falsely high value due to the preferential irradiation of the lymphocytes as they entered the tissues with high concentrations of ^{131}I and passed within range of the β-radiation. With 200 mCi this discrepancy was not noted, and comparable dose estimates, which ranged from 30 to 70 rad, were obtained with both techniques. Here the isotope was distributed much more uniformly around the body, as practically all the iodine remained as plasma inorganic iodide with just a transitory concentration in the bladder prior to elimination in the urine.

Conclusions

The effects of partial-body or non-uniform exposures on cytogenetic dose estimations are important factors, which need to be considered in the minority of accidental over-exposures to radiation involving large ($>$ 100 rad) local doses. *In vivo* and *in vitro* studies have indicated that selective reductions in the numbers of cells that come through the cycle for analysis at metaphase by 48–52 h are caused by both interphase death and mitotic delay. More experimental work needs to be undertaken to calibrate these effects, which appear to be dose-dependent.

The mixed culture technique seems to be a useful *in vitro* method for examining some problems of partial-body irradiation, but it urgently needs supplementing with *in vivo* human data on how radiation affects the normal pooling and recycling of lymphocytes and their uneven distribution through the body. The recently developed immunological techniques for discriminating between various types of lymphocytes have prompted an immense research interest in this field, and it is hoped that many of the problems of the *in vivo* and *in vitro* kinetics of lymphocyte populations will soon be resolved.

The development of computer-assisted cell finding and karyotyping may enable large numbers of cells to be scored economically. This would facilitate an analysis of the distribution of aberrations amongst the scored cells to determine its deviation from a Poisson distribution, and thus making possible an estimate of the irradiated volume of tissue. Such an approach might be combined with the analysis of sufficiently large numbers of skin fibroblasts, to permit the estimation of local skin doses based on dose-effect curves for stable aberrations.

Cytogenetic analysis after internal radioactive contamination is still a major problem, and can give misleading estimates of whole-body dose, possibly due to the selective irradiation of lymphocytes. More work in this field is clearly indicated, but it must be supplemented by accurate physical and bioassay data. Initially, therefore, work should be confined to those nuclides for which physical dosimetric techniques are reliable and where there is a good understanding of the metabolism of the nuclide in man.

Summary

Most accidental over-exposures to radiation involve partial or non-uniform irradiation of the body. This raises problems in interpreting chromosome aberration yields from peripheral blood lymphocytes, because both *in vivo* and *in vitro* dose-dependent selection against irradiated cells may occur and may affect the eventual dose estimate. Recent experimental work with human and animal lymphocytes indicates that interphase death and mitotic delay are the principal causes of the selective loss of cells, but no significant effect on the aberration yield is observed in cases involving localised doses of 100 rad or less. The uneven distribution of lymphocytes in the body may also result in variations in the aberration yield, depending upon which parts of the body are exposed. The possibility that fibroblast examinations may enable estimates of local skin doses is discussed. This might also assist in defining the limits of the exposure field, and could be combined with an analysis of the distribution of aberrations in lymphocytes to determine the dose to that volume. Non-uniform exposures also arise from the accidental intake of many radionuclides. In these cases cytogenetic analyses can give misleading estimates of the whole-body dose, possibly due to the selective irradiation of lymphocytes.

References

Bender, M.A. (1969) Human radiation cytogenetics, in *Advances in Radiation Research*, vol.3 (eds L.G.Augenstein, R.Mason & M.Zelle) pp.215–73. London: Academic Press.

Bigger, T.R.L., J.R.K.Savage & G.E.Watson (1972) A scheme for characterising ASG banding and an illustration of its use in identifying complex chromosomal rearrangements in irradiated human skin. *Chromosoma (Berl.) 39*, 297–309.

Blomgren, H., U.Glas, B.Melen & J.Wasserman (1974a) Blood lymphocytes after radiation therapy of mammary carcinoma. *Acta Radiol. Ther. Phys. Biol. 13*, 185–200.

Blomgren, H., J.Wasserman & B.Littbrand (1974b) Blood lymphocytes after radiation therapy of carcinoma of prostate and urinary bladder. *Acta Radiol. Ther. Phys. Biol. 13*, 357–64.

Chee, C.A., P.L.T.Ilbery & A.B.Rickinson (1974) Depression of lymphocyte replicating ability in radiotherapy patients. *Br. J. Radiol. 47*, 37–43.

Dolphin, G.W. (1969) Biological dosimetry with particular reference to chromosome aberration analysis, in *Handling of Radiation Accidents*, pp.215–24. IAEA-SM-119/4. Vienna: IAEA.

Evans, H.J. (1972) Action of radiations on human chromosomes. *Phys. Med. Biol. 17*, 1–13.

Evans, H.J., W.M.Court Brown & A.S.McLean, eds (1967) *Human Radiation Cytogenetics*. Amsterdam: North-Holland.

Lloyd, D.C., G.W.Dolphin, R.J.Purrott & P.A.Tipper (1977) The effect of X-ray-induced mitotic delay on chromosome aberration yields in human lymphocytes. *Mutat. Res. 42*, 401–12.

Lloyd, D.C., R.J.Purrott & G.W.Dolphin (1973) Chromosome aberration dosimetry using human lymphocytes in simulated partial body irradiation. *Phys. Med. Biol. 18*, 421–31.

Lloyd, D.C., R.J.Purrott, G.W.Dolphin, D.Bolton, A.A.Edwards & M.J.Corp (1975) The relationship between chromosome aberrations and low LET radiation dose to human lymphocytes. *Int. J. Radiat. Biol. 28*, 75–90.

Lloyd, D.C., R.J.Purrott, G.W.Dolphin, P.W.Horton, K.E.Halnan, J.S.Scott & G.Mair (1976) A comparison of physical and cytogenetic estimates of radiation dose in patients treated with iodine-131 for thyroid carcinoma. *Int. J. Radiat. Biol. 30*, 473–85.

Lloyd, D.C., R.J.Purrott, J.S.Prosser, G.W.Dolphin, P.A.Tipper, E.J.Reeder, C.M. White, S.J.Cooper & B.D.Stephenson (1977) Doses in radiation accidents investigated by chromosome aberration analysis. VII. A review of cases investigated: 1976. *National Radiological Protection Board Report*, *NRPB-R57*. Harwell: NRPB.

Matsubara, S., M.S.Sasaki & T.Adachi (1974) Dose response relationship of lymphocyte chromosome aberrations in locally irradiated persons. *J. Radiat. Res. 15*, 189–96.

McFee, A.F. (1977) Chromosome aberrations in the leukocytes of pigs after half body or whole body irradiation. *Mutat. Res. 42*, 395–400.

McFee, A.F., M.W.Banner & M.N.Sherrill (1974) Chromosome aberrations in the leukocytes of partial body and whole body irradiated swine. *Radiat. Res. 60*, 165–72.

Osgood, E.E. (1954) Number and distribution of human hemic cells. *Blood 9*, 1141–54.

Prosser, J.S. (1977) Survival of human T and B lymphocytes after X-irradiation. *Int. J. Radiat. Biol. 30*, 459–65.

Sharpe, H.B.A. (1969) Pitfalls in the use of chromosome aberration analysis for biological radiation dosimetry. *Br. J. Radiol. 42*, 943–4.

Sharpe, H.B.A., G.W.Dolphin, K.B.Dawson & E.O.Field (1968) Methods for computing lymphocyte kinetics in man by analysis of chromosomal aberrations sustained during extracorporeal irradiation of the blood. *Cell Tissue Kinet. 1*, 263–71.

Stevenson, A.C., J.Bedford, G.W.Dolphin, R.J.Purrott, D.C.Lloyd, A.G.S.Hill, H.F.H.Hill, J.M.Gumpel, D.Williams, J.T.Scott, N.W.Ramsey, F.E.Bruckner & C.B.D'A. Fearn (1973) Cytogenetic and scanning study of patients receiving intra-articular injections of gold-198 and yttrium-90. *Ann. Rheum. Dis. 32*, 112–23.

Stjernswärd, J., M.Jondal, F.Vanky, H.Wigzell & R.Sealy (1972) Lymphopenia and changes in distribution of human B and T lymphocytes in peripheral blood induced by irradiation for mammary carcinoma. *Lancet 1*, 1352–6.

Tamura, H., Y.Sugiyama & T.Sugahara (1974) Changes in the number of circulating lymphocytes with chromosomal aberrations following a single exposure of the pelvis to γ-irradiation in cancer patients. *Radiat. Res. 59*, 653–7.

Vodopick, H. & G.A.Andrews (1974) Accidental radiation exposure. *Arch. Environ. Health 28*, 53–6.

J. A. STEFFEN, K. SWIERKOWSKA,
A. MICHALOWSKI, E. KLING and A. NOWAKOWSKA

In vitro Kinetics of Human Lymphocytes activated by Mitogens

When human peripheral blood lymphocytes in cultures stimulated by mitogens are grown in the presence of amethopterin or some other agents that inhibit reversibly the synthesis of DNA, the responding cells can be blocked in their passage through the first *in vitro* cell cycle at the G_1/S boundary. The amethopterin (MTX)-imposed G_1/S block can be released within minutes by exogenous thymidine. Thus, when ^{3}H-thymidine is added at time intervals to replicate cultures prior to harvest, the kinetics of G_1/S accumulation can be evaluated by means of autoradiography. In addition, when the G_1/S block is released at the same time in several replicate cultures, which are then harvested at time intervals following incubation with colcemid, the variability of duration as well as median time of the $S + G_2$ phases can be directly evaluated from the patterns of the first generation mitotic wave and by estimating the ratios of the cells that arrive up to different times after G_1/S block release at mitosis to the cells that were accumulated at the G_1/S block (Steffen and Stolzmann 1969).

While using this approach in studies on the *in vitro* kinetics of the human lymphocyte response to PHA, we have established the following (Jasinska, Steffen and Michalowski 1970; Steffen and Stolzmann 1969); (a) under conditions of optimum PHA stimulation about 25–40% of cells die during the first 40 h of culture, but further cell loss up to the end of the fifth day of culture is negligible; (b) accumulation of cells at the G_1/S boundary starts during the second day of culture but may continue up to the fifth day; (c) the kinetics of G_1/S accumulation are frequently biphasic; (d) in cultures of cells from healthy donors, stimulated with an optimum concentration of PHA, the fraction of cells accumulated at the G_1/S boundary at the end of the fourth or fifth day of culture very seldom exceeds 0.6; (e) following release of the G_1/S block cells first start to appear at mitosis after about 6 h; the first generation mitotic wave is, however, always asymmetric and frequently bi-modal; (f) occasionally more than 75% of G_1/S cells reach the first mitosis during a 24-h collection period following release of the block. The above results imply a considerable heterogeneity, and possibly the involvement of different subpopulations or functional subsets in the response to PHA.

At least two lines of evidence suggest some correlations between the *in vitro* kinetics and *in vitro* chromosomal radiosensitivity of lymphocytes. Thus Bender and Brewen (1969) reported higher yields of unstable radiation-induced chromosomal aberrations in cultures harvested during the third day of cultivation, despite the presence of many more second-generation cells, than were obtained with cultures harvested at the end of the second day. In our laboratory considerable differences in the yields of unstable aberrations were found in first generation mitoses in different collection periods following the release of an amethopterin-imposed block in irradiated lymphocyte cultures (Steffen and Michalowski 1973).

In the present report we summarise the results of several series of experiments to evaluate further the heterogeneity of peripheral lymphocyte populations that are responsive to phytohaemagglutinin (PHA), concanavalin A (Con A) and pokeweed mitogen (PWM). These studies may be pertinent to the problem of the comparative radiosensitivity of subpopulations or functional subsets of lymphocytes circulating in the peripheral blood. In addition, preliminary results of studies on the *in vitro* radiosensitivity of purified T lymphocytes responsive to PHA and Con A will be reported.

Materials and Methods

Most of the experiments were performed on lymphocytes from blood of healthy donors. 100 ml or, exceptionally, 450 ml of blood was mixed, in a ratio of 6 to 1, with Hank's solution containing 10–40 IU of phenol-free heparin and 180 μg of kanamycin per ml. Leucocytes were separated from red blood cells together with the plasma either by 2 h of gravity sedimentation at 37°C, or by centrifugation with ficoll-isopaque or ficoll-uropoline solutions as described by Boyum (1968). In the majority of the experiments leucocytes were subsequently suspended in Hank's solution and carbonyl-iron (Koch-Light) was added at 1 mg per 10^6 cells. Following 30 min incubation at 37°C in a constant shaking water bath, cells remaining in suspension after the application of a strong magnet were separated from those that had sedimented. Suspensions containing 94–97% of mononuclear cells were usually obtained.

Separation of T from non-T lymphocytes relied on the ability of the thymus-dependent cells to form so-called 'spontaneous' (or 'non-immunologic') rosettes with erythrocytes of sheep (Jondal, Holm and Wigzell 1972). These E-rosette-forming cells were separated from those that did not bind sheep red blood cells (SRBC) by a modification of the method of Wybran *et al.* (1973). In brief, following three washes in Hank's EDTA solution (0.04 M for the first and 0.0027 M for the subsequent two washes) lymphocytes were suspended at 10^7 cells ml^{-1} in calf serum preabsorbed with SRBC. Aliquots (0.8 ml) were mixed with equal volumes of a freshly prepared 2% suspension of SRBC in Hank's solution, in tubes of 16 mm diameter. A two-step purification procedure was usually employed: (a) Following 5 min of centrifugation at 50*g*, the mixtures of lymphocytes and SRBC were kept for 20 min at room temperature, layered after gentle resuspension on ficoll-uropoline (sp. w. 1078) and centrifuged for 20 min at 950*g*. Cells from the interface of ficoll-uropoline and the supernatant layer were separated from those that bound SRBC and sedimented to the bottom of tubes. Red blood cells in the bottom (T cell) fraction were lysed by a 12 min exposure to 0.83% ammonium chloride solution and washed three times in Hank's solution. (b) Interface cells were washed, resuspended in calf serum, again mixed with a 2% suspension of SRBC and centrifuged for 5 min at 50*g*. Following 60 min at 0–4°C and gentle resuspension, this lymphocyte-SRBC mixture was layered on ficoll-uropoline and centrifuged at 0–4°C for 20 min at 950*g*. Only interface cells were collected following this second step of purification of non-T lymphocytes.

The purity of the sedimented (T) and the second interface (non-T) populations was evaluated by counting the fraction of cells that bound spontaneously to SRBC after 2 h at 0–4°C and, often, by scoring cells displaying a ring-shaped fluorescence following incubation with fluorescein-conjugated polyvalent rabbit-anti-human-Ig serum according to standard methods (Jondal, Holm and Wigzell 1972). The mean percentage of E-rosette-forming cells in the population prior to T cell separation was 73%. Usually more than 90% and, occasionally, more than 95% of E-rosette cells were found in the bottom (T cell) fraction. The interface populations, following two-step purification, usually contained less than 5% of E-rosette-forming cells.

Lymphocyte cultures with a cell density of 0.75×10^6 cells ml^{-1} were established, if not stated otherwise, in 2 or 4 ml aliquots of medium containing 80% of Parker's TC-199 solution, freshly supplemented with glutamine to a final concentration of 0.3 mg ml^{-1} and 20% of autologous plasma. Mitogens were added at the onset of cultivation to final concentrations that in preliminary experiments were found to elicit the highest mitotic response in 72-h cultures. These concentrations corresponded for PHA (a purified preparation from Wellcome Laboratories, containing 140 mitogenic units in 10 mg) to 5 μg per ml, for Con A (Sigma) to 17.5 μg of protein per ml, and for PWM (Gibco) to 0.025 ml per ml of culture.

In the majority of experiments freshly dissolved amethopterin (methotrexate, MTX, Lederle) was added to cultures giving a final concentration of 2.2×10^{-7} M. The original medium was usually replaced by fresh culture medium containing both MTX and mitogen, after two days. Cultures were harvested either: (a) at different time intervals, following 1 to 3 h of incubation with ^{3}H-methyl-thymidine (Amersham, spec. activ. 6 Ci $mmol^{-1}$) or (b) following release of the G_1/S block, usually after three days of culture, by 'cold' thymidine (Sigma, 50 μg per ml of culture), at 3-hour intervals, following 3 h incubation with colcemid (Steffen and Stolzmann 1969, Steffen and Michalowski 1973). Three to five preparations made from each culture, according to the method described by Moorhead *et al.* (1960), were processed by standard methods for autoradiography and/or evaluation of the mitotic index.

In some experiments designed to study cell cooperation in 'mixed' cultures of T and non-T lymphocytes, either the T or non-T cells were suspended prior to culture in Hank's solution with freshly dissolved mitomycin C (50 μg

ml^{-1}) and incubated for 30 min at 37°C. These cells were subsequently washed three or four times in Hank's solution to remove the excess inhibitor.

For evaluation of the response of some 'asynchronous' cultures ^{3}H-methyl-thymidine (Amersham), diluted by 'cold' thymidine to a specific activity corresponding to 1 μCi per ml of medium, was added 4 or 16 h prior to harvest. The cell pellets, prior to counting in a beta-scintillation counter, were processed, as described by Geha, Roser and Marler (1973).

In preparations from *in vitro* irradiated cultures, the yield of unstable aberrations was scored according to criteria outlined by Sharpe, Scott and Dolphin (1969). However, acentric fragments and minutes were grouped together, as were dicentrics and rings.

Samples of blood (12–40 ml) were obtained from cancer patients, following therapeutic partial-body irradiation. Procedures for separating lymphocytes from red blood cells, and for purifying T and non-T populations, were essentially the same as described above. In several experiments, due to low yields of leucocytes, the carbonyl-iron procedure for removing phagocytic cells was omitted. Standard microcultures with the medium containing MTX were established in experiments aimed at evaluating the yields of metaphases carrying unstable aberrations, in relation to the position of mitoses within the first-generation mitotic wave. Microcultures, which were usually employed for evaluation of aberration yields in 'asynchronous' cultures of whole blood, were established by a method described by Evans (1965). Such cultures were usually harvested at 4-h intervals starting from the 36th or 40th hour of culture, and aberration-carrying cells were scored at the onset of mitotic activity.

Results

Comparative Kinetics

When the kinetics of the initiation of the proliferative response in replicate cultures grown under conditions of amethopterin-imposed G_1/S block, stimulated respectively with optimum doses of PHA, Con A and PWM, are compared (table 1), the following features are apparent: (a) The overall pattern of G_1/S accumulation in cultures stimulated by each of these mitogens is the same, that is, cells continue to accumulate at the G_1/S boundary for at least 48 h. (b) As compared with responses to PHA, relatively few Con A-responsive cells appear at the G_1/S boundary following 48 and 72 h of cultivation. However in some of the experiments the fraction of G_1/S accumulated cells in Con A-stimulated cultures increases considerably during the fourth day of incubation, while a similar increase is not seen in parallel PHA cultures. (c) The fraction of responder cells in the initial population is always the highest in PHA-stimulated cultures. (d) No relationships are apparent between the relative magnitude of the response (e.g. as evaluated by the size of the G_1/S fraction following three or four days of culture) to each of the mitogens tested, in replicate cultures from the same donors.

To assess whether the same or different subsets of lymphocytes are

Table 1. Comparative kinetics of G_1/S accumulation of lymphocytes in cultures stimulated by PHA, Con A and PWM under conditions of MTX-imposed reversible DNA synthesis block. All results are means from three replicate cultures.

		Cells incorporating ^{3}H-TdR after G_1/S block release (%)			
Exp. no.	Mitogen	48 h	72 h	96 h	120 h
1	PHA	17.7	25.2	28.4	n.d.
	Con A	3.6	8.5	16.9	n.d.
	PWM	2.5	2.1	7.7	n.d.
2	PHA	10.0	20.3	21.1	n.d.
	PWM	2.4	6.0	6.2	n.d.
3	Con A	2.3	6.1	19.9	n.d.
	PWM	5.1	14.1	13.8	n.d.
4	PHA	n.d.	28.0	37.2	n.d.
	Con A	n.d.	3.3	4.2	n.d.
	PWM	n.d.	8.4	13.3	n.d.
5	PHA	n.d.	13.3	n.d.	36.0
	Con A	n.d.	3.5	16.3	17.4
	PWM	n.d.	1.6	n.d.	5.5

involved in the proliferative response elicited respectively by PHA, Con A and PWM, experiments were performed in which the fractions of G_1/S accumulated cells in replicate cultures stimulated by each of these mitogens were compared with G_1/S fractions of cultures stimulated by two mitogens together. Figure 1 summarises the results of three typical experiments from this series. It is apparent that when cultures were stimulated by PHA and Con A together the response is somewhat higher at 72 h as compared with cultures stimulated by PHA alone. However, following a further 24 h, fewer cells are accumulated at the G_1/S boundary in cultures stimulated by these two mitogens as compared with those stimulated by PHA alone. Similar results were obtained in cultures stimulated by both PHA and PWM. The results of stimulation by PHA and Con A, or PHA and PWM, together are thus very similar to those obtained with higher than optimum concentrations of PHA (Jasinska, Steffen and Michalowski 1969). That is, while the kinetics of G_1/S response seem to appear somewhat 'faster', a lower than optimum final response is obtained. Hence, most, if not all, Con A- and PWM-responsive cells are also potential PHA-responders. On the other hand, following stimulation with optimum doses of Con A and PWM added together, the fraction of G_1/S accumulated cells, both after 72 and 96 h of culture, approaches the values expected on the assumption that the effects of the two mitogens are additive. Thus, the majority of cells responding respectively to Con A and PWM seem to belong to different lymphocyte subsets.

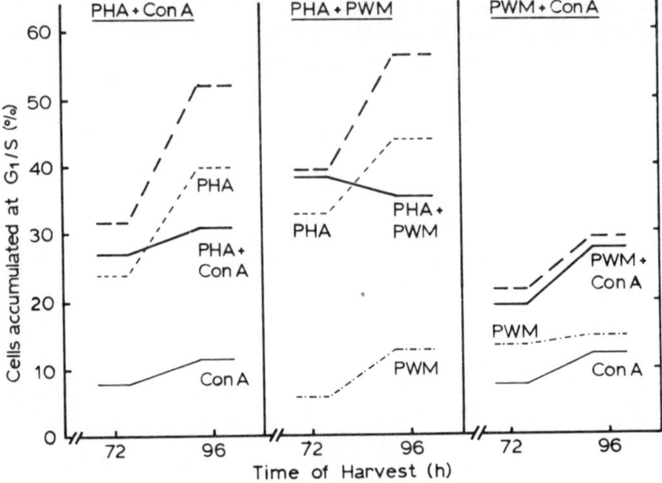

Figure 1. Percentages of G_1/S accumulated cells in 72- and 96-h cultures, grown in the presence of amethopterin, as evaluated by autoradiography 3 h after ^{3}H-thymidine was added. Replicate cultures were stimulated either with one or two mitogens in optimum concentration. Upper broken lines correspond to values expected under the assumption that each of the two mitogens stimulates different sub-classes of lymphocytes. All results are means from triplicate cultures. Left side, centre and right side of figure correspond to different experiments.

Table 2. Kinetics of accumulation at G_1/S of human peripheral T and non-T lymphocytes grown in presence of MTX in PHA-stimulated cultures. All results are means from three replicate cultures.

Exp. no.	Time of harvest (h)	G_1/S accumulated cells (%)		
		N-S	T	non-T
1	40	14.5	13.8	0.4
	64	44.9	47.0	3.5
	88	54.5	55.3	12.0
2	64	35.5	39.2	2.0
	88	45.0	44.5	n.d.
	112	58.1	59.2	0.5
3	40	28.1	7.2	1.4
	64	58.3	43.4	5.4
	94	47.7	38.7	16.8

N-S = non-separated lymphocytes. E-rosette-forming cells in consecutive experiments (%):
(1) N-S, 60.0; T, 89.0; non-T, 3.8
(2) N-S, 66.5; T, 92.0; non-T, 0.5
(3) N-S, 70.0; T, 94.5; non-T, 6.6

Comparative in vitro Kinetics and Cell Cooperation

The comparative kinetics of accumulation at the G_1/S boundary of the first cell cycle of PHA-stimulated, purified T and non-T lymphocytes is shown in table 2. It is apparent that the proliferative response of non-T cells is much smaller and delayed in time as compared with the purified T, as well as the non-separated lymphocytes. While a small proportion of T cells was always present in the non-T lymphocyte cultures, the fraction of responding cells in two out of three experiments clearly exceeded the percentage of E-rosette-forming cells present in the initial population. This implies a true involvement of the non-T lymphocytes in the proliferative response to PHA. The kinetics of G_1/S accumulation of purified T vs. non-separated lymphocytes was found to be quite similar in these experiments. It may also be noted that the size of the fraction of responding cells in the purified T cell population, at least up to the 96th hour of incubation, was seldom, if ever, higher than for cultures of non-separated lymphocytes.

Table 3. Cooperation of human peripheral T and non-T lymphocytes in the proliferative response to PHA, Con A and PWM in culture. Means from three replicate cultures. E-rosette-forming cells in the non-T population were less than 2% in all experiments.

Exp. no.	Mitogen	Time of harvest (h)	G_1/S accumulated cells (%)		
			T	non-T	T + non-T (1:1)
1	PHA	64	64.0	23.7	56.3
	Con A	64	13.0	31.7	43.4
2	PHA	64	14.4	1.7	25.2
	Con A	64	0.6	2.0	8.1
3	PHA	64	34.9	20.5	43.0
	PWM	64	6.7	2.5	8.2
4	Con A	64	4.1	0.0	16.5
	PWM	88	9.3	0.2	55.2
	PWM	64	9.7	0.0	8.5
		88	11.7	0.0	19.8

Essentially similar results were obtained when the response of T and non-T lymphocytes to PHA, Con A and PWM was compared (table 3). However, a considerable non-T cell response was found in some experiments in which the proportion of E-rosette-forming cells was negligible. When T and non-T lymphocytes, following separation, were mixed in cultures in a 1:1 ratio the resulting response was frequently much higher than the expected mean from T and non-T cells grown separately. This finding, which is particularly evident in some of the Con A-stimulated cultures, implies that T and non-T cell cooperation is occurring.

Direct evidence for cooperation of T and non-T lymphocytes was obtained in several experiments by pretreatment of the purified subpopulations

Table 4. Cooperation of human peripheral T and non-T lymphocytes in the proliferative response to mitogens: 'One-way' responses, PHA stimulation. Means from three replicate cultures. 0.5 and 1.5% of E-SRBC-forming cells were present in the non-T cell population in expts 1 and 2, respectively.

	^{3}H-thymidine incorporation/culture (c.p.m.)		
	Exp. 1		Exp. 2
Cells	64 h	88 h	72 h
T	16060	30570	77600
non-T	300	600	1180
T + non-T (1:1)	31430	34740	n.d.
$T_{mit.C}$	1370	930	2333
$T_{mit.C}$ + non-T (1:1)	9130	13500	3680
non-$T_{mit.C}$ + T (1:1)	7410	11470	41600

Table 5. Cooperation of human peripheral T and non-T lymphocytes in the proliferative response to mitogens: 'One-way' responses, Con A stimulation. Means from three replicate cultures. 0.5 and 7.0% of E-SRBC-forming cells were present in the non-T population in expts 1 and 2, respectively.

	^{3}H-thymidine incorporation/culture (c.p.m.)	
Cells	Exp. 1 (64 h)	Exp. 2 (80 h)
T	1360	19790
non-T	740	3410
T + non-T (1:1)	7780	15970
$T_{mit.C}$	900	860
$T_{mit.C}$ + non-T (1:1)	840	12430
non-$T_{mit.C}$ + T (1:1)	790	6950

with mitomycin C prior to culture. Mitomycin C-pretreated, purified T and non-T cells were mixed 1:1 with the other non-pretreated population, and cultures with PHA, Con A and PWM were established. The results of these experiments (tables 4–6) provide clear-cut evidence for a 'one-way' cooperation of T and non-T cells in the proliferative response to mitogens. Thus, it was demonstrated that T cells can help in the response of non-T, but the reverse does not occur. It may be noted, however, that either the 'helper' function of T cells or, alternatively, the ability of non-T lymphocytes to respond in the presence of T cells, was subject to considerable donor to donor variability.

The speed with which partially purified T and non-T lymphocytes moved

Table 6. Cooperation of human peripheral T and non-T lymphocytes in the proliferative response to mitogens: 'One-way' responses, PWM stimulation. Means from three replicate cultures. 5.5 and 1.0% of E-SRBC-forming cells were present in the non-T population in expts 1 and 2, respectively.

Cells	^{3}H-thymidine incorporation/culture (c.p.m.)	
	Exp. 1 (70 h)	Exp. 2 (80 h)
T	40 760	9 080
non-T	5 260	170
T + non-T (1:1)	33 360	n.d.
$T_{mit.C}$	5 130	380
$T_{mit.C}$ + non-T	9 460	580
non-$T_{mit.C}$ + T	23 300	4 120

through the S + G_2 phases of the cell cycle, was compared in experiments in which the amethopterin-imposed G_1/S block was released in replicate cultures at the onset of the 4th day of incubation. Replicate cultures were harvested at 3-h intervals following 3 h for incubation with colcemid, during a 24 h collection period (figure 2 and table 7). The ratio of cells that arrived at mitosis during this period to the fraction of cells accumulated at the G_1/S boundary prior to block release was always highest in cultures of non-T

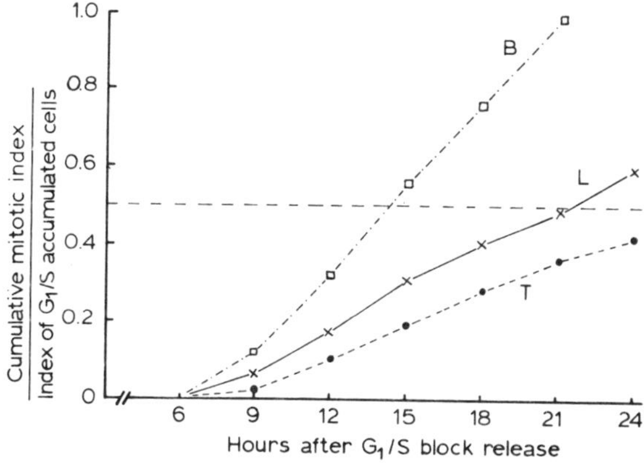

Figure 2. Comparative velocities of passage through the S + G_2 phases of the first *in vitro* cell cycle following release of amethopterin-imposed G_1/S block in cultures of whole (non-separated) peripheral lymphocyte population (L), partially purified T (T) and partially purified non-T (B) lymphocytes, stimulated by PHA. The percentage of G_1/S accumulated cells was estimated for the time corresponding to the 9–12 h collection period. Cumulative mitotic index corresponds to the fraction of cells that reached mitosis up to different time intervals following release of G_1/S block.

Table 7. Comparative $S + G_2$ kinetics of T and non-T cell enriched human peripheral lymphocytes stimulated by PHA

Exp. no.	Cells	% of E-SRBC-forming	L.C.I.[1]	C.M.I.[2]	C.M.I./L.C.I.[3]
1	N-S	70	40.4	21.5	0.52
	T	86	48.3	20.2	0.42
	non-T	18	11.1	11.5	1.04
2	N-S	71	46.5	21.8	0.47
	T	90	51.0	24.7	0.49
	non-T	19	26.0	18.0	0.69
3	N-S	69	48.5	23.8	0.59
	T	91	34.4	16.2	0.47
	non-T	39	46.3	30.8	0.66

[1] Percent of G_1/S accumulated (^{3}H-TdR incorporating) cells at time of block release.
[2] Percent of cells that arrived at mitosis during a 24 h period after release of G_1/S block.
[3] Calculated fraction of G_1/S cells that reached mitosis up to 24 h after block release.

lymphocytes. Due to the presence of a considerable proportion of T cells in the non-T populations (highly purified non-T cells responded poorly in these experiments), the results cannot be considered as quantitative. They indicate, however, a much shorter duration of the $S + G_2$ phases of the non-T cells as compared with the T and non-separated lymphocytes.

In vitro Cell Cycle Kinetics and Comparative Responsiveness

To establish relationships between the *in vivo* life span and the *in vitro* cell cycle kinetics, the relative frequencies of metaphases carrying unstable chromosomal aberrations were measured at different times following the release of the G_1/S block in lymphocyte cultures from several patients, 2–13 months after partial-body irradiation.

Most of the patients were treated by irradiation for carcinoma of the uterine cervix or corpus. Their treatment consisted of local X-irradiation (250 kV, 1.5 mm Cu filtration) delivered to four gynaecological fields in split doses every second day through a period of 3–5 weeks; the sum of the total skin dose ranging from 14400 to 16800 rad. In addition, these patients obtained both intravaginal and intrauterine ^{226}Ra treatment; the total dose thus delivered amounted to 6385–12895 mg h^{-1}. Also included in the study were patients with breast carcinoma who had received X-irradiation after surgery. The radiation was delivered to three fields in split doses, each second day for three weeks, resulting in a total skin dose of about 14500 rad.

Since a detailed account of this study will be published elsewhere, only some typical experiments will be described here (figure 3 and 4, tables 8 and 9). Following the release of the amethopterin-imposed block after three days

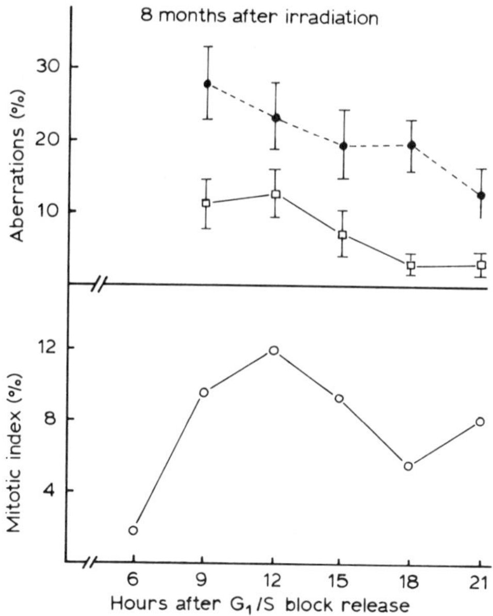

Figure 3. Percentages of metaphases carrying dicentrics and rings (solid line, upper graph) and acentric fragments (broken line, upper graph) in relation to their position within the first-generation mitotic wave in PHA-stimulated replicate cultures of lymphocytes from a patient 8 months after partial-body irradiation. The pattern of the first-generation mitotic wave, following release of the amethopterin-imposed G_1/S block, is shown in the lower part of the graph.

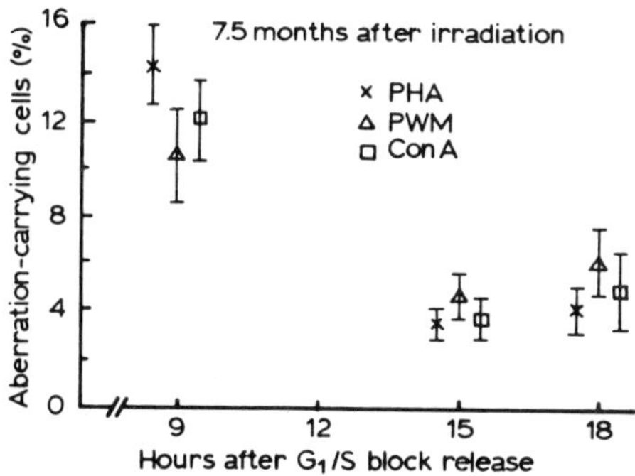

Figure 4. Percentages of metaphases carrying dicentrics and rings in the 6–9, 12–15 and 15–18 h collection periods, following release of G_1/S block, in replicate PHA-, Con A- and PWM-stimulated cultures of lymphocytes from a patient 7.5 months after partial-body irradiation.

of culture, metaphases carrying unstable aberrations were most frequently encountered in collection periods corresponding to 6–9 h and 9–12 h after block release. The frequencies of aberration-carrying metaphases in consecutive collection periods following G_1/S block release tended to increase with time after *in vivo* irradiation. In cultures from patients 9–13 months after irradiation, metaphases carrying asymmetric exchanges were encountered up to ten times as frequently in the 6–9 h or 9–12 h collection periods after the G_1/S block release, as compared with cultures harvested 6 or 9 h later. Some second generation mitoses, as visualised by the increase of frequencies of metaphases with dicentrics not accompanied by acentric fragments, started to appear in most experiments during the 18–21 h collection period following the block release, but were encountered in higher frequencies only 3 h later.

A very similar pattern of frequencies of aberration-carrying metaphases in relation to time of harvest after G_1/S block release was encountered in replicate cultures stimulated respectively by PHA, Con A and PWM (figure 4).

The relative overall response of long-lived lymphocytes appeared, however, to be much higher in Con A- and PWM-stimulated cultures compared with those stimulated by PHA. This was apparent from the comparison of frequencies of aberration-carrying metaphases in 'asynchronous' whole blood microcultures harvested after 48 or 54 h of cultivation, at the beginning of mitotic activity (table 8).

Table 8. Percentages of metaphases carrying asymmetric chromosomal exchanges in replicate cultures of PHA-, Con A- and PWM-stimulated lymphocytes from patients 7–11 months after therapeutic irradiation.

		Percent of metaphases with asymmetric exchanges		
Exp. no.	Mitogen	A[1]	B[2]	A/B
1	PHA	14.2 ± 1.9	2.9 ± 0.7	4.9
	PWM	10.5 ± 1.5	4.7 ± 0.9	2.2
	Con A	12.0 ± 2.0	9.5 ± 1.4	1.3
2	PHA	6.4 ± 1.2	2.7 ± 0.7	2.4
	Con A	7.1 ± 1.3	5.7 ± 0.7	1.3
3	PHA	4.5 ± 1.2	2.0 ± 0.9	2.3
	PWM	5.7 ± 0.7	3.3 ± 3.0	1.7
	Con A	9.9 ± 2.1	7.8 ± 2.7	1.3

[1] In synchronous cultures, 6–9 h after release of G_1/S block.
[2] Non-synchronous, 42–46 h cultures.

Finally, it was found that long-lived lymphocytes are present both in the T and non-T populations of PHA-responsive lymphocytes. This is evident from experiments in which lymphocytes from patients after partial-body irradiation were separated on the basis of their ability to form E-rosettes. They were then grown with mitomycin C-pretreated autologous T cells, to ensure the optimum response in the non-T population, and harvested at the beginning of the third day of culture (table 9).

The results described here can be thus summarised as follows: (a) A

Table 9. **Relative frequencies of aberration-carrying metaphases in PHA-stimulated cultures of partially purified T and non-T lymphocytes from patients with carcinoma of the uterine cervix after radiation therapy.**

Exp. no.	Months after irradiation	Type of aberration	Percentages of aberration-carrying metaphases		
			N-S	T	non-T
1	9	AE	9.8 ± 2.2	9.4 ± 1.5	11.6 ± 2.2
		F	11.3 ± 2.4	14.9 ± 1.9	15.3 ± 2.5
2	10	AE	5.4 ± 1.6	3.5 ± 1.0	3.3 ± 1.6
		F	7.4 ± 1.9	4.4 ± 1.2	4.0 ± 1.8
3	9	AE	12.4 ± 1.7	8.7 ± 1.6	6.2 ± 1.3
		F	16.9 ± 2.0	12.6 ± 1.9	10.0 ± 1.7
4	12	AE	3.3 ± 1.2	1.6 ± 0.9	4.8 ± 1.8
		F	5.8 ± 1.6	3.2 ± 1.3	7.5 ± 2.3
5	12	AE	5.5 ± 2.1	2.0 ± 0.7	3.0 ± 1.1
		F	5.5 ± 2.1	2.4 ± 0.8	4.8 ± 1.4

AE = asymmetric exchanges, F = fragments, N-S = non-separated.
Percentages of E-SRBC-forming cells: N-S, 52–69%; T, 80–87%; non-T, 1.8–9.5%.

clear-cut relationship seems to exist between the *in vivo* life span and the *in vitro* cell cycle kinetics of lymphocytes; the *in vivo* long-lived cells displaying a shorter mean duration of the $S + G_2$ period of the cell cycle, as compared with the whole lymphocyte population. (b) A relatively larger fraction of long-lived lymphocytes appears to participate in the response to Con A as well as to PWM, as compared with the PHA-responsive population. (c) Long-lived lymphocytes are present both in the T and non-T fractions of PHA-responsive cells.

Comparative in vitro Chromosomal Radiosensitivity

In two series of preliminary experiments, populations of non-separated T plus non-T lymphocytes and purified T cells, obtained from the same samples of blood from healthy donors, were irradiated at room temperature with 100 rad of γ-rays from a ^{60}Co source, at a dose rate of 53 rad min^{-1}.

In the first series of experiments (table 10) PHA and Con A respectively were added to replicate macrocultures, which were harvested after 54 h of incubation when first mitoses started to appear. The mean frequencies of aberrations per metaphase in this series, in cultures stimulated with PHA, were higher in the purified T cells as compared with the non-separated whole lymphocyte populations. In addition, aberration yields in PHA-stimulated cultures of purified T lymphocytes in both experiments were considerably higher as compared with parallel cultures stimulated with Con A.

In the second series of experiments (figure 5) aberration yields were scored in replicate cultures of non-separated lymphocytes and purified T cells harvested at different time intervals after the release of the amethopterin-imposed G_1/S block at the beginning of the 4th day of culture. The patterns

Table 10. Yields of unstable chromosomal aberrations in 54 h cultures of γ-irradiated (100 rad) non-separated and purified T lymphocytes stimulated by PHA vs. Con A.

			Aberrations/metaphase	
Exp. no.	Cells	Mitogen	Asymmetric exchanges	Fragments
1	N-S	PHA	0.061 ± 0.014	0.174 ± 0.024
		Con A	0.088 ± 0.022	0.160 ± 0.030
	T	PHA	0.129 ± 0.031	0.303 ± 0.048
		Con A	0.053 ± 0.019	0.205 ± 0.037
2	N-S	PHA	0.095 ± 0.018	0.237 ± 0.029
		Con A	0.084 ± 0.025	0.145 ± 0.033
	T	PHA	0.144 ± 0.032	0.247 ± 0.041
		Con A	0.091 ± 0.023	0.177 ± 0.032

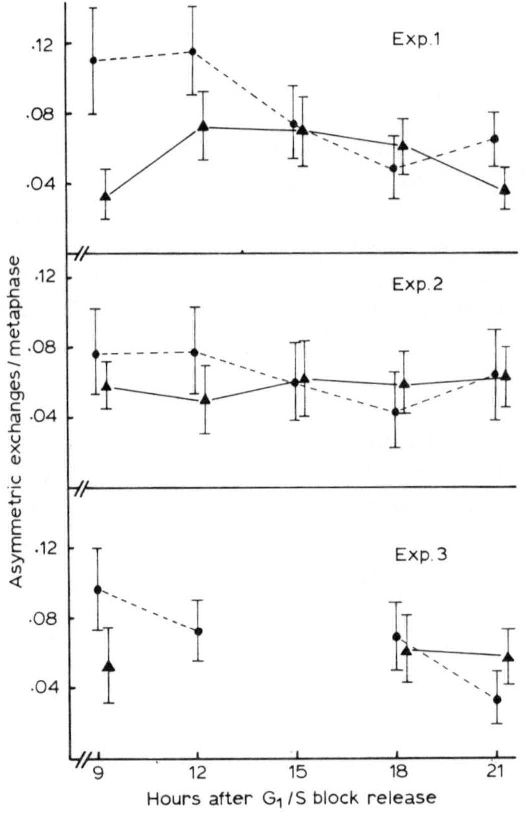

Figure 5. Frequency of asymmetric exchanges per metaphase in consecutive collection periods following release of G_1/S block at the beginning of the 4th day of cultivation in PHA-stimulated cultures of whole (non-separated) lymphocyte populations (solid line) vs. purified T cells (broken line), irradiated *in vitro* with 100 rad of γ-rays. The percentages of E-SRBC rosette-forming cells in consecutive experiments were as follows: non-separated lymphocytes 79.0, 85.0 75.5; purified T lymphocytes 93.5, 97.0, 94.0.

of aberration frequencies in metaphases in relation to the time after release of the block in cultures of non-separated lymphocytes displayed donor to donor variation. On the other hand, more reproducible patterns, implying a reverse correlation between radiosensitivity and the duration of the $S + G_2$ period of the cell cycle, emerged from the evaluation of the parallel series of purified T cell cultures.

While direct studies on radiosensitivity of non-T cells are needed, and the present experiments need confirmation, these preliminary results may imply differences of T and non-T lymphocyte radiosensitivity and heterogeneous radiosensitivity of the cell population, probably related to differences of *in vitro* cell cycle kinetics.

Discussion

The population of mammalian peripheral lymphocytes consists of two developmentally-independent compartments, of thymus-dependent (T) and thymus-independent (non-T) cells, within each of which distinct subsets of cells can be distinguished by surface markers, life-span and immunological potential. These subsets may reflect either different developmental pathways or functional diversity related to the degree of maturity (Greaves, Owen and Raff 1974; Raff and Cantor 1971; Shiku *et al.* 1976; UICC 1974; Wigzell 1973). It seems very likely that the heterogeneity of the human peripheral lymphocyte population is reflected by the patterns of *in vitro* kinetics of cells responsive to PHA and other non-specific mitogens (Steffen and Stolzmann 1969; Jasinska, Steffen and Michalowski 1970).

In recent years a large body of experimental work concerning relationships between the functional diversity and mitogen responsiveness of lymphocytes both from mice and man has been published. In particular, earlier assumptions (Greaves and Janossy 1972) that some of the plant lectins, most notably Con A and PWM, can elicit a selective response of T and non-T lymphocytes respectively, could not be sustained by further work (Ling and Kay 1975). Following studies on separated human T and non-T lymphocytes it has been established that PHA, Con A and PWM can elicit the response of thymus-dependent lymphocytes, whereas the response of non-T cells to these mitogens is largely T-dependent (Lohrmann, Novikovs and Grow 1974; Steffen 1974). The results of experiments reported here indicate, however, that only part of the circulating T lymphocytes will respond to PHA, Con A and PWM under conditions of optimum stimulation, and that the magnitude of this response is subject to considerable donor-to-donor variability. In addition, while the response of non-T cells is to a large degree T-cell-dependent, the primary responsiveness of a fraction of non-T cells from some donors cannot be excluded. Finally, the T-cell-dependent response of non-T cells in cultures stimulated by non-specific mitogens differs considerably from donor to donor. Presumably this is either due to a different 'helper' activity of the T cells, or a different ability of non-T lymphocytes to respond to T cell 'help'.

Some of the relationships between the presence of surface markers and immunological potential as well as maturity and life span both for the T and non-T compartments have been established in mice (Raff and Cantor 1971, Shiku *et al.* 1976, UICC 1974, Wigzell 1973), but little similar information is yet available for man. Some of the specific features of lymphocytes in cultures stimulated respectively by PWM (Douglas *et al.* 1967; Parkhouse, Janossy and Greaves 1972) and Con A (Ling and Kay 1975, Janossy and Greaves 1972) may imply a degree of selectivity of stimulation of different lymphocyte subsets by these mitogens. Experiments performed in cultures grown under conditions of amethopterin-imposed G_1/S block support this view, since: (a) in replicate cultures from the same donor no correlation was found between the magnitude of response to each of the mitogens tested; and (b) the magnitude of response in cultures stimulated with Con A and PWM together correspond to values expected on the assumption of simple additive effects of these two mitogens. However, the latter conclusion must be considered as preliminary, until these experiments are repeated on purified T and non-T lymphocyte subpopulations.

Relatively little is known about relationships between *in vivo* life span and immunological potential of lymphocytes (Greaves, Owen and Raff 1974). While long-lived cells are linked to immunological memory, this has yet to be proven in direct experiments. Long-lived lymphocytes are present both in the T and non-T compartments in mice (Iversen 1973) and seem to predominate in the more mature subclass of T lymphocytes (Raff and Cantor 1971). Both short- and long-lived lymphocytes are among the PHA-responsive cells in man. This is apparent from studies on frequencies of metaphases carrying unstable chromosomal aberrations in relation to time following partial-body irradiation (Buckton, Court Brown and Smith 1967). Experiments reported here provide evidence that cells with a minimum life span of about ten months are present both in the fractions of PHA-responsive T and non-T cells of man. In addition, our studies on relative frequencies of metaphases carrying unstable chromosomal aberrations in cultures from patients after partial-body irradiation, stimulated respectively by PHA, PWM and Con A, indicate different relative involvement of these cells in the response to the mitogens.

Only some of the relationships between the functional heterogeneity and *in vitro* kinetics of lymphocytes in cultures stimulated with plant lectins have been established so far. In particular: (a) the majority of non-T cells seem to need more time to arrive at the G_1/S boundary of the first *in vitro* cell cycle as compared with T lymphocytes. This may reflect the need of T cell 'help' in the proliferative response of the former; (b) non-T cells seem to display a shorter mean $S + G_2$ phase duration as compared with T lymphocytes, although a considerable variability in the speed of passage through the proliferation cycle of the T cells may also be noted; (c) Lymphocytes with a long life span *in vivo* have a distinctly shorter $S + G_2$ period *in vitro* than *in vivo* short-lived cells. However, it remains to be established whether this correlation holds for long-lived lymphocytes present both in the T and non-T compartments.

The heterogeneity of the PHA-responsive population may express itself in different sensitivities to *in vitro* irradiation, as postulated by Bender and Brewen (1969) and Steffen and Michalowski (1973) and suggested by the results of some preliminary experiments reported here. While the composition of the responding cell population may perhaps account for some donor to donor variability of aberration yields following *in vitro* irradiation of lymphocytes, these effects can probably be minimised by standardisation of culture conditions and time of harvest of cultures of lymphocytes from healthy donors. They may, however, considerably influence the results in diseases associated with changes in the composition of the population of circulating lymphocytes. Moreover, the functional heterogeneity, and, in particular, variability of the size of the fraction of long-lived cells involved in the response to mitogenic stimuli, may be a critical factor to be taken into account in comparative studies of *in vitro* and *in vivo* effects of ionising radiations.

Summary

Human peripheral lymphocytes stimulated in culture with PHA, Con A and PWM under conditions of amethopterin-imposed reversible block of DNA synthesis, continue to accumulate at the G_1/S boundary of the first proliferation cycle for at least four days. Evidence is presented that the majority of Con A- and PWM-responsive lymphocytes are also potentially PHA-responsive, but Con A and PWM seem to stimulate different subsets of peripheral lymphocytes.

Highly purified non-T lymphocytes respond poorly to PHA, Con A and PWM stimulation, but a considerable response of these cells can be elicited in the presence of T lymphocytes pre-incubated with mitomycin C. The accumulation of non-T lymphocytes at the G_1/S boundary is delayed as compared with T lymphocytes, but the mean duration of $S+G_2$ cell cycle phases of the former is considerably shorter as compared with the T cells.

Metaphases carrying non-stable chromosomal aberrations in cultures of patients 2–13 months after partial-body irradiation are found predominantly in the first part of the mitotic wave after G_1/S block release. This implies that *in vitro* long-lived lymphocytes display a shorter mean duration of $S+G_2$ period as compared with short-lived cells. Relatively more long-lived cells are found in the Con A- and PWM-responsive population as compared with PHA-responsive cells. Long-lived lymphocytes are found both in the T and non-T compartments of cells that respond to PHA.

Results of preliminary experiments imply differences of *in vitro* radiosensitivity of PHA- vs. Con A-responsive T lymphocytes, and a heterogeneous radiosensitivity of T cells that seems to be related to their cell cycle kinetics.

Acknowledgements

This work was supported by the Polish Government's Programme of Cancer Research, PR-6 (grant no. 1302), and by the Polish Academy of Sciences (grant no. R/00145).

References

Bender, M.A. & J.G.Brewen (1969) Factors influencing chromosome aberration yields in the human peripheral leukocyte system. *Mutat. Res. 8*, 383–99.

Boyum, A. (1968) A one-stage procedure for isolation of granulocytes and lymphocytes from human blood. *Scand. J. Clin. Lab. Invest. 21*, 21–9.

Buckton, K.E., W.M.Court Brown & P.G.Smith (1967) Lymphocyte survival in men treated with X-rays for ankylosing spondylitis. *Nature 214*, 470–3.

Douglas, S.D., P.F.Hoffman, J.Borjeson & L.M.Chessin (1967) Studies on human peripheral lymphocytes *in vitro*. III. Fine structural features of lymphocyte transformation by pokeweed mitogen. *J. Immunol. 98*, 17–30.

Evans, H.J. (1965) A simple microtechnique for obtaining human chromosome preparations with some comments on DNA replication in sex chromosomes of goat, cow and pig. *Exp. Cell Res. 38*, 511–16.

Everett, N.B. & R.W.Tyler (1967) Lymphopoiesis in the thymus and other tissues: Functional implications. *Int. Rev. Cytol. 22*, 205–37.

Geha, R.S., F.S.Rosen & E.Marler (1973) Identification and characterization of subpopulations of lymphocytes in the human peripheral blood after fractionation in discontinuous gradients of albumin. The cellular defect in X-linked agammaglobulinemia. *J. Clin. Invest. 52*, 1726–34.

Greaves, M.F. & G.Janossy (1972) Elicitation of selective T and B lymphocyte responses by cell surface binding ligands. *Transplant. Rev. 11*, 87–130.

Greaves, M.F., J.J.T.Owen & M.C.Raff (1974) *T and B Lymphocytes: Origins, Properties and Roles in the Immune Responses*, pp.62–4. Amsterdam: Excerpta Medica.

Iversen, J.G. (1973) Long-lived B memory cells separated on antigen-coated bead columns. *Nature 243*, 23–4.

Janossy, G. & M.F.Greaves (1972) Lymphocyte activation: discriminating stimulation of lymphocyte subpopulations by phytomitogens and heterologous anti-lymphocyte sera. *Clin. Exp. Immunol. 10*, 525–35.

Jasińska, J., J.A. Steffen & A.Michalowski (1970) Studies on *in vitro* lymphocyte proliferation in cultures synchronized by the inhibition of DNA synthesis. II. Kinetics of the initiation of the proliferative response. *Exp. Cell Res. 61*, 333–41.

Jondal, M., G.Holm & H.Wigzell (1972) Surface markers of human T and B lymphocytes. I. A large population of lymphocytes forming non-immune rosettes with sheep red cells. *J. Exp. Med. 136*, 207–15.

Ling, N.R. & J.E.Kay (1975) *Lymphocyte Stimulation*, pp.237–52. Amsterdam: North Holland.

Lohrmann, H.P., L.Novikovs & R.G.Grow jr (1974) Cellular interactions in the proliferative response of human T and B lymphocytes to phytomitogens and allogeneic lymphocytes. *J. Exp. Med. 139*, 1553–67.

Moorhead, P.S., P.C.Nowell, W.J.Mellman, D.M.Batipps & D.A.Hungerford (1960) Chromosome preparations of leukocytes cultured from human peripheral blood. *Exp. Cell Res. 20*, 613–16.

Parkhouse, R.M.E., G.Janossy & M.F.Greaves (1972) Selective stimulation of Ig M synthesis in mouse B lymphocytes by pokeweed mitogen. *Nature 235*, 21–3.

Raff, M.C. & H.Cantor (1971) Subpopulations of thymus cells and thymus derived lymphocytes, in *Progress in Immunology* (First International Congress of Immunology) pp.83–93. New York: Academic Press.

Sharpe, H.B.A., D.Scott & G.W.Dolphin (1969) Chromosome aberrations induced in human lymphocytes by X-irradiation *in vitro*: the effect of culture techniques and blood donors on aberration yield. *Mutat. Res. 7*, 453–61.

Shiku, H., P.Kisielow, E.A.Boyse & H.F.Oettgen (1976) Immunogenetic identification of functional T cell subsets. *Transplant. Proc. 8*, 381–91.

Steffen, J.A. (1974) Studies on the heterogeneity of human non-specific mitogen responsive lymphocytes. *Proc. 1st Meeting of the Polish Immunological Society*, Wroclaw, pp.34–6.

Steffen, J.A. & A.Michalowski (1973) Heterogeneous chromosomal radio-sensitivity of phytohaemagglutinin-stimulated human blood lymphocytes in culture. *Mutat. Res. 17*, 367–76.

Steffen, J.A. & W.M.Stolzmann (1969) Studies on *in vitro* lymphocyte proliferation in cultures synchronized by the inhibition of DNA synthesis. I. Variability of S plus G2 periods of first generation cells. *Exp. Cell Res. 56*, 453–60.

UICC (1974) *Multiple Myeloma and Related Immunoglobulin-Producing Neoplasms.* International Union Against Cancer Technical Report, pp.3–17. Geneva: UICC.

Wigzell, H. (1973) The functional significance of surface structures on lymphocytes. *Ann. Immunol. (Inst. Pasteur) 124C*, 7–15.

Wybran, J., S.Chantler & H.H.Fudenberg (1973) Isolation of normal T cells in chronic lymphocytic leukaemia. *Lancet 1*, 126–9.

T. SOFUNI, H. SHIMBA, K. OHTAKI and A. A. AWA

A Cytogenetic Study of Hiroshima Atomic-Bomb Survivors

In our previous observations on the somatic chromosomes of atomic bomb survivors of Hiroshima and Nagasaki, we demonstrated that cells with radiation-induced chromosome aberrations have persisted among the circulating lymphocytes for at least three decades after exposure to A-bomb irradiation, and that the frequency of the aberrant cells is in general proportional to the radiation dose. The majority of such aberrant cells were identified as having symmetrical exchanges (i.e. stable-type, reciprocal translocations and pericentric inversions), while those with asymmetrical aberrations, such as dicentrics and rings (also referred to as unstable-type aberrations), were an order of magnitude less frequent (Awa *et al.* 1971, Awa 1975). These findings prompted us to examine the radiation-induced chromosome aberrations in the somatic cells of A-bomb survivors in further detail, by comparing the results derived from a conventional staining and a trypsin-Giemsa banding method, in order to obtain more accurate information about the type and frequency of induced symmetric aberrations. The purpose of this study was also to evaluate the microscopic criteria presently in use for the detection of symmetric exchanges observed in A-bomb survivors.

Materials and Methods

Twenty-three atomic bomb survivors of Hiroshima, participants in the Adult Health Study Program, receiving biennial physical examinations at the Radiation Effects Research Foundation (Beebe and Usagawa 1968) were selected for study, inasmuch as they had had a previous cytogenetic examination in our laboratory. In each instance, more than 10% of cells from the lymphocyte culture showed radiation-induced chromosome aberrations. Their estimated exposure doses ranged between 101 and 850 rad of mixed gamma and neutrons (Auxier *et al.* 1966, Milton and Shohoji 1968). Past radiation history other than A-bomb exposure was carefully reviewed for each individual, and it was found that none of them had received either radiation therapy, including radioisotope treatment, at any time in the past, or any extensive diagnostic irradiation within the year preceding blood sampling.

For all of these cases, lymphocyte cultures were successful using a whole blood culture method (Hungerford 1965). The culture incubation time for this study was 52 h, during the last two of which colchicine was present. The cells were first treated with a hypotonic solution consisting of a mixture of KCl and sodium citrate, then fixed with an acetic acid-methanol solution, and finally were spread and dried on slides by mild flaming. The slides thus prepared were stained for 20 min with 5% Giemsa solution diluted by a phosphate buffer at pH 6.8, and then washed and dried.

Well-spread metaphases were selected at random and photographed under the microscope, without mounting the coverslip on the slide. The exact location of each metaphase on the slide was recorded, for subsequent re-examination. After routine observations, the slides were soaked with tetrachloroethylene to remove immersion oil from the slide, then in an acetic acid-methanol mixture to decolorise the slide, washed in tap water and dried.

The trypsin-G-banding technique employed here was essentially the same as the original method of Seabright (1971): the slides were treated for 1 to 15 s in 0.2% trypsin solution (1:250, Difco) dissolved in a Ca-, Mg-free balanced salt solution, followed by washing in running water, and re-stained with a 5% Giemsa solution at pH 6.8.

The banded metaphases, which had been previously photographed, were re-located, re-examined and again photographed. Karyotype analyses from the printed photographs of the same metaphases derived from the two different staining methods were carried out separately, so that the types and frequencies of radiation-induced chromosome aberrations were evaluated independently.

The ordinary staining method will be abbreviated hereafter as O-method, and G-banding as G-method.

Results

Table 1 summarises our previous cytogenetic findings among 386 Hiroshima A-bomb survivors (Awa *et al.*, cited from unpublished data). It is clear from this table that there is an increase in the frequency of exchange

Table 1. Type and frequency of chromosome aberrations by dose (Hiroshima). (From Awa *et al.*, unpublished data.)

Dose group (rad)	No. of cases	No. of cells examined	A dic + r (%)	B t + inv (%)	C Total (%)	A/C
1– 99	70	6458	31 (0.48)	123 (1.90)	154 (2.38)	0.20
100–199	137	12632	64 (0.51)	509 (4.03)	573 (4.54)	0.11
200–299	72	6482	49 (0.76)	577 (8.90)	626 (9.66)	0.08
300–399	43	3896	41 (1.05)	495 (12.71)	536 (13.76)	0.08
400–499	30	2866	17 (0.60)	426 (14.86)	443 (15.46)	0.04
500 +	34	3219	40 (1.24)	587 (18.24)	627 (19.48)	0.06
Total	386	35553	242 (0.68)	2717 (7.64)	2959 (8.32)	0.08

dic = dicentric, r = centric ring, t = reciprocal translocation, inv = pericentric inversion.

aberrations with increasing dose. Symmetric aberrations, such as reciprocal translocations and pericentric inversions, were found to predominate over asymmetric ones, and they were the major components contributing to the dose-aberration relationship. In contrast to the predominance of symmetric aberrations, asymmetric exchanges (dicentrics and rings) were far less frequent, being less than 10% of the total number of aberrant cells analysed, although a dose-response relation was nonetheless observed.

The results of karyotype analysis, using both O- and G-methods, of 896 metaphases derived from 23 Hiroshima survivors are shown in tables 2 and 3. Though the number of metaphases, as well as of cases, was small, there was in general a dose-dependent increase in the frequency of cells with chromosome aberrations for either method. The frequency of aberrant cells was clearly higher in the G- than in the O-series for every dose group (table 2). The ratio of aberrant cells detected by O to that detected by G was 0.86 (293 for O vs. 342 for G).

Table 2. Frequency of aberrant cells by dose in 23 heavily exposed A-bomb survivors of Hiroshima.

Dose group (rad)	Mean	No. of cases	No. of cells analysed	No. of aberrant cells O (%)	No. of aberrant cells G (%)
100–199	157.8	8	295	68 (23.1)	78 (26.4)
200–299	258.2	6	257	80 (31.1)	92 (35.8)
300–399	331.7	3	137	52 (38.0)	63 (46.0)
400–499	405.7	3	85	25 (29.4)	29 (34.1)
500 +	710.7	3	122	68 (55.7)	80 (65.6)
Exposed total	311.1	23	896	293 (32.7)	342 (38.2)
Total no. of cells photographed				1758	1114

O = ordinary stain, G = G-banding stain.

Of the 896 metaphases analysable by both methods, 548 cells were identified as normal, while aberrations were observed by either one or both methods in the remaining 348 cells (table 3). Since the frequency of asymmetric aberrations (dicentrics, rings, acentric rings, terminal and interstitial deletions) was extremely low, there being only 17 in 348 cells (or 5% of the total number of aberrant cells), the present analysis was confined to symmetric aberrations.

Further comparison of the karyotypes from the 348 cells with aberrations showed that the aberrations were identical in 151 cells (43%), while they were not identical in the remaining 197 cells. The latter group of aberrant cells was further divided into three subgroups as shown in table 3; (a) though abnormalities were identified by both methods, the types of aberrations were not identical (for example, a pericentric inversion by O was in fact determined to be a reciprocal translocation or an even more complicated exchange by G);

Table 3. Comparison of karyotypes of the same metaphases between ordinary (O) and G-banding methods (G).

O		G	Karyotype	No. of karyotypes
N	:	N	identical	548
AB	:	AB	identical	151
			non-identical	136
N	:	AB	non-identical	55
AB	:	N	non-identical	6
Total				896
No. of non-identical karyotypes				197

N = normal karyotype, AB = abnormal karyotype.

this sub-group consisted of 136 cells; (b) abnormalities were detected only by the G- but not by the O-method (55 cells); and (c) the G-method failed to detect any abnormality in cells that had already been identified as abnormal by O (6 cells). For the 6 cells in (c), there were 4 in which one chromosome by O-stain appeared to be abnormal, since it was either overcontracted or unusually twisted, but by G-stain its banding pattern proved to be normal. In the remaining two cells, an exchange could conceivably have occurred at the negative regions of the distal part of certain chromosomes, indicating that even the banding technique may fail to detect exchanges of this type.

In order to further simplify the results observed from the G-method, the 197 cells with chromosome aberrations were classified into three groups, according to the types of aberrations, by scoring the number of breaks involved in each exchange; i.e. S-type cells as having simple aberrations caused

Table 4. Analysis of G-band metaphases with non-identical karyotypes between two methods.

O		G	No. of metaphases			Total
			S	C	U	
AB	:	AB	74	35	27	136
N	:	AB	50	1	4	55
AB	:	N	6	—	—	6
Total			130	36	31	197

N = normal karyotype, AB = abnormal karyotype. S = simple-type with 1 or 2 breaks, C = complex-type with 3 or more breaks, U = unidentifiable-type (for detail, see text).

by one or two breaks, C-type as complex interchanges with three or more breaks involved, and U-type in which the types of exchanges could not be determined. Table 4 shows that there were 31 U-type cells, in which aberrations were so complicated that even the banding method could not specify the types of exchanges.

Of the 130 S-type cells with an aberrant chromosome(s) due to a simple deletion or an interchange in the G-series, 50 cells were previously identified as normal in the O-series. There were 36 C-type cells, which showed various types of complex exchanges.

Although the types of exchanges detected by the G- but not by the O-method among the present cases have not yet been fully characterised, preliminary analysis showed that these aberrations consisted of paracentric inversions (in 12 chromosomes so far confirmed), intra- and interchanges of chromosome parts with equal lengths, or even with unequal lengths that were judged nonetheless to be within the normal limits of variation by ordinary staining. As for complex exchanges, insertions of chromosome segments of both inversion- and translocation-types, and sequential translocations involving more than three chromosomes were frequently noted. These exchanges undoubtedly contributed to an elevated frequency of aberrant cells in the G-series.

In the O-series, aberrations designated as deletions were frequently observed, in which only one of the chromosomes in the complement was shorter than the partner, but about a half of them were classified as reciprocal translocations by the G-method because the corresponding abnormal counterpart could be identified.

Discussion

The majority of exchange aberrations observed in the peripheral blood lymphocytes of Hiroshima and Nagasaki A-bomb survivors are the symmetric type, while cells with asymmetric aberrations are very low in frequency (Awa *et al.* 1971, Awa 1975). It seems likely that in these individuals, irradiated more than 20 years ago, the cells with asymmetric exchanges would have been eliminated from the lymphocyte population over this period, perhaps because of mitotic disturbances. In this connection the symmetric exchange, therefore, would seem to be a useful indicator for evaluating the dose-aberration relationship, especially in those exposed to large doses of radiation long before the cytogenetic examination.

Unfortunately, microscopic detection of symmetric exchanges is extremely limited in the conventional staining method, since the criteria for the identification of individual chromosomes are based only on relative lengths and arm ratios (or indices) including a certain amount of variability for each chromosome. A large fraction of such aberrations is thus likely to be overlooked (Evans 1974, Sasaki 1974). For example, Sasaki (1974) reported from his *in vitro* irradiation experiments with human lymphocytes that the frequency of symmetric aberrations was only about 20% of asymmetric exchanges. Since the probability of both symmetric and asymmetric aberrations being induced by irradiation is assumed to be equal, about 80% of the symmetric exchanges remain undetected by the ordinary staining method.

By the use of banding techniques, it was possible to analyse the same metaphases in two different ways: first by an ordinary staining method, and then by the trypsin-G-banding method, so that the frequencies of chromosome aberrations derived from these two methods can be compared directly with each other. The present results showed that the frequency of aberrant cells was definitely higher in the G- than in the O-series. The O-method detected 86% of the aberrant cells identified by the G-method, which was higher than hitherto believed by radiation cytogeneticists (see Evans 1974, Sasaki 1974, Seabright 1971).

Some types of exchanges were observed in the G-series that could not be identified by the O-staining method. Paracentric inversion is a typical example, the frequency of which was close to that of the total number of asymmetric aberrations observed in the sample. In addition to paracentric inversions, several types of insertions and interchanges of chromosome segments with both equal and unequal lengths were also detected frequently.

Another noteworthy feature was that some of the translocations, which were identified as simple reciprocal types by the O-method, were in fact more complicated, involving more than three breaks, and often showed sequential interchanges akin to 'musical chairs'. Further analyses by the G-banding method are in progress for the types of aberrations, numbers of breaks per cell, and the sites of break-points in the somatic cells of A-bomb survivors.

Summary

A total of 896 metaphases obtained from 2-d-cultures of peripheral blood lymphocytes of 23 heavily exposed A-bomb survivors of Hiroshima were examined first after ordinary staining, and then re-examined after trypsin-G-band staining. The frequencies of cells with radiation-induced chromosome aberrations, mainly of the symmetric type, were compared for the two methods. There were 348 metaphases identified as having abnormal karyotypes by either one or both methods. Of these aberrant cells, 293 were found to have chromosome aberrations by the ordinary stain. There were 55 metaphases in which abnormalities were detected only by G-banding, while 6 cells were identified as abnormal by ordinary stain but as normal by G-banding, 4 of which were misjudged in the ordinary preparation due to the presence of partially distorted chromosomes. Further G-banding analysis identified various exchanges, including several types of insertions and paracentric inversions, which could not be detected by the ordinary staining technique.

Acknowledgements

The authors wish to thank Dr Howard B. Hamilton, Chief of the Department of Clinical Laboratories, Radiation Effects Research Foundation, for his continued encouragement during the present study.

This study was supported in part by a Grant-in-aid for Cancer Research from the Ministry of Health and Welfare, Japan.

References

Auxier, J.A., J.S.Cheka, F.F.Haywood, T.D.Jones & J.H.Thorngate (1966) Free-field radiation-dose distributions from the Hiroshima and Nagasaki bombings. *Health Phys. 12*, 425–9.

Awa, A.A., S.Neriishi, T.Honda, M.C.Yoshida, T.Sofuni & T.Matsui (1971) Chromosome-aberration frequency in cultured blood-cells in relation to radiation dose of A-bomb survivors. *Lancet 2*, 903–5.

Awa, A.A. (1975) Review of thirty years study of Hiroshima and Nagasaki atomic bomb survivors. II. Biological effects, G chromosome aberrations in somatic cells. *J. Radiat. Res. Suppl.*, 122–31.

Beebe, G.W. & M.Usagawa (1968) The major ABCC samples. *ABCC Tech Rep.*, pp.12–68.

Evans, H.J. (1974) Effects of ionizing radiation on mammalian chromosomes, in *Chromosomes and Cancer* (ed. J.German), pp.191–237. New York: John Wiley.

Hungerford, D.A. (1965) Leukocytes cultured from small inocula of whole blood and the preparation of metaphase chromosomes by treatment with hypotonic KCl. *Stain Tech. 40*, 333–8.

Milton, R.C. & T.Shohoji (1968) Tentative 1965 radiation dose (T65D) estimation for atomic bomb survivors, Hiroshima and Nagasaki. *ABCC Tech. Rep.*, pp.1–68.

Sasaki, M.S. (1974) Chromosome aberrations by ionizing radiation. *Tokyo J. Med. Sci. 82*, 208–22.

Seabright, M. (1971) A rapid banding technique for human chromosomes. *Lancet 2*, 971–2.

UNSCEAR (1969) *Radiation-induced Chromosome Aberrations in Human Cells.* UNSCEAR report, General Assembly Official Records, 24th Session, Suppl. 13 (A/7613). New York: United Nations.

W. KEMMER, W. SCHMUTZLER and A. STEINSTRÄSSER

Radiation Dose and Chromosome Aberrations in Radiotherapy Patients

Statistically significant and reproducible calibration curves relating radiation dose and chromosome aberrations can be obtained if blood is irradiated *in vitro* under constant conditions. Using these curves for dose estimation in accidentally irradiated persons can be helpful if physical dosimetry is unavailable (Purrott *et al.* 1975, Kemmer 1976). In some cases, however, the aberration yield does not agree with the expected or calculated absorbed radiation dose. A good example is Thorotrast patients (Buckton *et al.* 1973, Kemmer *et al.* 1973), where biological dosimetry fails even when knowledge of the microdistribution of Thorotrast permits calculations of the radiation doses to lymph nodes or lymphocytes (Steinsträsser, Kemmer and Muth 1977). The major difficulty in obtaining a significant dose-effect relationship in such cases is a consequence of the inhomogeneous distribution of radiation in the body. To attempt to resolve this we searched for a group of persons exposed to homogeneous partial-body irradiation under controlled conditions. A preliminary account is given of a study of patients treated with ^{60}Co γ-rays for gynaecological tumours or Hodgkin's disease.

Materials and Methods

All patients were irradiated with fractionated doses of ^{60}Co γ-rays produced by a Siemens Gammatron. They were treated with doses of 200 rad d^{-1}, equivalent to a weekly accumulated dose of 1000 rad. Up to 5000 rad were given. In no cases were cytostatic drugs administered. One patient with Hodgkin's disease, irradiated on the supradiaphragmatic field, and three patients treated for gynaecological tumours were studied. In all patients control blood was drawn immediately prior to the first irradiation. A second blood sample was taken a short time (up to ten minutes) after exposure, and in all cases the samples were set up in culture two hours later.

Standard lymphocyte culturing methods and techniques of metaphase preparation were employed (Kemmer 1978). The chromosomes were scored by a TV-linked microscope, and only dicentric chromosomes with acentric fragments were noted for biological dosimetry purposes. The lymphocytes in the samples from patients given 5000 rad grew poorly in culture, and no results are given.

Results and Discussion

The results of chromosome aberration analysis are presented in table 1, where it may be seen that no dicentrics or other chromosome aberrations were found in control blood samples.

Table 1. Dicentrics per 100 cells and total radiation dose.

		Dose (rad)			
Case no.	Control	1000	2000	3000	4000
1 G	0	7	11	—	—
2 G	0	10	11	11	21
3 G	0	2	12	—	—
4 H	0	12	38	33	30
	Mean ± S.D.	7.8 ± 4.3	18 ± 13.3	22 ± 15.6	25.5 ± 6.4

G = gynecological tumour, H = Hodgkin's disease. Dashes indicate that the lymphocytes grew poorly in culture.

Chromosome Aberrations at High Radiation Doses

The results obtained to date show that even after high partial-body irradiation, the absolute number of chromosome aberrations is very low. At these high doses a higher chromosome aberration yield might have been expected, but it must be remembered that the irradiation was to part of the body and that the total dose was accumulated from doses of 200 rad d^{-1}. We are nevertheless surprised that the highest aberration level was between 0.3 and 0.4 per cell, a result not anticipated from *in vitro* calibration curves. A one-way analysis of variance of the aberration rate shows no significant differences between the values within and between the patients except for case 4 at 2000 rad (table 1). There is also no significant increase of aberration yield in relation to increasing accumulated dose. These results are similar to those obtained from our Thorotrast patients. The mean absorbed dose from α-irradiation of lymphocytes calculated from the microdistribution of Thorotrast in single lymph nodes was up to 375 rads. On the assumption that the RBE is 15 for thorium α-particles, the absorbed dose is equivalent to about 5700 rad of X-rays (Kemmer, Steinsträsser and Muth 1977; Steinsträsser, Kemmer and Muth 1977). The highest number of chromosome aberrations in our Thorotrast patients was 0.25 per cell, and the results are shown in figure 1.

Chromosome Aberrations at Low Radiation Doses

Buckton *et al.* (1971) and Silberstein *et al.* (1974) have published data on aberration frequencies in peripheral blood lymphocytes from patients

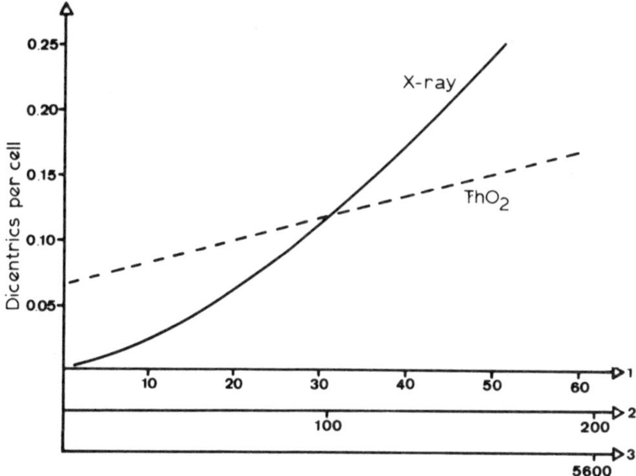

Figure 1. Comparison of calculated dose-effect relationship in Thorotrast patients (dotted line, ThO_2) with an X-ray calibration curve. Abscissae: (1) ^{224}Ra equivalent value; (2) X-ray doses in rad; (3) calculated α-dose in lymphocytes of Thorotrast patients.

receiving low doses (25 to 200 rad) of X-rays to the whole body, and their data are shown in table 2. By a one-way analysis of variance Silberstein *et al.* (1974) showed that there was no significant correlation between the induced chromosome aberration frequency and the radiation dose. This led the authors to conclude that 'this result must call into question the use of radiation-induced chromosome aberrations in man' for dosimetry.

Table 2. Mean value of chromosome aberrations and whole-body dose in therapy patients.

Dose (rad)	No. of patients	Mean (D + R)/C
25[1]	6	0.028
50[1]	7	0.045
100[2]	3	0.043
200[2]	4	0.098

[1] after Buckton *et al.* 1971.
[2] after Silberstein *et al.* 1974.
(D + R)/C = dicentrics and rings per cell.

We are of the opinion that the wide variation in results of chromosome aberration analysis is caused by differences in the immunological status of individuals and hence the distribution of several lymphocyte subpopulations

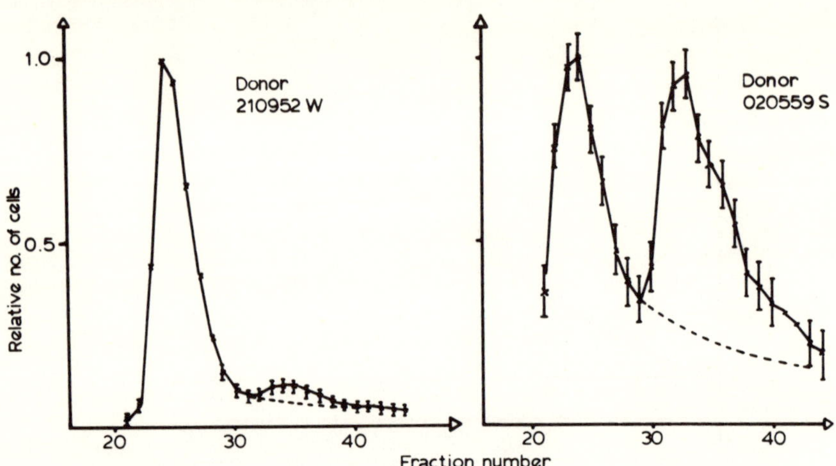

Figure 2. Distribution of sub-populations of lymphocytes, separated by free-flow electrophoresis. Left peak, T-lymphocytes; right peak, B-lymphocytes.

in peripheral blood. Figure 2 shows the distribution of T and B subpopulations of lymphocytes in blood samples of two female blood donors. The T and B subpopulations were separated by free flow electrophoresis (Steinsträsser, pers. comm.). The number of different cell types in these subpopulations is different, both between several persons and within a person sampled on different dates.

Conclusion

Before it is possible to use chromosome aberration yield as a biological dosimeter, fundamental studies on the relationship between various groups of radiosensitive lymphocytes in blood, and their distribution in blood samples, must be undertaken. Moreover, further studies on the relationship between radiosensitivity and the response to PHA-stimulation are necessary.

Summary

The first results of chromosome analysis of patients irradiated for Hodgkin's disease or gynaecological tumours are presented. The patients underwent an irradiation of the supradiaphragmatic lymph node chain or the abdominal region with ^{60}Co γ-rays, generated by a Siemens Gammatron. The accumulated radiation doses were up to 5000 rad, fractionated in doses of 200 rad d^{-1}. The chromosome aberration frequency was lower than expected and the individual values changed within wide limits. The uses of chromosome aberration frequencies as a biological dosimeter are discussed.

Acknowledgements

The authors wish to thank Dr Dietz (Radiologische Universitätsklinik der Universität des Saarlandes) for his help in selecting patients for this study. The dosimetry was made in cooperation with the irradiation treatment planning group, under Prof. Dr H-K. Leetz.

References

Buckton, K.E., A.O.Langlands, P.G.Smith, G.E.Woodcock & P.C.Looby (1971) Studies on chromosome aberration production after whole-body irradiation in man. *Radiat. Biol. 19*, 369–78.

Kemmer, W. (1976) Chromosomenaberrationen als biologisches Dosimeter, in *Die Überwachung der Strahlenexposition der Arbeitskräfte. Zielsetzung, Durchführung, Ergebnisse und Bewertung*, Fachverband für Strahlenschutz, pp.326–43.

—— (1978) Chromosome aberrations as a biological dosimeter in radiation protection (manuscript in preparation).

Kemmer, W., A.Steinsträsser & H.Muth (1977) Chromosome aberrations as a biological dosimeter in Thorotrast patients: dosimetric problems, given at the International Meeting on the Toxicity of Thorotrast and other Alpha-Emitting Heavy Elements (Lisboa, 28 June–2 July). *Environ. Res.* (in press).

Kemmer, W., H.Muth, F.Tranekjer & U.Borkenhagen (1973) Chromosome aberration caused by Thorotrast, in *Third International Meeting on the Toxicity of Thorotrast* (Copenhagen, 25–27 April) Risø Report 294, pp.104–13. Copenhagen.

Purrott, R.J., D.C.Lloyd, J.S.Prosser, G.W.Dolphin, P.A.Tipper, E.J.Reeder, C.M.White, S.J.Cooper & B.D.Stephenson (1975) *The Study of Chromosome Aberration Yield in Human Lymphocytes as an Indicator of Radiation Dose.* Harwell: NRPB R–35.

Silberstein, E.B., C.J.Ewing, G.K.Bahr & J.G.Kereiakes (1974) The human lymphocytes as a radiobiological dosimeter after total body irradiation. *Radiat. Res. 59*, 658–64.

Steinsträsser, A., W.Kemmer & H.Muth (1977) Microdistribution of thorotrast conglomerates in lymph nodes and radiation dose of single lymphocytes, given at the International Meeting on the Toxicity of Thorotrast and other Alpha-Emitting Heavy Elements (Lisboa, 28 June–2 July 1977). *Environ. Res.* (in press).

J.G.BURR, N.WALD, S.PAN and K.PRESTON JR

The Synergistic Effect of Ultrasound and Ionising Radiation on Human Lymphocytes

The recently increasing use of ultrasound for both diagnostic and therapeutic medical purposes has stimulated research on the biological effects of ultrasonic radiation. An important area of this research concerns the possible synergistic effect of ultrasound and ionising radiation. Such an effect might be detrimental to human health, as, for example in the developing fetus, which has a growing likelihood of receiving the combination of exposures in diagnostic studies. On the other hand, such synergism might lead to an improvement in the treatment of neoplastic disease. To evaluate both possibilities, the reports of pertinent research were reviewed.

A synergism between ultrasonic radiation and ionising radiation has been reported by numerous investigators using various levels and types of radiation and different types of biological end-points. The noted synergistic effects have been quite variable and have included increases in cell reproductive death (Craig and Tyler 1977; Dharkar 1964; Javish 1966; Martins 1971; Spring, Rytila and Blomquist 1970; Todd and Schroy 1974); increases in chromosome aberrations (Kim 1968, Kunze-Muhl 1975); reduction in ionising radiation dose to achieve tumour remission (Lehmann and Krusen 1955, Paal *et al.* 1963, Pydorich 1966, Woeber 1959); increases in cellular membrane effects (Martins 1971, Rapacholi 1970); increases in mutation rates (Conger 1948); decrease in plant growth (Spring 1969); and congestion in Conchae (Fujita and Sakuma 1974). Some studies, however, have shown no synergistic effects (Clark, Hill and Adams 1970; Hering and Shepstone 1976). The majority of the studies appear to support the hypothesis that some type of synergism occurs when ultrasound and ionising radiation are administered together; however, it is also apparent that quantitation, comparison and interpretation of the results of these studies is made extremely difficult by the lack of consistent methods for determining the amount of radiation producing the effects.

This paper is a report on several preliminary experiments that were conducted in our laboratory in an attempt to verify the existence of a synergistic effect between ultrasound and ionising radiation on the chromosomes of human lymphocytes, and to learn what parameters of the radiation exposures might require control and quantitation in future studies. Lymphocytes were chosen because of their synchronous cell population, their well quantitated response to ionising radiation (Buckton and Evans 1973, Lloyd *et al.* 1975), their previous utilisation in ultrasound studies (Braeman, Coakley and Gould 1974; MacIntosh, Brown and Coakley 1975; Mermut *et al.* 1973; Thacker 1972), and the known correlation between their *in vitro* and *in vivo* responses (Brewen, Preston and Littlefield 1972). It must be emphasised that these were preliminary experiments to ascertain the need for improved experimental design and dosimetry in the more detailed studies that are presently being undertaken.

Materials and Methods

Biological Test System

The biological test system used was the cytogenetic change in lymphocytes in human peripheral blood. Blood was drawn from the same healthy male donor into heparinised vacutainers no earlier than 2 h before each series of experiments. The blood was held at room temperature (22°C) at all times except during ultrasound exposures, in which the temperature rose but the rise was not measured, and during culturing. One ml samples of the blood were exposed to various sequences of ultrasound and ionising radiation and the blood cultured for 48 h (first metaphase) at 37°C. Cells were cultured and harvested using the method described by Buckton and Evans (1973) with minor modifications. All exposures were scored on a blind basis to remove observer bias. Only those metaphase spreads that contained 46 centromeres were scored for one-hit aberrations (gaps, minutes, fragments) and two-hit aberrations (dicentrics, rings). Comparisons were based on the totals for each aberration category.

Radiation Exposure System

The exposure system used for simultaneous exposure of the biological test system to ultrasound and ionising radiation is shown in figure 1. The basic components of the system are an ionising radiation source, a sonation system, and a biological sample holder.

The ionising radiation source was an Atomic Energy of Canada Limited, Theratron 80, ^{60}Co irradiator of 5648 Ci as of October 1970. Exposures were made with the biological sample holder at a depth in water of 10.5 cm at 80 cm SSD and 10×10 cm field size, with a resultant dose rate of 45.32 rad min^{-1} delivered to the samples. No correction was made for the 3 mm-thick glass in the exposure tank wall. All exposures were conducted using identical geometry.

The sonation system included a sonation source, a 40 l glass-walled tank ($50 \times 26.5 \times 30$ cm), sonation medium (degassed water), a cone style sound absorber (rubberised wool), and an ultrasound calibration system. The ultrasound source was a Medico Products, Mark V-B, Physical Therapy device of 1 MHz frequency, CW, with an effective radiation area of 8.7 cm^2. The output of the ultrasound unit was calibrated using a hanging ball radiometer (Dunn

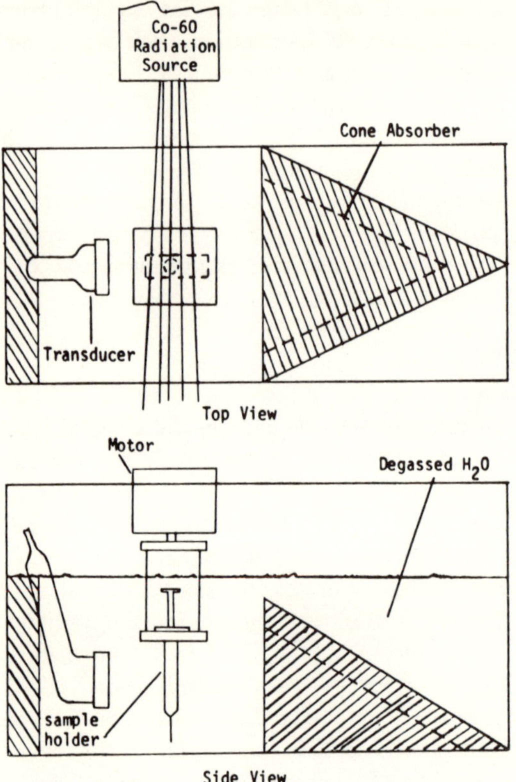

Figure 1. Schematic diagram of irradiation and sonation apparatus.

and Fry 1971; Hill 1970). The intensity was measured at the location where the biological sample holder was to be placed and the output was monitored during all exposures using an oscilloscope connected directly to the ultrasound transducer.

The biological sample holder was a 3 ml Plastipak Disposable syringe made of thin-walled polypropylene, which is non-toxic and sterile. The syringe was held a fixed distance (4 cm) from the ultrasound transducer and rotated at 45 rpm with an electric motor. One ml of blood was exposed and the needle hub was cleaned before placing the sample in the culture medium.

Experimental Design

Various sequences of exposures were made, using combinations of ultrasound at 2 W cm^{-2} for 30 min and γ-radiation at 45.3 rad min^{-1} for a total of 200 rad. Specific sequences included:

1. Ultrasound alone
2. γ-ray alone, incubated immediately (control for 5,6,7,8)
3. γ-ray alone, incubated after 2 h delay at 22°C (control for 10)
4. γ-ray alone, incubated after 6 h delay at 22°C (control for 9)
5. Ultrasound 6 h before γ-ray

6. Ultrasound immediately before γ-ray
7. Ultrasound and γ-ray simultaneously (γ-ray given at mid-point of ultrasound exposure)
8. Ultrasound immediately after γ-ray
9. Ultrasound 2 h after γ-ray
10. Ultrasound 6 h after γ-ray

Results

The results of the various exposure sequences are shown in table 1 and figure 2. The results of sequence 1 (ultrasound alone) are consistent with the literature (Braeman, Coakley and Gould 1974; MacIntosh, Brown and Coakley 1975; Mermut *et al.* 1973; Thacker 1972), and show no chromosome aberrations (for the exposure system used). Sequences 2, 3 and 4 (γ-ray alone) show aberration yields that are also consistent with the data of Buckton and Evans (1973) and Lloyd *et al.* (1975).

Table 1. Aberration frequencies per cell for sequences of ultrasound and ionising radiation exposures (see text).

Sequence	1	2	3	4	5	6	7	8	9	10
One-hit										
Gaps	0	.015	.010	.010	.020	.015	.014	.019	.019	.005
Fragments	0	.213	.240	.308	.221	.223	.603	.458	.309	.346
Minutes	0	.025	.077	.038	.029	.064	.014	.028	.034	.063
Totals	0	.252	.327	.356	.270	.302	.630	.505	.362	.415
Two-hit										
Dicentrics	0	.149	.154	.125	.103	.099	.301	.196	.121	.205
Rings	0	.000	.000	.000	.005	.015	.014	.014	.010	.015
Totals	0	.149	.154	.125	.108	.114	.315	.210	.130	.220
Total	0	.401	.481	.481	.377	.416	.945	.715	.493	.634
Metaphases scored	200	202	104	104	204	202	73	214	207	205

Figure 2 shows the effect of the different combinations of ultrasound and γ-rays used in sequences 5 to 10. There is no statistically significant difference ($p < 0.05$ using a one-sided Student's t-test; Armitage 1971) between the control (sequence 2) and combined sequences 5 and 6 for all types of chromosome aberrations. This suggests that the ultrasound has no radiosensitising effect on the lymphocytes. There is, also, no statistically significant difference between the control (sequence 3) and combined sequence 9 or between the control (sequence 4) and combined sequence 10 for all types of aberrations. This suggests that the synergistic effect does not occur if ultrasound is not given within 2 h after γ-rays.

There is, however, a highly significant difference between control (sequence 2) and combined sequences 7 and 8 for all types of aberrations except two-hit aberrations in sequence 8. It appears that the ultrasound must

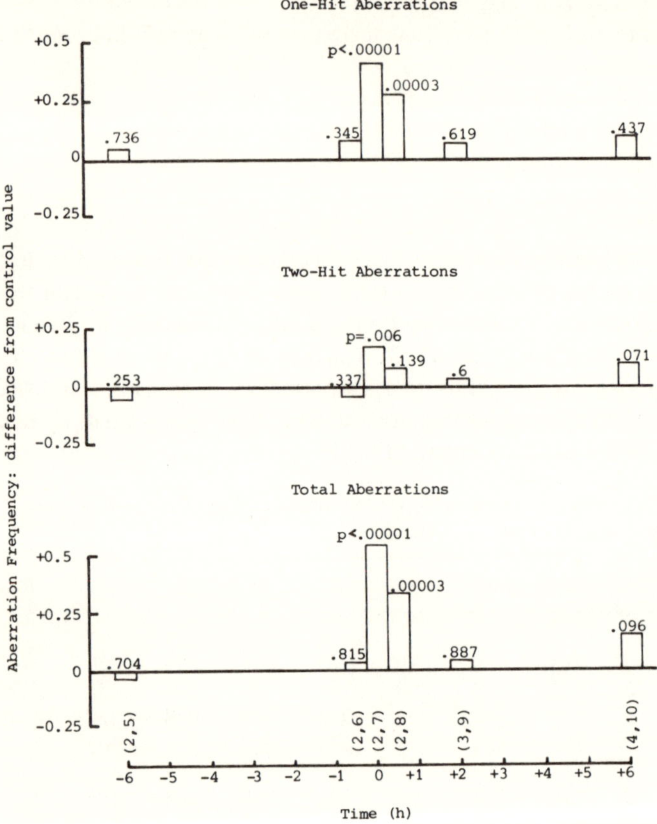

Figure 2. Effect of exposure sequence on chromosome aberration frequencies. Values plotted are the difference in aberration frequency between the control value (γ-ray alone) and the combined sequence (ultrasound exposure before, simultaneously with, or after γ-ray) vs. the time of γ-irradiation. The time scale shows the time at which ultrasound was applied, relative to irradiation. The specific sequence numbers for each set of bars are given just above the time scale, and show the control sequence number followed by the combined sequence number. The P-values are the probabilities that the control and combined sequences are equal, using a one-sided Student's t-test.

follow the γ-rays or be administered simultaneously to produce a synergistic effect. The effect noted was a substantial increase in acentric fragments and a minor increase in dicentrics when the two forms of radiation are given simultaneously.

The low number of metaphases scored (73) in sequence 7 (simultaneous ultrasound and γ-ray) was due to poor staining of the slides for that exposure. Only those metaphase spreads that were clearly visible were scored to avoid observer error. The low number scored does not affect the overall result bcause the number of metaphases scored is considered in the statistical analysis (one-sided Student's t-test).

Discussion

The results of these experiments suggest that a synergism exists in the production of chromosome damage in human small lymphocytes exposed *in vitro* to combined sequences of ultrasound and ionising radiation. This synergism appears only when ultrasound is applied simultaneously or shortly after ionising radiation, and diminishes to zero after 2 h delay. These results confirm and extend the findings of Kunze-Muhl (1975) and Kim (1968). Kunze-Muhl showed a synergistic effect when ultrasound (3 W cm^{-2}cw) was used immediately after irradiation (250 kVp X-rays) but no experiments were reported with ionising radiation given simultaneously, 2 h or 6 h after ultrasound exposures.

Whether the synergistic effect is caused by heating, cavitation or shear stresses is not clear from the data, but possible mechanisms for this effect can be hypothesised. The fact that the ultrasound given immediately before γ-irradiation has no effect suggests that the chromosomes must be broken initially by the ionising radiation before the ultrasound can influence the aberration yield. The possibility that the synergistic effect is due to some interference with the mechanism for repair of radiation-induced chromosome damage, as first postulated by Conger (1948), is supported by the findings that synergism was diminished by even several minutes delay in following γ-irradiation with ultrasound, and was not observed when the delay exceeded 2 h. This is compatible with several studies showing that radiation damage to chromosomes is repaired within 30 to 90 minutes, with a large portion completed in the first few minutes (Dewey, Miller and Leeper 1971; Prempree and Merz 1969; Prempree, Michelson and Merz 1969; Winans, Dewey and Dettor 1972).

If one accepts the initial hypothesis that the synergism is due to some interference with the repair mechanism, then some further speculation may be stimulated by examining the chromosome aberration yields. The data show that both dicentrics as well as acentric fragments are increased. If the synergistic effect is due to an 'inhibition' of the repair mechanism one might expect to see an increase in acentric fragments and a decrease in dicentrics, which are a form of misrepair. Since there is an increase in dicentrics as well as in acentric fragments, other possible mechanisms must be considered.

Three such possible mechanisms are: (a) The ultrasound, either by temperature increase or mechanical motion, causes the chromosomes that were broken by ionising radiation to move about, increasing the probability of interaction with other broken chromosomes and thereby increasing the probability of misrepair. (b) Ultrasound used with ionising radiation increases the total number of double-strand DNA fractures, which are responsible for the chromosome aberrations seen at metaphase. Ultrasound, which is not capable of double-strand DNA fractures when used alone, may be capable of fracturing single-strand DNA. Since ionising radiation caused many more single-strand DNA fractures than double-strand DNA fractures (Burrell, Feldschreiber and

Dean, 1971; Lehmann and Ormerod 1970), there would be many single strands of DNA available for action by the ultrasound. (c) Ultrasound causes some as yet undetermined change in the kinetics and/or chemical nature of the repair mechanism.

These preliminary experiments, although limited in design and dosimetry, do confirm the existence of a synergism between ultrasound and ionising radiation on human lymphocytes. The results have helped to define the scope of further investigations. Our laboratory is currently developing a comprehensive ultrasound and ionising radiation dosimetry system to accomplish future detailed and carefully controlled investigations. This system will control and monitor the temperature of the biological samples at all times, control and monitor the ultrasound exposures at all times, and control the ionising radiation dose and dose rate. In addition, heat and ionising radiation sequences will be used to simulate the temperature changes caused by the ultrasound. In this way it may be possible to obtain further information about the mechanism of the synergistic effects of ultrasound and ionising radiation.

Summary

Small lymphocytes in human peripheral blood were exposed to various sequences of ultrasound (2 W cm^{-2} for 30 min) and ^{60}Co γ-irradiation (45.32 rad min^{-1} for 200 rad total) to determine if a synergistic effect exists in producing chromosome damage. The results showed a statistically significant difference between controls and combined sequences where ultrasound was given either simultaneously with γ-rays or immediately after γ-rays. Both acentric fragments and dicentric chromosomes were significantly increased in the simultaneous sequences, but only fragments were significantly increased by ultrasound given immediately after γ-ray sequence. No significant differences were noted when ultrasound was used before γ-ray or when ultrasound was given later than 2 h after γ-ray. Several hypotheses that could explain this apparent synergism are discussed and a general description of further investigation is presented.

References

Armitage, P. (1971) *Statistical Methods in Medical Research*, pp.101–7. New York: John Wiley.

Braeman, J., W.T.Coakley & R.K.Gould (1974) Human lymphocyte chromosome and ultrasonic cavitation. *Br. J. Radiol. 47*, 158–61.

Brewen, J.G., R.J.Preston & L.G.Littlefield (1972) Radiation-induced human chromosome aberration yields following an accidental whole-body exposure to Co-60 γ-rays. *Radiat. Res. 49*, 647–56.

Buckton, K.E. & H.J.Evans (1973) *Methods for the Analysis of Human Chromosome Aberrations*. Geneva: World Health Organization.

Burrell, A.D., P.Feldschreiber & J.Dean (1971) DNA-membrane association and the repair of double breaks in X-irradiated *Micrococcus radiodurans*. *Biochim. Biophys. Acta 247*, 38–53.

Clark, P.R., C.R.Hill & K.Adams (1970) Synergism between ultrasound and X-rays in tumour therapy. *Br. J. Radiol. 43*, 97–9.

Conger, A.D. (1948) The cytogenetic effect of sonic energy applied simultaneously with X-rays. *Proc. Nat. Acad. Sci. USA 34*, 470–4.

Craig, A.G. & J.M.R. Tyler (1977) Synergism between gamma and ultrasonic irradiation of the bacterium *E. Coli B*, in *Proc. IVth Int. Congr. IRPA* (Paris, France, April 24–30) pp.229–32.

Dewey, W.C., H.H.Miller & D.B.Leeper (1971) Chromosome aberrations and mortality of X-irradiated mammalian cells: emphasis on repair. *Proc. Nat. Acad. Sci. USA 68*, 667–71.

Dharkar, S.P. (1964) Sensitization of microorganism to radiation by previous ultrasonic treatment. *J. Food Sci. 29*, 641–3.

Dunn, F. & F.J.Fry (1971) Ultrasonic field measurement using the suspended ball radiometer and thermocouple probe, in *Interaction of Ultrasound and Biological Tissues*, pp.173–6. Workshop Proceedings, 8–11 Nov. DHEW Publication 73–8008.

Fujita, S. & S.Sakuma (1974) The influence of ultrasound on ionizing radiation effects. *Radiat. Res. 58*, 311.

Hering, E.R. & B.J.Shepstone (1976) Observations on the combined effects of ultrasound and X-rays on the growth of the roots of *Zea Mays. Phys. Med. Biol. 21*, 263–71.

Hill, C.R. (1970) Calibration of ultrasonic beams used in biomedical research. *Phys. Med. Biol. 15*, 241.

Javish, J. (1966) Distribution of chromosome damage in irradiated Chinese hamster cell cultures. *Radiat. Res. 27*, 493 (abst).

Kim, A.M. (1968) *Chromosomenanalysen nach Behandlung von Meschlichem Venenblut mit Röntgenstrahlen und Ultraschall*, PhD thesis, Univ. of Vienna.

Kunze-Muhl, E. (1975) Chromosome damage in human lymphocytes after different combinations of X-ray and ultrasonic treatment, in *Proc. Second European Congr. Ultrasonics in Medicine* (Munich, May 12–16) pp.1–9. Amsterdam: Excerpta Medica.

Lehmann, A.R. & M.G.Ormerod (1970) Double-strand breaks in the DNA of a mammalian cell after X-irradiation. *Biochim. Biophys. Acta 217*, 268–77.

Lehmann, J.F. & F.H.Krusen (1955) Biophysical effects of ultrasonic energy on carcinoma and their possible significance. *Arch. Phys. Med. Rehabil. 36*, 452–9.

Lloyd, D.C., R.J.Purrott, G.W.Dolphin, D.Bolton, A.A.Edwards & M.J.Corp (1975) The relationship between chromosome aberrations and low LET radiation dose to human lymphocytes. *Int. J. Radiat. Biol. 28*, 75–90.

MacIntosh, I.J.C., R.C.Brown & W.T.Coakley (1975) Ultrasound and *in vitro* chromosome aberrations. *Br. J. Radiol. 48*, 230–2.

Martins, B.K. (1971) *A Study of the Effects of Ultrasonic Waves on the Reproductive Integrity of Mammalian Cells Cultured In Vitro.* PhD thesis, Lawrence Berkeley Laboratory, Univ. of California, Donner Lab, Berkeley, Cal., LBL-37.

Mermut, S., K.P.Katayama, R.D.Castillo & H.W.Jones (1973) The effect of ultrasound on human chromosomes *in vitro. Obstet. Gynecol. 41*, 4–6.

Paal, M., J.Koever, S.Coemor & R.Humka (1963) Histochemical studies of the action of ultrasound and X-irradiation on Ehrlich ascites tumor in mice. *Rehabil. (Bonn) 16*, 129–32.

Prempree, T. & T.Merz (1969) Radiosensitivity and repair time: the repair time of chromosome breaks produced during different stages of the cell cycle. *Mutat. Res. 7*, 433–40.

Prempree, T., A.Michelson & T.Merz (1969) The repair of chromosome breaks induced by pulsed X-rays of ultra-high dose rate. *Int. J. Radiat. Biol. 15*, 571–4.

Pydorich, M.S. (1966) Complex treatment of eyelid epithelioma. *Oftal'mol. Zh. 21*, 281.

Rapacholi, M.H. (1970) Electrophoretic mobility of tumor cells exposed to ultrasound and ionizing radiation. *Nature 227*, 166–7.

Spring, E. (1969) Increased radiosensitivity following simultaneous ultrasonic and gamma ray irradiation. *Radiol. 93*, 175–6.

Spring, E., A.Rytila & K.Blomquist (1970) The effect of simultaneous gamma and ultrasonic radiation upon the growth of He-La cells, in *Proc. First Nordic Meeting of Medical and Biological Engineering* (Helsinki) pp.95–7.

Thacker, J. (1972) The possibility of genetic hazards from ultrasonic radiation. *Curr. Top. Radiat. Res. Q. 8*, 235–58.

Todd, P. & C.B.Schroy (1974) X-ray inactivation of cultured mammalian cells: enhancement by ultrasound. *Radiol. 113*, 445–7.

Winans, L.F., W.C.Dewey & C.M.Dettor (1972) Repair of sublethal and potentially lethal X-ray damage in synchronous Chinese hamster cells. *Radiat. Res. 52*, 333–51.

Woeber, K. (1959) Combination of ultrasound and X-ray radiation in the treatment of cancer. *Int. J. Phys. Med. 4*, 10.

D. SCOTT and R. R. MARSHALL

An Investigation of UV-Induced Chromosome Changes in relation to Lethality in Normal and UV-Sensitive Human Cells

The fact that xeroderma pigmentosum (XP) cells, defective in DNA excision repair, are more sensitive to ultraviolet (UV)-induced chromosome structural aberrations than normal human cells was first demonstrated by Parrington, Delhanty and Baden (1971). We have confirmed and extended these observations to investigate the role of chromosome damage in UV-induced cell lethality and to study other possible mechanisms of cell death. We have also performed cytogenetic studies (quantitation of chromosome structural aberrations, sister chromatid exchanges and multipolar mitoses) on UV-sensitive cells from an individual (designated 11961) who is not an XP patient and who has no detectable DNA repair defects (Lehmann *et al.* 1977) to see if the UV-sensitivity resulted from increased susceptibility to chromosome changes.

Materials and Methods

Cell lines. The XP cells (XP25RO, de Sanctis Caccione syndrome, complementation group A) were kindly supplied by Dr D. Bootsma (Erasmus University, Rotterdam), and the normal cells (designated 1BR) and 11961 cells by Dr C. F. Arlett (MRC Cell Mutation Unit, Sussex University). These three cell lines are from male donors and were used for experiments between the 8th and 26th passage, being routinely grown at 37°C in an atmosphere of 5% CO_2 and 95% air in Eagle's MEM medium plus 15% fetal calf serum and subcultured 1:2 every 4–5 days. Cells were always seeded into petri dishes 2 days before irradiation.

Irradiation was with a low pressure germicidal UV lamp of wavelength predominantly (> 90%) 254 nm (Fox and Fox 1971) at an incident dose rate of 0.25 $J\,m^{-2}s^{-1}$ after growth medium had been completely removed from the petri dishes.

Cell survival (colony forming ability) was determined by the method of Arlett, Harcourt and Broughton (1975) using a feeder layer technique. Plating efficiencies of controls were between 12 and 52%.

Chromosome structural aberration and sister chromatid exchange

(SCE) frequencies were analysed in metaphase preparations after pre-fixation treatment with vinblastine sulphate (0.1 μg ml^{-1}) and exposure to hypotonic KCl (0.075M) for 10 min or a 1:1 mixture of 0.075M KCl and 1% sodium citrate for 30 min. Cells were either fixed (aceto-alcohol 1:3) *in situ* whilst attached to cover-slips placed in petri dishes, or as a pellet in a centrifuge tube having first been removed from the petri dishes by trypsinisation. For analysis of SCEs in control and irradiated cells, BUdR at 5 μg ml^{-1} was added to the growth medium for 44–48 h before *in situ* fixation and cells were stained with 10 μg ml^{-1} Hoescht 33258 and 2% Giemsa according to the method of Perry and Wolff (1974). For analysis of conventional structural aberrations cells were stained with aceto-orcein and aberrations classified according to the system described by Savage (1975).

Multipolar mitoses and bridges and fragments were scored in anaphase and telophase cells fixed *in situ* without any vinblastine or hypotonic treatment.

Cell cycle times of control and irradiated normal and XP cells were derived from labelled mitosis curves after pulse labelling of cells growing on cover slips with ^{3}H-thymidine (^{3}HTdR, 0.1 μCi ml^{-1}, 1.0 Ci mmol^{-1}) for 15 min, washing with phosphate buffered saline (PBS) containing 4 μg ml^{-1} unlabelled TdR, completely removing the PBS and giving a dose of 0 or 0.5 J m^{-2} UV. After irradiation, fresh medium containing 4 μg ml^{-1} unlabelled TdR was added to the cells, which were fixed at approximately 3-h intervals up to 37 h post-treatment. At each fixation time a mitotic arresting agent, vinblastine sulphate (0.1 μg ml^{-1}), was added to those cultures that were to be fixed at the next sampling time, in order to obtain a cumulative mitotic index over the whole post-treatment period. Before *in situ* fixation, the culture medium was gently replaced with PBS and then 2–3 ml of fixative was slowly

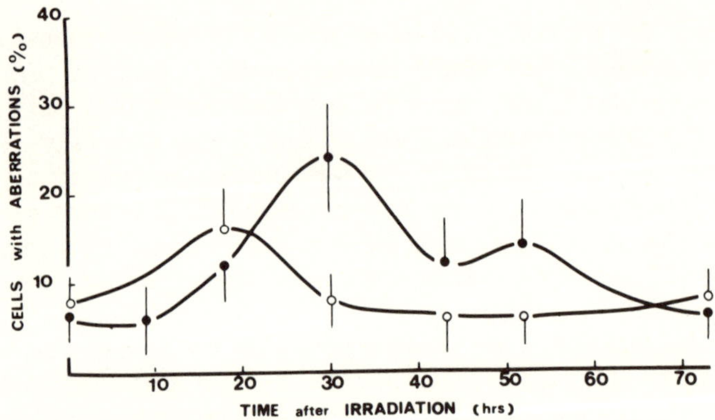

Figure 1. Frequency of metaphase cells with chromosomal aberrations after exposure of normal (open circles) and XP cells (solid circles) to 0.5 J m^{-2} UV. Control frequencies are given at time zero. Fifty cells were scored at each treatment/fixation time. Survival values for normal and XP cells determined at the same time were 65 ± 2 and 13 ± 0.5% respectively, the former being atypically low.

added. Cells had a further two changes of fixative before being left to air dry, and were then stained with aceto-orcein, dipped in Ilford L4 autoradiographic emulsion and exposed for two weeks.

Results

Comparison of XP and Normal Cells

Chromosome structural aberrations at metaphase were more frequent in XP than in normal cells after 0.5 (figure 1) or 1.0 J m^{-2} UV (figure 2); in normal cells the frequencies barely increased above the control levels shown at time zero in figures 1 and 2. Details of the aberration types observed are given elsewhere (Marshall and Scott, 1976; Marshall, 1976) and consisted mainly of chromatid deletions and a few exchanges.

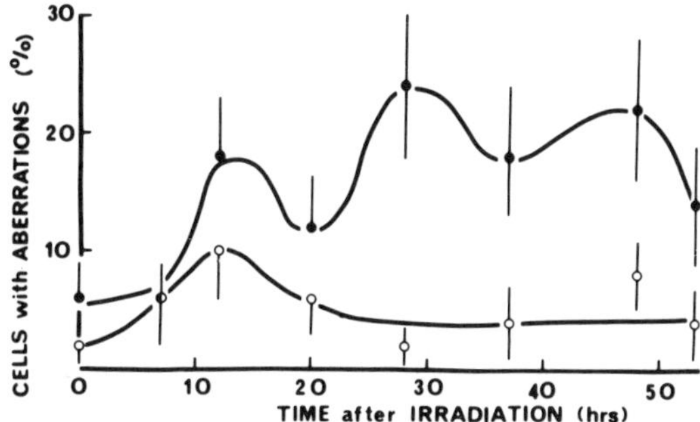

Figure 2. Frequency of metaphase cells with chromosomal aberrations after exposure of normal (open circles) and XP cells (solid circles) to 1.0 J m^{-2} UV. Other details as for figure 1, except that survival values for normal and XP cells were 100 ± 12 and 4 ± 0.5% respectively.

Cell cycle times were estimated after a dose of 0.5 J m^{-2} in order to calculate the frequency of chromosomally aberrant cells at their first, second and later cell cycles. The total cycle times of control and irradiated normal and XP cells, derived from the time interval between the half maximum value

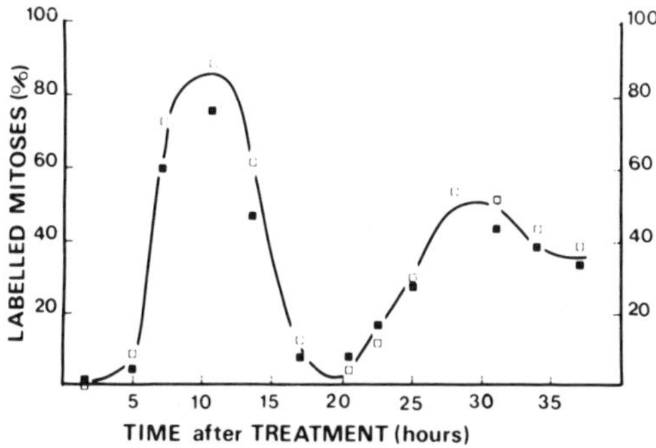

Figure 3. Labelled mitosis curve for normal human fibroblasts: unirradiated controls (open squares) and UV-irradiated (0.5 J m^{-2}) cells (solid squares). 100–200 mitoses were examined for labelling at each treatment/fixation time.

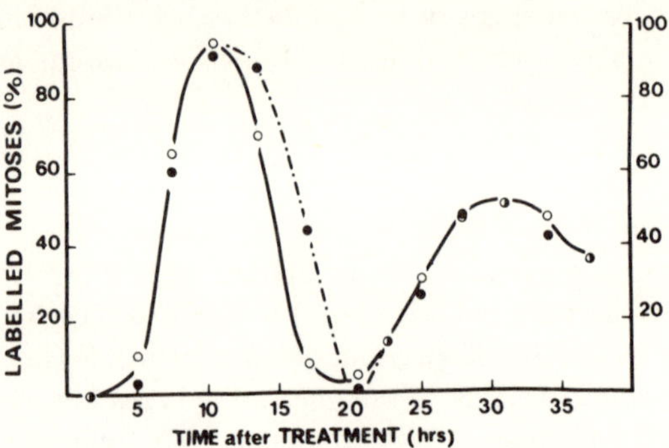

Figure 4. Labelled mitosis curve for XP fibroblasts: unirradiated controls (open circles) and UV-irradiated (0.5 J m^{-2}) cells (solid circles). 100–200 mitoses were examined at each treatment/fixation time.

on the first and second ascending limbs of the labelled mitosis curves (Quastler and Sherman 1959) was 20 h (figures 3 and 4). The only effect of UV was on the duration of the S phase of XP cells, which increased from 7.75 to 10.25 h. However, there appears to have been no delay, and possibly an acceleration, in the progression of cells that were in G_1 at the time of irradiation to their first mitosis (or of G_2 cells to their second mitosis), since the timing of the second wave of S cells was very similar for control and irradiated cells; thus the total cell cycle time was the same as in controls, at 20 h.

Metaphase aberration frequencies in relation to cell death. A comparison of the labelled mitosis curves after 0.5 J m^{-2} with that for metaphase aberration frequencies after the same dose (figures 5 and 6), indicates that only at 18 h post-irradiation of normal cells is the frequency of aberrations above the

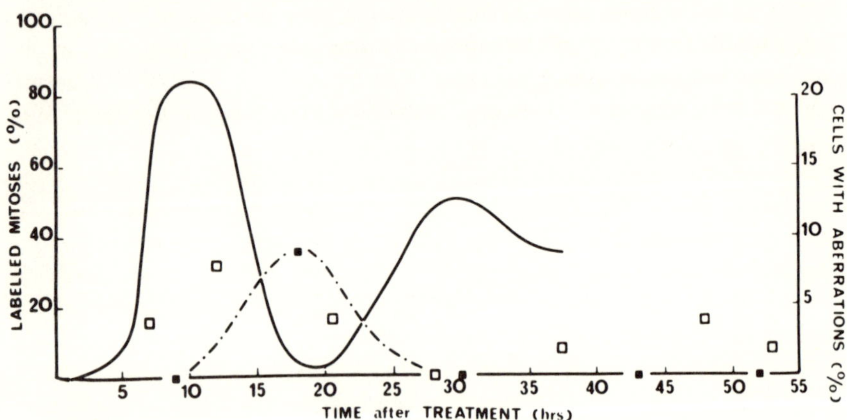

Figure 5. Relationship between the labelled mitosis curve after 0.5 J m^{-2} UV (see figure 3 for data) and the frequency of metaphase cells with chromosomal aberrations after 0.5 (solid squares) or 1.0 J m^{-2} (open squares) UV-irradiation of normal human fibroblasts. Aberration data as in figures 1 and 2 but with control frequencies subtracted.

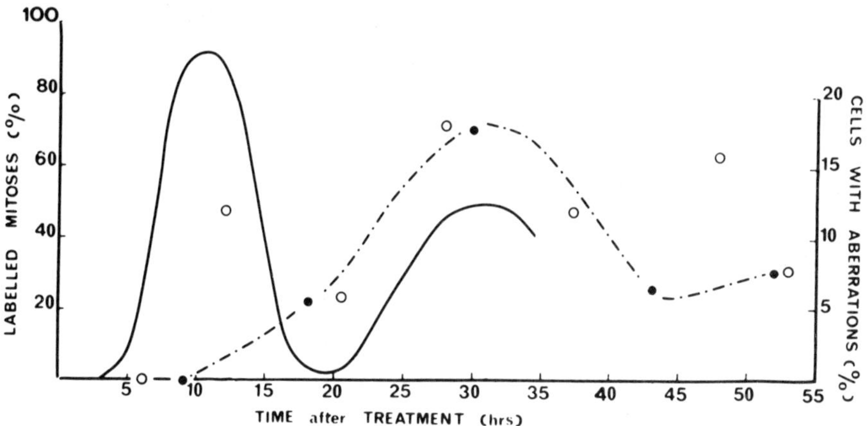

Figure 6. Relationship between the labelled mitosis curve after 0.5 J m^{-2} UV (see figure 4 for data) and the frequency of metaphase cells with chromosomal aberrations after 0.5 (solid circles) or 1.0 J m^{-2} (open circles) UV-irradiation of XP cells. Aberration data as in figures 1 and 2 but with control frequencies subtracted.

control level. This means that these aberrations were mainly in first division cells, whereas in XP cells aberrations are found primarily in cells at their second and third mitosis.

On the assumption that the metaphase aberration frequencies can reasonably be represented by the line drawn in figure 6, in spite of the limited number of sampling times, and that the cycle time is 20 h for XP cells after 0.5 J m^{-2} UV, we can calculate the frequencies of aberrations in each cell cycle and compare these values with the proportion of cells killed by this dose (87%), to see if the former can account for the latter. Too few aberrations were induced in normal cells to make this calculation meaningful.

Our basic assumption, in trying to compare aberration frequencies with cell killing, is that the loss of any chromosome material in the form of an acentric fragment will be *cell lethal.* However, because of the method of assessing cell killing by loss of colony-forming ability (CFA), only those aberrations that result in fragment loss from both daughter cells after mitosis (e.g. chromosome-type deletions) will be *colony-lethal*; those aberrations resulting in one normal and one abnormal (deleted) cell (e.g. chromatid deletions) will not cause loss of CFA. The proportion of cells carrying aberrations that are expected to be lethal to one or both daughter cells after the first mitosis is very low at 1.1 and 3.3% respectively (table 1). The majority (95.6%) of cell pairs derived from this first mitosis will therefore be chromosomally normal, and in order to prevent this cell pair from forming a colony *both* must sustain chromosome damage of the type lethal to both their progeny.

The average frequency of this type of chromosome damage over the second cell cycle (20 to 40 h) was 5.1% (table 1), so the probability of a pair

Table 1. Fractions of XP25RO metaphase cells containing chromosome aberrations likely to lead to the death of both, one or neither daughter cell after mitosis. Cells received 0 or 0.5 J m^{-2} UV.

Time after UV (h)	Percentage of cells with: aberrations lethal to both daughters[1]	aberrations lethal to one daughter[2]	no lethal aberrations[3]
9	3	3	94
18	4.7	7.3	88.0
30	6.3	16.3	77.4
43	2	10	88
52	6	8	86
72	2	4	94
Control	0	6.7	93.3
Mean frequencies for first two cycles after UV (above control)			
1st cycle	3.3	1.1	95.6
2nd cycle	5.1	6.3	88.6

[1] Chromosome-type dicentrics (with accompanying acentric fragments), centric rings plus fragments, deletions and acentric rings. Also isochromatid deletions, triradials and 50% of asymmetrical chromatid exchanges.
[2] Chromatid-type deletions, duplication/deletions, minutes, incomplete symmetrical exchanges and 50% of asymmetrical chromatid exchanges.
[3] No chromosomal aberrations or just gaps or inversions.

of normal cells, arising from first mitosis, both sustaining lethal damage is $0.956 \times 0.051 \times 0.051 = 0.25\%$. At first mitosis 1.1% of cells sustained damage lethal to one daughter cell (table 1); thus 5.1% of these potential colony-forming units will sustain damage that will lead to loss of CFA (i.e. $0.051 \times 0.911 = 0.056\%$). Clearly, therefore, only aberrations that affect both daughter cells and that arise during the first cell cycle will have any significant effect on CFA, and the frequency of these (3.3%) in UV-irradiated XP cells is extremely small compared with the degree of cell killing (87%). Chromosome structural aberrations must therefore play only a very minor role in UV-induced cell killing.

Interphase death. It has been shown, in cell lines derived from the Chinese hamster and the mouse, that after UV-irradiation a significant proportion of cells fail to undergo mitosis (Chu 1965, Todd *et al.* 1969, Thompson and Humphrey 1970), a phenomenon that we shall refer to as interphase death. Domon and Rauth (1968, 1969) claim that in mouse L cells the fraction that fails to undergo mitosis is equal to the fraction that loses CFA.

The extent of interphase death was investigated in our normal and XP cells after 0.5 J m^{-2} UV in conjunction with the cell cycle determinations described earlier, by successive accumulation of metaphases with vinblastine to obtain a cumulative mitotic index (MI) value for the whole post-irradiation sampling period (figures 7 and 8). If all irradiated cells undergo their first

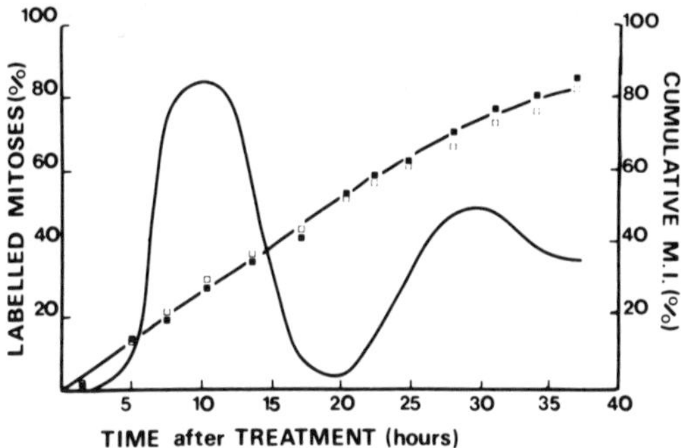

Figure 7. Relationship between the labelled mitosis curve (see figure 3 for data) and the cumulative mitotic index (CMI) of normal human fibroblasts: unirradiated controls (open squares) and UV-irradiated (0.5 J m^{-2}) cells (solid squares). 800–1000 cells scored for MI (meta- and pro-metaphases) at each treatment/fixation time. The failure of the CMI to reach 100% after a cycle time (20 h) may be due to (a) some cells not being in cycle, (b) a lag-time between adding vinblastine and metaphase accumulation beginning, (c) detachment of mitotic cells, although this was not evident from phase-contrast observations, (d) probably a combination of a, b and c. Survival level measured simultaneously was $80 \pm 11\%$ at 0.5 J m^{-2} UV.

post-treatment mitosis, the cumulative MI over the first post-irradiation cell cycle (20 h) should be similar for control and irradiated cells. This is the case for the normal cells exposed to 0.5 J m^{-2} (figure 7), a dose that induced 20% cell death. In irradiated XP cells about 20% less cells were accumulated over the 20 h post-irradiation period than in controls (figure 8), but this amount of interphase death is quite inadequate to explain the 80% loss of CFA. With interphase death at 20% and an aberration frequency affecting both daughter cells of approximately 4% (table 1) in those cells (80%) that do undergo their first mitosis, we can only account for about 23% cell death in terms of interphase death and chromosome aberrations.

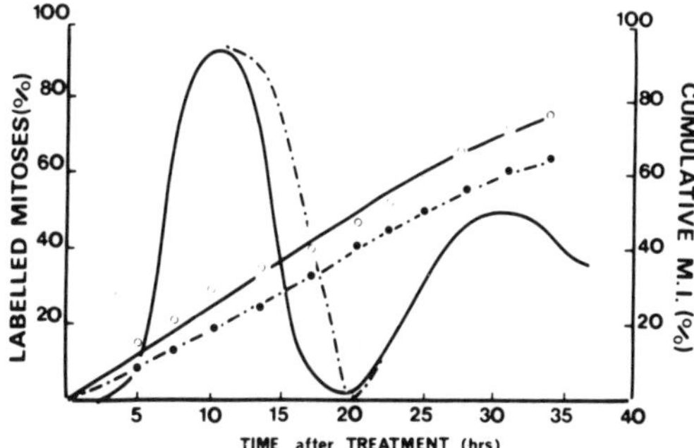

Figure 8. Relationship between the labelled mitosis curve (see figure 4 for data) and the CMI of XP cells: unirradiated controls (open circles) and UV-irradiated (0.5 J m^{-2}) cells (solid circles). Other details as for figure 7, except that survival was $18 \pm 4\%$.

UV-Induced Aberrations and Lethality

The death of XP cells exposed to UV occurs mainly after their second cell cycle, since the cumulative MI over the second cell cycle was similar for control and irradiated cells (figure 8), and from cell counts we have found that after 0.75 J m^{-2}, which gave 85% cell death, exponential growth was not resumed until after 140 h post-irradiation.

Multipolar mitoses. In time-lapse photography studies of UV-irradiated mouse L cells, Thompson and Humphrey (1970) showed that a significant proportion of cells (about 20% after 7.5 J m^{-2}) had abnormal mitoses that 'failed to complete cytokinesis and/or were multipolar'. We have examined approximately 75 ana/telophases in fixed preparations at 12, 18 and 30 h after 0.5 J m^{-2} UV in normal and XP cells, and have seen no abnormalities except occasional anaphase fragments, but no bridges or multipolar divisions.

Sister-chromatid exchange (SCE) frequencies have been compared in UV-irradiated normal and XP cells by de Weerd-Kastelein, Keijzer and Rainaldi (1977), who found that the XP strain that we have used (XP25RO) was extremely sensitive to UV-induced SCEs; a dose of only 0.3 J m^{-2} induced an average of 10 SCEs per cell above the spontaneous frequency of 7 per cell. This fact, coupled with the observation that cultured fibroblasts of Bloom's syndrome patients, which have a high spontaneous frequency of SCEs (Chaganti, Schonberg and German 1974), also have a low plating efficiency and poor growth (Giannelli *et al.* 1977), suggested to us that high frequencies of SCE might be cell-lethal. If there is a probability of the process of sister chromatid exchange being error-prone, such that exchange does not occur at exactly homologous loci on the two sister chromatids, this would lead to duplication of chromatin in one chromatid and loss (deletion) in the other, and the latter (and perhaps even the former) could be lethal. Obviously the higher the SCE frequency the greater would be the probability of error. This hypothesis, that SCE induction is a significant mechanism of cell killing after UV, was tested in the UV-sensitive 11961 cells (see below).

Comparison of 11961 and Normal Cells

Chromosome structural aberrations at metaphase were not significantly

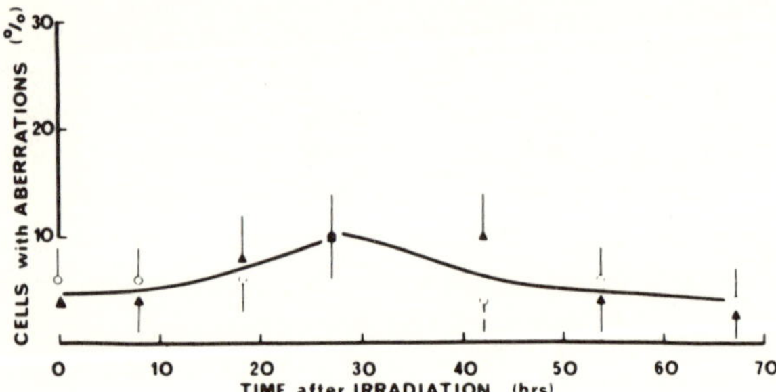

Figure 9. Frequency of metaphase cells with chromosomal aberrations after exposure of normal (circles) and 11961 cells (triangles) to 1.5 J m^{-2} UV. Control frequencies are given at time zero. 50 cells were scored at each treatment/fixation time.

increased above control levels after a 1.5 J m^{-2} UV exposure of normal and 11961 cells (figure 9, table 2), although this dose killed about 20% and 90% of the cells, respectively. This was also the case in a separate experiment (table 2) at the higher dose of 2.0 J m^{-2}, although at both doses there was a small increase in aberrations in irradiated 11961 cells above controls, which was not statistically significant. These data provide a very clear demonstration of cell killing by UV in the absence of chromosome structural aberrations.

Anaphase observations. Multipolar mitoses were virtually absent from control and irradiated (1.5 or 2.0 J m^{-2}) normal and 11961 cells (table 2). A few bridges and fragments were observed, but their frequency did not significantly increase after irradiation.

Table 2. Cytogenetic observations in control and UV-irradiated normal and 11961 cells. After 1.5 J m^{-2} UV, metaphase aberrations were analysed at 8, 19, 27, 42, 53 and 67 h after irradiation, anaphase aberrations at 12, 18, 24 and 30 h, and SCEs at 44 and 48 h. After 2.0 J m^{-2}, metaphase aberrations were analysed at 18, 24 and 30 h, anaphases at 12, 18, 24 and 30 h, and SCEs at 44 and 48 h. At each treatment/fixation time 50 cells were analysed except for SCEs, for which 25 cells were scored.

UV dose:	1.5 J m^{-2}		2.0 J m^{-2}	
Cell type:	Normal	11961	Normal	11961
% cells with chromosome structural aberrations (metaphase)				
Control	6.00 ± 0.48	4.00 ± 0.39	2.00 ± 0.09	2.00 ± 0.09
Irradiated	6.00 ± 0.08	7.00 ± 0.09	2.00 ± 0.07	3.30 ± 0.15
% cells with bridges and fragments (anaphases)				
Control	1.27 ± 0.07	1.53 ± 0.06	4.30 ± 0.10	3.50 ± 0.09
Irradiated	3.12 ± 0.14	1.01 ± 0.05	7.00 ± 0.13	3.00 ± 0.09
% multipolar anaphases				
Control	0.63 ± 0.05	0	0	0
Irradiated	1.04 ± 0.05	0	0	0
SCEs per cell				
Control	12.0 ± 0.7	15.3 ± 0.8	7.7 ± 0.5	9.4 ± 0.7
Irradiated	11.4 ± 0.8	18.1 ± 0.7	10.2 ± 0.9	11.5 ± 1.2

SCEs. Since there is no significant induction of conventional chromosome structural aberrations or multipolar mitoses in 11961 cells after UV doses that kill a significant proportion of cells, this cell line provided us with an opportunity to test the hypothesis that cell killing by UV is mediated through the induction of high frequencies of SCEs (see above). However, unlike the situation with XP25RO cells (de Weerd-Kastelein, Keijzer and Rainaldi 1977), UV-irradiation did not markedly increase the frequency of SCEs in 11961 cells (table 2). In our experiments a dose of 0.3 J m^{-2} on XP25RO cells kills about 60% of them, and this dose induced an average of 10 SCEs per cell in the experiments of de Weerd-Kastelein *et al.* At UV doses of 1.5 and 2.0 J m^{-2}, which kill over 90% of 11961 cells, less than 3 SCEs were induced above control levels. It seems unlikely therefore that death results from SCE induction in these cells. Although at 1.5 J m^{-2} SCEs increased above controls in 11961 cells but not in normal cells, at 2.0 J m^{-2} SCE frequencies increased in both cell lines.

The spontaneous frequency of SCEs was different in the two experiments whose results are summarised in table 2, perhaps because of differential incorporation of BUdR, but in both experiments the SCE frequency in unirradiated 11961 cells was higher by 2–3 per cell than in normal controls, in experiment 1 significantly so ($P=0.001$). To establish whether or not 11961 cells have a consistently higher spontaneous frequency of SCEs than normal cells will require parallel studies with other normal cell lines.

Interphase death. We have not investigated the amount of UV-induced interphase death in 11961 cells as we did for XP cells, but the fact that UV irradiation did not markedly reduce the mitotic index in the samples used for our chromosome studies suggests that this phenomenon will not account for a major fraction of the cell death observed.

Discussion

Chromosome changes and cell death. There is now good evidence that a major fraction of the cell death induced by ionising radiation at relatively low doses is attributable to induced chromosome structural aberrations (Marshall and Scott 1976). A positive correlation between UV-induced chromosome damage and cell killing has also been demonstrated by, e.g., Parrington, Delhanty and Baden (1971), Griggs and Bender (1972, 1973), Kato (1974), and Nilsson and Lehmann (1975), which led us to investigate their quantitative relationship in human cells.

It is clear from our investigations that conventional chromosome structural aberrations play little (XP cells) or no (11961 cells) part in UV-induced cell lethality. Neither can we explain the cell death in terms of other visible cytogenetic changes (i.e. multipolar mitoses or SCEs) or as a failure to undergo first post-irradiation mitosis. Whatever mechanism is responsible for most of the cell death, at least of XP cells, several cell cycles are required for its complete expression since chromosomally normal cells pass through several cell cycles before dying. Perhaps UV can inactivate gene expression without the actual loss of DNA in the form of chromosome fragments that appears to be necessary after ionising radiation (Grote and Revell 1972). The UV-irradiated cells may be able to pass through a number of cell cycles, using residual gene products present before irradiation, before death ensues.

The mechanism of structural aberration induction by UV in normal and XP cells. The fact that there are more UV-induced aberrations in XP than in normal cells clearly implicates unexcised pyrimidine dimers (Cleaver 1974) as the lesions from which the aberrations are derived. Our observation that whereas aberrations are primarily confined to first division in normal human fibroblasts, but are more frequent at second and even third mitosis of XP cells than at first division, can be explained on a model proposed by Bender, Griggs and Walker (1973), which assumes that the unreplicated G_1 chromosome is uninemic and that a chromatid break is equivalent to an unrepaired DNA double-strand break.

We suggest that in normal cells excision of pyrimidine dimers without insertion of new bases in the DNA leaves single-strand breaks that are replicated in S to give chromatid aberrations at first mitosis, as proposed by Bender *et al.* (1973). In XP cells very little excision takes place, so cells will pass into their first S phase with many unexcised dimers, and a gap will be formed in the newly-synthesised strand opposite the dimer (Lehmann 1972). If a small number of these gaps persists until the next S phase they could then be replicated, giving double-strand breaks leading to chromatid aberrations at the subsequent (later than first) mitosis.

Summary

Cultured fibroblasts of excision-repair defective xeroderma pigmentosum (XP) cells sustain more conventional chromosome structural aberrations than normal human fibroblasts when exposed to UV. By measuring cell cycle times after UV-irradiation we find that whereas in normal cells such aberrations are primarily confined to cells at their first post-irradiation mitosis, XP cells aberrations occur mainly in later cell cycles.

The frequency of cells with chromosomal aberrations is totally inadequate to explain the degree of cell killing. Neither can we explain the mechanism of cell killing in terms of abnormal segregation of chromosomes at mitosis (multipolarity) or failure to undergo first post-irradiation mitosis (interphase death), the frequency of the latter being small compared with the proportion of cells that die. Chromosomally normal cells pass through several cell cycles before dying.

UV-sensitive cells from an individual (designated 11961), who is not an XP and has no detectable DNA repair defects (Lehmann *et al.* 1977), exhibit no chromosome aberrations or multipolar mitoses above control values even after UV doses that kill over 90% of cells. A small increase in sister chromatid exchange (SCE) frequencies is observed after these UV doses and the spontaneous frequency of SCEs in 11961 cells is a little higher than in normal cells studied in parallel. Cell death cannot be attributed to visible UV-induced chromosome changes.

Acknowledgements

This work was supported by grants from the Cancer Research Campaign and the Medical Research Council. We would like to thank Mrs Christine Blease for expert technical assistance and Miss Ann R. Currie, Dr R. Rowley, Mr A. I. Galbraith and Mr G. M. Wolfe for microscope work.

References

Arlett, C.F., S.A.Harcourt & B.C.Broughton (1975) The influence of caffeine on cell survival in excision-proficient and excision-deficient xeroderma pigmentosum and normal human cell strains following ultraviolet light irradiation. *Mutat. Res. 33*, 341–6.

Bender, M.A., H.G.Griggs & P.L.Walker (1973) Mechanisms of chromosomal aberration production. I. Aberration induction by UV light, *Mutat. Res. 20*, 387–402.

Chaganti, R.S.K., S.Schonberg & J.German (1974) A manyfold increase in sister chromatid exchanges in Bloom's syndrome lymphocytes. *Proc. Nat. Acad. Sci. USA 71*, 4508–12.

Chu, E.H.Y. (1965) Effects of ultraviolet radiation on mammalian cells: I. Induction of chromosome aberrations. *Mutat. Res. 2*, 75–94.

Cleaver, J.E. (1974) Repair processes for photochemical damage in mammalian cells. *Adv. Radiat. Biol. 4*, 1–75.

de Weerd-Kastelein, E.A., W.Keijzer & P.Rainaldi (1977) Induction of sister chromatid exchanges in xeroderma pigmentosum cells following UV exposure. *Mutat. Res. 46*, 163 abs.

Domon, M. & A.M.Rauth (1968) Ultraviolet irradiation of mouse L cells: effects on DNA synthesis and progression of cells through the cell cycle. *Radiat. Res. 35*, 350–68.

——(1969) Effects of caffeine on ultraviolet irradiated mouse L cells. *Radiat. Res. 39*, 207–21.

Fox, M. & B.W.Fox (1971–72) The establishment of cloned cell lines from Yoshida sarcomas having differential sensitivities to methylene dimethane sulphonate *in vivo* and their cross-sensitivity to X-rays, UV and other alkylating agents. *Chem.-Biol. Interactions 4*, 363–75.

Giannelli, F., P.F.Benson, S.A.Pawsey & P.E.Polani (1977) A contribution to the study of Bloom's syndrome. *Mutat. Res. 46*, 121–2.

Griggs, H.G. & M.A.Bender (1972) Ultraviolet and gamma-ray induced reproductive death and photoreactivation in a Xenopus tissue culture cell line. *Photochem. Photobiol. 15*, 517–26.

——(1973) Photoreactivation of ultraviolet induced chromosomal aberrations. *Science 179*, 86–8.

Grote, S.J. & S.H.Revell (1972) A correlation of chromosome damage and colony forming ability in Syrian hamster cells in culture irradiated in G_1. *Curr. Top. Radiat. Res. Q. 7*, 303–9.

Kato, H. (1974) Differential responses of several aneusomic cell clones to UV light. *Exp. Cell Res. 83*, 55–62.

Lehmann, A.R. (1972) Post-replication repair of DNA in ultraviolet-irradiated mammalian cells. *J. Mol. Biol. 66*, 319–37.

Lehmann, A.R., S.Kirk-Bell, C.F.Arlett, S.A.Harcourt, E.A. de Weerd-Kastelein, W. Keijzer & P.Hall-Smith (1977) Repair of ultraviolet light damage in a variety of human fibroblast cell strains. *Cancer Res. 37*, 904–10.

Marshall, R.R. (1976) *Cytogenetic studies in drug and radiation resistance.* PhD thesis, Victoria University of Manchester, England.

Marshall, R.R. & D.Scott (1976) The relationship between chromosome damage and cell killing in UV-irradiated normal and xeroderma pigmentosum cells. *Mutat. Res. 36*, 397–400.

Nilsson, K. & A.R.Lehmann (1975) The effect of methylated oxypurines on the size of newly-synthesised DNA and on the production of chromosome aberrations after UV irradiation on Chinese hamster cells. *Mutat. Res. 30*, 255–66.

Parrington, J.M., J.D.A.Delhanty & H.P.Baden (1971) Unscheduled DNA synthesis, UV-induced chromosome aberrations and SV40 transformation in cultured cells from xeroderma pigmentosum. *Ann. Hum. Genet. 35*, 149–60.

Perry, P. & S.Wolff (1974) New Giemsa method for the differential staining of sister chromatids. *Nature 251*, 156–8.

Quastler, H. & F.G.Sherman (1959) Cell population kinetics in the intestinal epithelium of the mouse. *Exp. Cell. Res. 17*, 420–38.

Savage, J.R.K. (1975) Classification and relationships of induced chromosomal structural changes. *J. Med. Genet. 12*, 103–22.

Thompson, L.H. & R.M.Humphrey (1970) Proliferation kinetics of mouse LP59 cells irradiated with ultraviolet light, a time lapse photographic study. *Radiat. Res. 41*, 183–201.

Todd, P., T.P.Coohill, A.B.Hellewell & J.A.Mahoney (1969) Post-irradiation properties of cultured Chinese hamster cells exposed to UV light. *Radiat. Res. 38*, 321–39.

K.E.BUCKTON, G.E.HAMILTON, L.PATON, and A.O.LANGLANDS

Chromosome Aberrations in irradiated Ankylosing Spondylitis Patients

At the previous International Symposium held in this Unit in 1966, entitled 'Human Radiation Cytogenetics', the cytogenetic information available from peripheral blood lymphocytes of patients who had undergone a course of radiotherapy was presented (Buckton *et al.* 1967a and b). Cytogenetic changes following partial-body irradiation (1500 rad in ten fractions to the whole spine and sacroiliac joints, which was and is the standard radiotherapy treatment in Edinburgh for ankylosing spondylitis) were seen to persist for up to at least ten years, the time interval of the study. We are now able to report on the fate of the radiation-damaged cells following this treatment regime for up to twenty years. Information is also available for up to thirty years post-treatment, but this includes patients that had higher doses, 2000 and 2500 rad to standard fields, and a variety of treatments.

Materials and Methods

All patients considered here have had only the standard course of treatment, described above, and the number of cells analysed per year and the frequency of aberrations seen is given in table 1.

Since the early 1970s, advances in cytogenetic techniques have allowed more detailed analysis of cells by the use of banding techniques. The techniques used were the ASG technique, which was developed in this laboratory (Sumner, Evans and Buckland 1971), and also an R-band technique described by Sehested (1974).

Results and Discussion

As the amount of information available from banded preparations is still quite small, all the data we have for cultures incubated for 48 h, when the majority of cells will be in their first division *in vitro*, have been included in figure 1. The shape of the curve for Cu cells, i.e. cells with unstable or asymmetrical aberrations, is very similar to the previous curve (Buckton *et al.* 1967a) and is seen to decrease still further after ten years. At twenty years post-treatment only one per cent of the cells carry a Cu abnormality. The use of banded chromosome preparations does not assist in the detection of asymmetrical exchanges. In fact the number of cells with dicentrics seen

Table 1. Frequency of aberrations.

ime post-eatment r)	Total cells analysed	Unstable (Cu) cells		Stable (Cs) cells		Dicentrics		Rings		Acentric fragments	
		no.	%	no.	%	no.	per cell	no.	per cell	no.	per cell
< 1/12	1375	517	37.6	135	9.8	446	0.32	57	0.042	230	0.17
1–6/12	930	305	32.7	103	11.1	263	0.28	39	0.042	169	0.18
1 ± 0.5	1097	206	18.8	113	10.3	168	0.15	13	0.012	116	0.11
2 "	550	60	10.9	49	8.9	46	0.84	6	0.011	24	0.044
3 "	393	28	7.1	34	8.6	24	0.61	3	0.007	15	0.038
4 "	990	54	5.4	89	8.9	38	0.38	8	0.008	30	0.030
5 "	1812	96	5.2	164	9.0	83	0.05	4	0.002	46	0.025
6 "	1336	53	4.0	170	12.7	44	0.033	10	0.007	23	0.017
7 "	2043	62	3.0	190	9.3	55	0.027	3	0.001	34	0.017
8 "	1422	45	3.1	124	8.7	39	0.027	11	0.008	19	0.013
9 "	1410	24	1.7	128	9.1	18	0.012	6	0.004	10	0.007
10 "	1575	38	2.4	155	9.8	26	0.016	2	0.001	25	0.016
11 "	1320	18	1.4	165	12.5	10	0.008	2	0.001	9	0.007
12 "	960	12	1.2	87	9.1	8	0.008	1	0.001	7	0.007
13 "	935	17	1.8	87	9.3	14	0.015	2	0.002	7	0.007
14 "	991	19	1.9	142	14.3	10	0.010	1	0.001	14	0.014
15 "	1350	9	0.6	115	8.5	4	0.003	1	0.001	5	0.004
16 "	1150	17	1.4	128	11.1	11	0.009	1	0.001	6	0.005
17 "	830	8	0.9	82	9.9	8	0.009	1	0.001	3	0.004
18 "	1340	21	1.6	197	14.7	8	0.006	2	0.001	15	0.011
19 "	920	9	1.0	117	12.7	4	0.004	1	0.001	4	0.004
20 "	980	6	0.6	86	8.7	5	0.005	0	0	3	0.003
21 "	550	4	0.7	81	14.7	3	0.005	0	0	3	0.005
22 "	380	2	0.5	24	6.3	0	0	1	0.002	0	0
23 "	810	13	1.6	45	5.5	4	0.005	2	0.002	7	0.009
24 "	510	5	1.0	36	7.1	3	0.006	0	0	1	0.002
25 "	325	1	0.3	33	10.0	0	0	0	0	1	0.003
26 "	225	4	1.7	21	9.3	2	0.008	0	0	0	0
27 "	300	1	0.3	23	7.6	0	0	0	0	0	0
28 "	200	0	0	10	10.5	0	0	0	0	0	0
> 29	260	5	1.9	18	6.9	5	0.02	0	0	1	0.004

in a banded preparation is often fewer, because (a) there is a tendency to select better quality cells for analysis than was necessary on a conventionally stained preparation, and (b) a centromere is not as obvious on a banded preparation as on a conventionally stained slide; because of the apparent constriction in all pale bands, this is particularly true of dicentrics involving an acrocentric, where the break is close to the centromere and the banding pattern is barely altered.

In contrast, the detection of Cs cells, or cells with a stable or symmetrical rearrangement only, is seen to be improved by the use of banded preparations. The level of these cells fluctuates about a mean of 10% for at least thirty years from exposure and probably will remain at this level for the rest

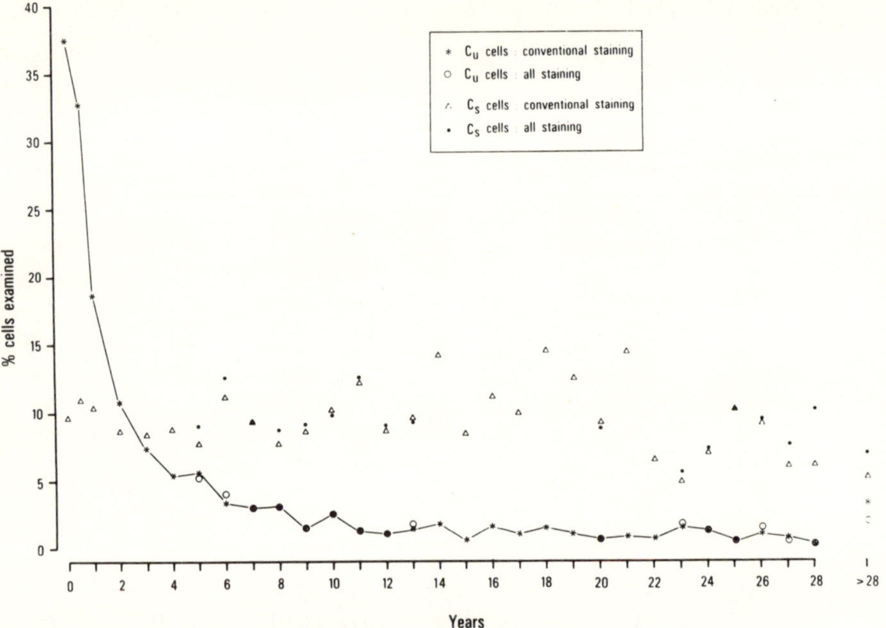

Figure 1. Frequency of cells with unstable (Cu) and stable (Cs) aberrations, with time in years, in the peripheral blood of patients treated with 250 kV X-ray therapy for ankylosing spondylitis.

of the patient's life. The cells with stable aberrations only, because they can duplicate themselves and are likely to remain in the body indefinitely, may give rise to clones of abnormal cells with an altered potential for growth. In all, approximately 200 patients treated for ankylosing spondylitis have been examined cytogenetically, many of them on several occasions. In only three of them has a definite clone been identified. Sixteen other patients have had two cells with the same rearrangement, but we have not been able to confirm the presence of a clone on subsequent examination. However, as already mentioned, the numbers of cells analysed using banded preparations is as yet quite small.

Although banded preparations do not improve the detection of unstable aberrations, a banded preparation is a great help in interpreting all the abnormalities that are present in a cell. For example, it can be determined whether an acentric fragment is a result of a true terminal deletion involving only one break, or a more complex rearrangement such as an incomplete exchange or interstitial deletion involving at least two breaks. Detection of any additional abnormalities that the cell might carry is also easier. Often one abnormality is quite obvious, but with careful analysis less obvious abnormalities in the cell become apparent (figure 2).

The frequency of dicentrics and rings per cell has been shown previously to fall with time post-irradiation; this is evident from table 1 and figure 3.

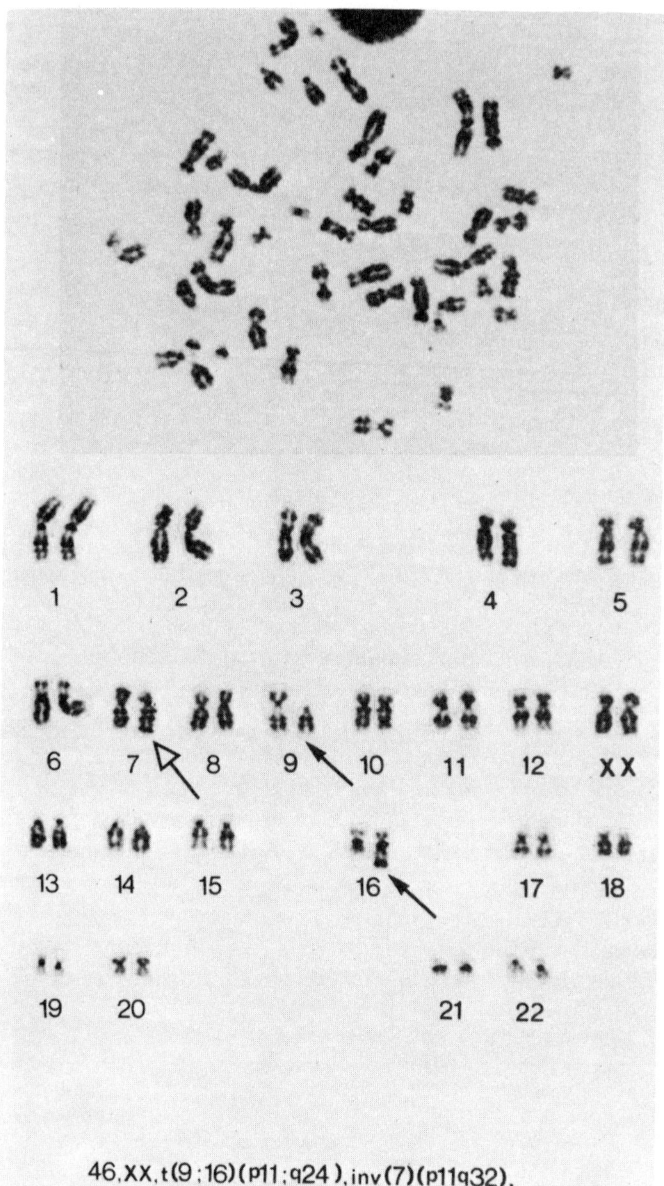

Figure 2. Radiation-damaged cell showing damage that would (closed arrowhead) and would not (open arrowhead) have been apparent with conventional staining.

Over the first four years post-treatment, there is a rapid drop in the number of dicentrics seen at an estimated rate of $42.7 \pm 7.5\%$ per annum. Over the first three or four years after treatment, there is a gradual restoration of the small lymphocyte population to approximately pre-treatment numbers

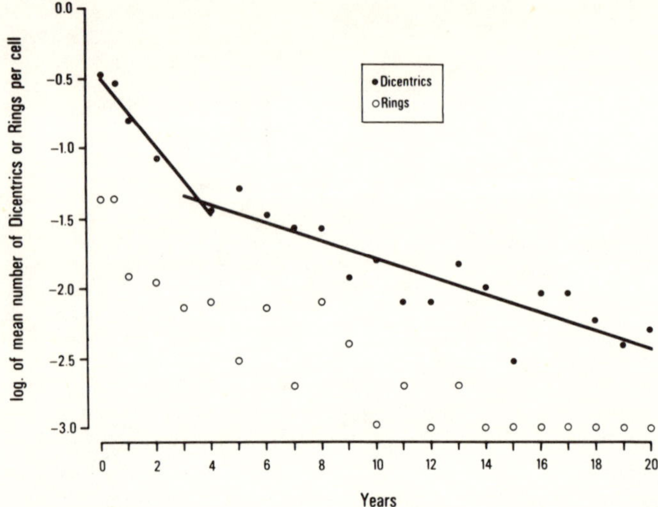

Figure 3. Frequency of dicentrics and rings, with time in years, in the peripheral blood of patients treated with 250 kV X-ray therapy for ankylosing spondylitis.

(Buckton *et al.* 1967a). This repopulation is probably largely from undamaged cells, which will cause a dilution effect of the abnormal cells. After four years and up to twenty years, the estimated decrement is 13.8 ± 3.3% per annum. No line has been plotted for the decrease in the frequency of ring chromosomes, because these are much fewer in number.

Our interest in the irradiated ankylosing spondylitic population is not only to monitor changes in the incidence of chromosome damage with time following radiation exposure, but to study the distribution of damage between cells and to establish whether certain chromosomes are selectively damaged or selectively survive.

In the last few years very few patients have undergone radiotherapy treatment for ankylosing spondylitis in Edinburgh. For this reason only one

Table 2. Numbers of cells with aberrations.

Time post-treatment	3 months			6 months	
Staining	banded G	banded R	conventional	banded	conventional
Total cells analysed	50	50	50	50	50
Cells without aberrations	14	16	12	16	22
Cells with asymmetrical exchanges only	13	9	16	8	10
Cells with asymmetrical and symmetrical exchanges	11	18	13	16	10
Cells with symmetrical exchanges only	12	7	9	10	8

patient has been studied shortly after treatment, using banded preparations. From table 2 it can be seen that the number of cells free of aberrations in this patient, who was examined three and six months post-treatment, is similar whether the cells were analysed using banding techniques or not; however, the complexity of the damage seen in the abnormal cells following banding is greater. In fact, the detection of symmetrical exchanges in the cells of this patient three months post-treatment is improved by one-third by the use of banding (table 3). This was also true of data derived from cells irradiated *in vitro* (Buckton 1976). From table 3 it can be seen that using banding techniques the ratio of asymmetrical exchanges (dicentrics) to symmetrical exchanges (reciprocal translocations) is approximately 1 : 1.

Table 3. Number of abnormalities seen in 100 cells.

	Asymmetrical		Symmetrical		
	banding	conventional	banding	conventional	
Dicentrics	31	36	27	21	Reciprocal translocations involving two chromosomes
Tricentrics	2	4	0	0	Reciprocal translocations involving three chromosomes
Rings	2	5	3	0	Pericentric inversions
Interstitial deletions	0	2	0	0	Paracentric inversions

In the cells from the patient treated three months previously, the point at which the chromosomes had been broken could be plotted for 328 breaks on the autosomes and 4 on the sex chromosomes (table 4). An expected number of breaks for each chromosome was calculated, assuming that the number of breaks would be proportional to length. From this it was found that chromosomes 1 and 2 had only two-thirds the number of breaks expected, and chromosomes 3 and 19 only half; 12 and 17 had one-third more than expected. These last two chromosomes were also favoured after irradiation *in vitro* (Buckton 1976). In contrast, the distribution of breaks in cells studied four or more years after exposure provided a rather different picture. With G-banding, 755 break points were identified, 728 on the autosomes and 27 on the sex chromosomes. The observed distribution is given in table 4, the expected number of breaks being calculated as above. Chromosomes 9, 10 and 11 have a particularly high number of breaks, and 3, 19 and 22 very few; this is contrary to published results for *in vitro* irradiation (Holmberg and Jonasson 1973, Seabright 1973, Buckton 1976) and *in vivo* irradiation (San Roman and Bobrow 1973).

When the breaks detected in cells from patients treated four or more years previously are plotted on a map of banded chromosomes (Paris Conference 1971), certain bands appear to be favoured (figure 4). This is particularly marked in certain chromosomes, such as chromosome 3, which

Table 4. Break points per chromosome.

3 & 6 months post-treatment				>4 yr. post-treatment		
No. of breaks obs	No. of breaks exp	Difference obs−exp	Chromosome number	No. of breaks obs	No. of breaks exp	Difference obs−exp
19	28	−9	1	62	61	+1
20	27	−7	2	56	58	−2
13	23	−10	3	36	50	−14
20	21	−1	4	38	46	−8
14	20	−6	5	42	44	−2
23	20	+3	6	38	43	−5
17	18	−1	7	42	39	+3
19	16	+3	8	35	36	−1
21	16	+5	9	49	35	+14
14	15	−1	10	45	33	+12
19	15	+4	11	52	34	+18
22	15	+7	12	39	34	+5
9	12	−3	13	20	27	−7
14	12	+2	14	31	26	+5
13	12	+1	15	26	25	+1
11	11	—	16	23	24	−1
19	11	+8	17	24	24	—
14	10	+4	18	22	21	+1
5	9	−4	19	9	19	−10
7	8	−1	20	14	19	−5
6	6	—	21	16	14	+2
9	7	+2	22	9	15	−6
2	8	−6	X	22	19	+3
2	3	−1	Y	5	8	−3

has less breaks than expected in proportion to its length, but 39% of those that do occur are in the two median pale bands on each arm. Also chromosome 21 is notable: it has 16 breaks on it, 10 of which are in band q22.

Conclusions

It would seem from the evidence we have so far, from *in vitro* irradiated blood and *in vivo* samples taken soon after treatment, that all chromosomes are approximately equally damaged by the irradiation, with the possible exception of those mentioned above. However, cells that have survived *in vivo* for more than four years appear to show that damage at certain sites, particularly at the centromeres, can be better tolerated by the cells. It may be that certain chromosomes or parts of chromosomes are less important to the function and proliferative capacity of lymphocytes and their precursors than others. We know that actively proliferating leukaemic cells are often characterised by a loss of a number 7 chromosome and a gain of a number 8, and also, to a lesser extent, a gain of chromosomes 9, 10 and 11 (Rowley 1977).

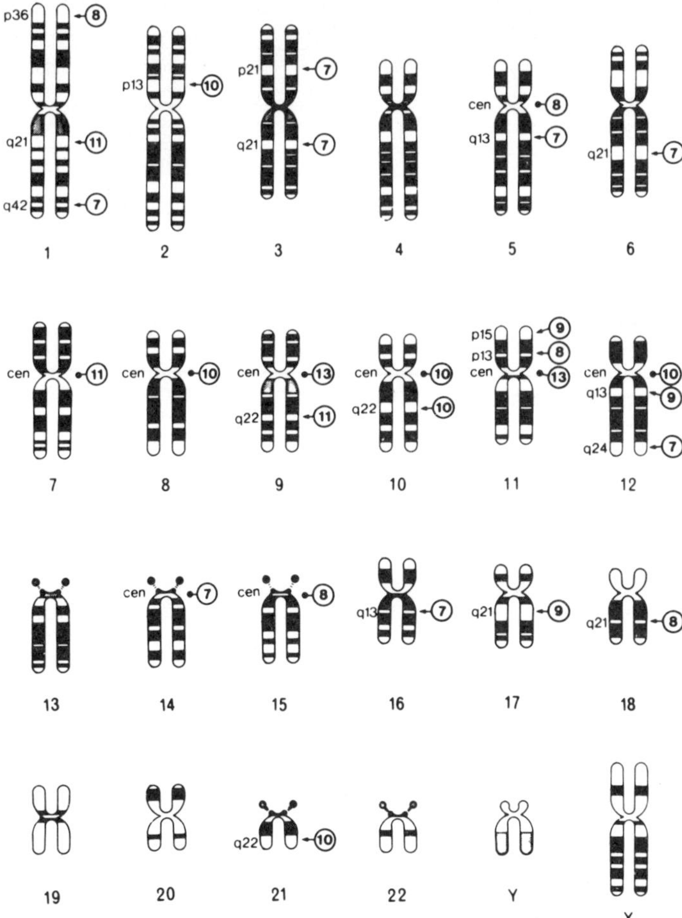

Figure 4. Human chromosome map with bands marked in which more than 6 breaks out of the 755 identified occurred. The numbers in circles record the actual number of breaks that occurred in any particular band. The centromere is taken to include bands p11 and q11.

At the present time, chromosome aberrations observed in the peripheral blood lymphocytes of an individual may be used as an indicator of radiation exposure. The fact that many of these cells can survive in the body of an irradiated individual for twenty and more years is now well established. What correlation may exist between these irradiation-damaged cells and possible late effects of the irradiation, such as the increased frequency of mid-line cancers and leukaemia, still remains an enigma. Further data, on the sites of radiation-induced chromosome exchange in cells that can survive *in vivo*, on the chromosome sites involved in cells that are observed to form clones, and on the cytogenetics of radiation-induced malignancies, may reveal a pattern of chromosome change from which it may be possible to get some indication of those changes that are 'benign' and the kinds of rearrangements that are associated with the initiation and development of malignancies.

Acknowledgements

This work is partly supported by a grant from the International Atomic Energy Agency.

References

Buckton, K.E. (1976) Identification with G and R-banding of the position of breakage points induced in human chromosomes by *in vitro* X-irradiation. *Int. J. Radiat. Biol. 29*, 475–88.

Buckton, K.E., P.G.Smith & W.M.Court Brown (1967a) The estimation of lymphocyte lifespan from studies on males treated with X-rays for ankylosing spondylitis, in *Human Radiation Cytogenetics* (ed. H.J.Evans, W.M.Court Brown & A.S.McLean) pp.106–14. Amsterdam: North Holland.

Buckton, K.E., A.O.Langlands, P.G.Smith & J.McLelland (1967b) Chromosome aberrations following partial and whole body X-irradiation in man: Dose-response relationships, in *Human Radiation Cytogenetics* (ed. H.J.Evans, W.M. Court Brown & A.S.McLean) pp.122–35. Amsterdam: North Holland.

Holmberg, M. & J.Jonasson (1973) Preferential location of X-ray-induced chromosome breakage in the R-bands of human chromosomes. *Hereditas 74*, 57–68.

Paris Conference (1971) *Standardization in Human Cytogenetics*. Birth Defects: Original Article Series vol. 8, no. 7, 1972. New York: National Foundation – March of Dimes.

Rowley, J.D. (1977) Chromosomes in malignancies, in *Proc. V International Congress of Human Genetics*. Amsterdam: Excerpta Medica (in press).

San Roman, C. & M.Bobrow (1973) The sites of radiation-induced breakage in human lymphocyte chromosomes, determined by quinacrine fluorescence. *Mutat. Res. 18*, 325–31.

Seabright, M. (1973) High resolution studies on the pattern of induced exchanges in the human karyotype. *Chromosoma 40*, 333–46.

Sehested, J. (1974) A simple method for R-banding of human chromosomes, showing a pH-dependent connection between R and G-bands. *Humangenetik 21*, 55–8.

Sumner, A.T., H.J.Evans & R.A.Buckland (1971) New technique for distinguishing between human chromosomes. *Nature New Biol. 232*, 31–2.

M. SEABRIGHT

Participation of Human Chromosomes in Induced Exchanges

The distribution of X-ray induced exchanges and other aberrations was studied in 1000 cells obtained from ten normal subjects and fifteen patients with abnormal chromosome complements. The material used was derived from venous blood samples, treated with X-rays at a dose of 300 rad. Standard cultures were set up, incubated for 48–52 hours and harvested. The analysis of the lesions was carried out from karyotypes of G-banded chromosomes (Seabright 1973a).

Results and Discussion

The main points that emerged from the study of the first group of ten normal individuals were: (a) all lesions occurred in regions that, at metaphase, appeared as light bands, and (b) there was a significant deficiency of exchanges involving the sex-chromosomes with the autosomes.

In view of the latter finding, nine patients with sex-chromosome abnormalities ranging from 45,X0 to 49,XXXXY and two Xi(X) were investigated in order to ascertain whether the deficiency of exchanges in the X and Y chromosomes would persist in abnormal complements. The data gathered from the normal subjects provided a set of expectations, which was used to test the results obtained from this second group.

The analysis showed that the frequency of breaks in the X chromosome was as expected in respect to its length, but the significant lack of breaks that resulted in exchanges with the autosomes was maintained in these abnormal individuals. Moreover, an excess of pericentric inversions in the X was noted. These inversions were associated with the presence of 'hot spots' in regions Xp22 and Xq13 (Seabright 1973b).

It was also observed that the total number of breaks in the autosomes of these individuals was significantly higher than that found in normal subjects. The hypothesis that the presence of supernumerary or rearranged chromosomes may increase their susceptibility to irradiation was therefore formulated. To test this hypothesis a third group of patients with abnormalities of the autosomes were studied. This group included a trisomy 21, a 21;21 Robertsonian translocation, two balanced reciprocal translocations and a father and his mentally retarded son. Both had a very large 1qh segment.

The results of the analysis indicated that the average number of breaks in the autosomes of these individuals was almost 50% higher than that found in normal subjects. These observations suggested that any type of anomaly of the chromosomes may be responsible either for an increase in their fragility or for a decreased ability to repair (Seabright 1976). Finally, the participation of the autosomes in the formation of dicentrics and translocations was examined.

It is generally agreed that the number of breaks per chromosome is dependent on chromosome size and perhaps the amount of light-band material present. Given this general limitation, the frequency with which the autosomes may exchange with other specific autosomes may vary. To investigate this possibility a record was made of the numbers of each type of dicentric and translocation that occurred. These data were analysed using a variant of the cluster analysis described by Lance and Williams (1967), which was carried out on the Reading University computer by Professor R. Curnow. Each dendrogram thus obtained demonstrated the patterns of relationship between autosomes in relation to exchanges of symmetrical and asymmetrical type respectively. A list of K-values was also obtained. These values are indices of the degree of clustering shown by the nearest neighbours and were used to derive the branched-chain diagrams in the dendrograms (figures 1 and 2). The X and Y chromosomes were not included because of the paucity of exchanges found in these chromosomes.

From these diagrams it is apparent that the pattern of clustering was different between the dicentrics and the translocations. For example, in the dicentric data, chromosomes 14 and 21 were involved with each other in the formation of a high number of dicentrics. Chromosomes 8 and 11 also showed a high degree of clustering for dicentrics. But these chromosomes did

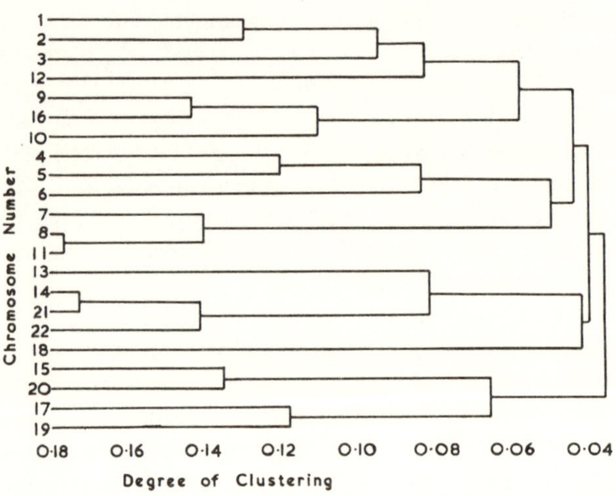

Figure 1. Dendrogram showing the clustering behaviour of autosomes in relation to dicentric formation.

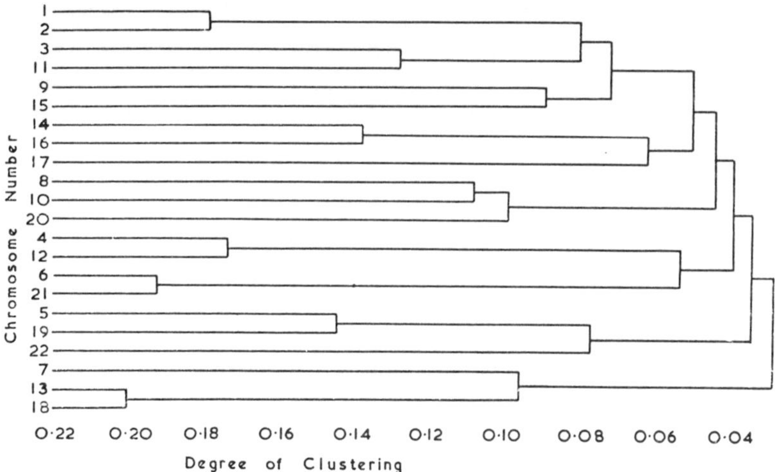

Figure 2. Dendrogram showing the clustering behaviour of autosomes in relation to translocation formation.

not have the same affinity for each other in forming translocations. This difference may be a reflection of the ratio of dicentrics to translocations obtained, which deviated significantly from the theoretical 1:1. In fact, there was an excess of dicentrics, 381:279 translocations. This deviation could perhaps be explained by failure to detect a number of symmetrical exchanges. For example, equal exchanges between homologues or small terminal translocations would not be recognised with any of the banding techniques in use at present. The fact that the short arms of the acrocentrics were involved more frequently in the formation of dicentrics was not surprising, but there was a significant excess of events involving chromosomes 14 and 21. Oddly, number 15 was more involved with 20 rather than with the other acrocentrics, and 13 was frequently involved with 18 in forming translocations.

Conclusion

In conclusion, some specificity in induced exchanges between autosomes has been found. For example, the clustering of 14 and 21 in the formation of dicentrics, which is consistent with the estimated frequencies of naturally occurring events of this type.

Summary

The patterns of breaks and exchanges induced *in vitro* by X-irradiation were determined from G-banded chromosomes in a sample of chromosomally normal individuals and in subjects with numerical and structural abnormalities.

The main findings of this study were: (a) all breaks appeared to have occurred in light-band regions; (b) there was a significant paucity of X/autosome exchanges; (c) the autosomes of abnormal subjects showed a generalised increase in the number of breaks, which was almost 50% higher than that of normal individuals; (d) significant deviations from expectation were found when the chromosome regions involved in the formation of dicentrics, translocations, inversions and rings were analysed separately; (e) the frequency with which autosomes may exchange with other specific autosomes to form dicentrics or translocations was examined using cluster analysis. The dendrograms thus derived showed differences of clustering between chromosomes. For example, 8 and 11 were apparently involved with each other in forming a high number of dicentrics, but not for translocations. A significant excess of events resulting in dicentrics was also found in chromosomes 14 and 21, which may be consistent with the estimated frequencies in naturally occurring events of this type.

References

Lance, G.N. & W.T.Williams (1967) A general theory of classification sorting stratagems. *Comp. J. 9*, 373–80.

Seabright, M. (1973a) High resolution studies on the patterns of induced exchanges in the human karyotype. *Chromosoma 40*, 333–46.

——(1973b) Non-involvement of the human X chromosome in X-ray-induced exchanges. *Cytogenet. Cell Genet. 12*, 342–56.

——(1976) Patterns of induced aberrations in humans with abnormal autosome complements, in *Chromosomes Today*, vol.5 (eds P.L.Pearson and K.R.Lewis) pp.293–8. New York: Wiley.

J.R.K.SAVAGE and T.R.L.BIGGER

Aberration Distribution and Chromosomally Marked Clones in X-Irradiated Skin

In contrast to epidermal cells, dermal fibroblasts do not undergo a rapid cycle of loss and replacement throughout the life of the organism. Nevertheless, these cells do divide *in vivo* in response to local injury, and will readily proliferate *in vitro* for a period of weeks or months. Because of this slow turnover, structural chromosomal changes induced *in vivo* by ionising radiation would be expected to persist for long periods, and these have been found in fibroblasts grown *in vitro* from biopsies of irradiated human skin (e.g. Visfeldt 1966). Real progress in the identification and quantitation of the surviving symmetrical aberration forms only became possible with the advent of banding, which allows the detection of a reasonable proportion of such changes.

This paper reports some of our findings in dermal fibroblasts cultured from biopsies of clinically normal skin removed from the irradiated area of patients that had received courses of radiotherapy. The biopsies were taken at intervals ranging from one week to 60 years following completion of the treatments.

Materials and Methods

Samples

Skin was obtained from patients who, having had a course of radiotherapy, had returned for surgery, either in connection with tumour recurrence or for the removal of damaged skin and cosmetic reconstruction. Small strips were removed from a region of the irradiated field judged to be clinically free from gross abnormality or tumorous tissue. 10 ml of venous blood was also donated by each patient.

Table 1 summarises the patient details and times of biopsy. The radiation doses quoted are taken from the case notes and are often approximate, generally referring to total applied dose.

Skin Culture

Depending on size, each biopsy was divided into three or more pieces, one for histology and the remainder for culture. For culture, each piece was subdivided into about ten small ($\sim$0.25 mm^2) bits and these were explanted

Table 1. Source of irradiated skin.

Patient (sex)	14a(M)	14b(M)	9(M)	6(F)	5(M)	16(F)	17(F)
Estimated X-ray dose (R)	2000	2000	4480	1000	2000	3400	?(large)
Time of skin biopsy post-treatment	1 h	1 wk	7 mo	14 mo	8.5 yr	13.7 yr	60 yr
Origin of skin	back of hand		cheek	neck	cheek	cheek	cheek
Age at biopsy	50	50	67	2	60	79	78
No. of cells examined	40	40	83	76	36	37	92
Range of cytologically normal cells in culture flasks	20–50	20–55	5–13	16–33	0	0	0–5
Evidence for *in vivo* division and migration	—	—	+	+	—	—	+

into chicken plasma clots within Falcon plastic flasks. The pieces of skin from different parts of the biopsy were therefore in separate flasks, and were kept separate through all subsequent preparative stages so that cells from different regions could be examined and compared. Within 3–4 d fibroblasts migrated from the explants, and by one week there were sufficient to subculture to fresh medium. The growth medium used was Eagles MEM supplemented with 10% human AB serum, and gassed with a 5% CO_2/air mixture.

Cytological preparations were made of cells from this first subculture at 14–18 d from primary explant. Prior to fixation and air-drying, the cultures were treated with colcemid for 3 h. One week later the chromosomes were banded using the ASG technique (Sumner, Evans and Buckland 1971).

It should be emphasised that all the cultures were of primary, untransformed cells, and may therefore be assumed to have become karyotypically stable once inviable aberrations were eliminated.

Blood Samples

Whole blood microcultures, using PHA stimulation, were made for each patient. Lymphocytes were sampled at 48 h and the preparations banded one week later. These cells provided a check on the normality of the patients' karyotype.

Analysis

A laboratory banding norm was established from normal, unirradiated skin and blood, and a numbering system developed to facilitate the rapid recording and location of aberration 'break points'. Details and a comparison with the Paris Conference (1971) system have been published elsewhere (Bigger, Savage and Watson 1972; Savage, Watson and Bigger 1973a).

For each cell examined, a complete analysis was made of the aberrations present and their presumptive break points, and a photographic record of the karyotyped cell kept for later reference. Additional photographs of the derivative and abnormal chromosomes from clonal cells were made to facilitate inter-cell comparisons.

Results

Unirradiated Material

Normal unirradiated skin from four subjects (two male, two female) was treated in exactly the same way as the irradiated biopsies, and 50 banded metaphases examined for each. Apart from 2–5% aneuploidy, part of which was attributable to technical procedures, a few (<1%) chromatid-type aberrations were observed. There were no gross asymmetrical or symmetrical chromosome-type changes.

Irradiated Material

Histology. Histological sections were examined. Tumour cells were not seen. Epidermis and dermis appeared normal, apart from occasional reduction in dermal cellularity. No typical radiation-induced 'giant nuclei' were seen.

Karotypes from blood lymphocytes. In none of the six subjects did banding reveal structural changes that could be attributed to radiation. Two patients, however, carried inherited abnormalities, present in all cells analysed from blood and skin:

Patient 6 was 46, XX, del(X)(pter→q26:)

Patient 9 was 46, XY, inv(9)(pter→p13::q22→p13::q22→qter)

All others were karyotypically normal.

Frequency of 'normal' cells in skin cultures. A minimum of two Falcon flasks (each representing a different region of the biopsy) was available for each patient. The frequency of cells with no detectable karyotypic changes ('normal cells') varied considerably between flasks, and the extreme ranges found are given in table 1. In two patients (5 and 16) no normal cells were found in any flask.

Types and frequencies of chromosomal structural change. The most frequent aberration observed in all patients was the symmetrical chromosome

Table 2. Distribution of translocations between cells: test of goodness-of-fit to a Poisson distribution by dispersion index.

	Cells with													
Patient	0	1	2	3	4	5	6	7	8	9	10	11	Unit normal deviate	
14a	13	13	6	4	3								1.08	
14b	16	7	11	4	1	1							1.53	
9	7	18	18	21	11	5	0	2					−0.24	2.60[1]
6	21	11	21	12	7	1	1						1.36	
5	0	9	5	9	7	6							−1.23	
16	0	2	7	16	9	0	0	0	3				−0.92	1.19
17	2	13	14	7	11	22	13	2	4	2	0	1	1.87	
Total	59	73	82	73	49	35	14	4	7	2	0	1	6.89[2]	

[1]P=0.01–0.001. [2]P<0.001.

interchange (reciprocal translocation; for terminology see Savage 1976). A total of over 1000 was scored from all samples, but a number of these were repeated forms in clonal cells. Just over 600 different (i.e. judged to be independent) types were available for subsequent break-point analysis. Almost all were simple and complete and, in contrast to data from primary aberrations, there were very few complexes. Frequencies in some cells reached very high values (table 2).

The next most frequent aberration was the terminal deletion, which results in the cell being monosomic for part (sometimes a large part, figure 1) of an arm. A total of 58 independent examples was found. Although all have been classified as terminal, the possibility that a small proportion of the observed shortened arms could have arisen from interstitial deletions of dark G-bands cannot be ruled out, for the band patterns on some arms do

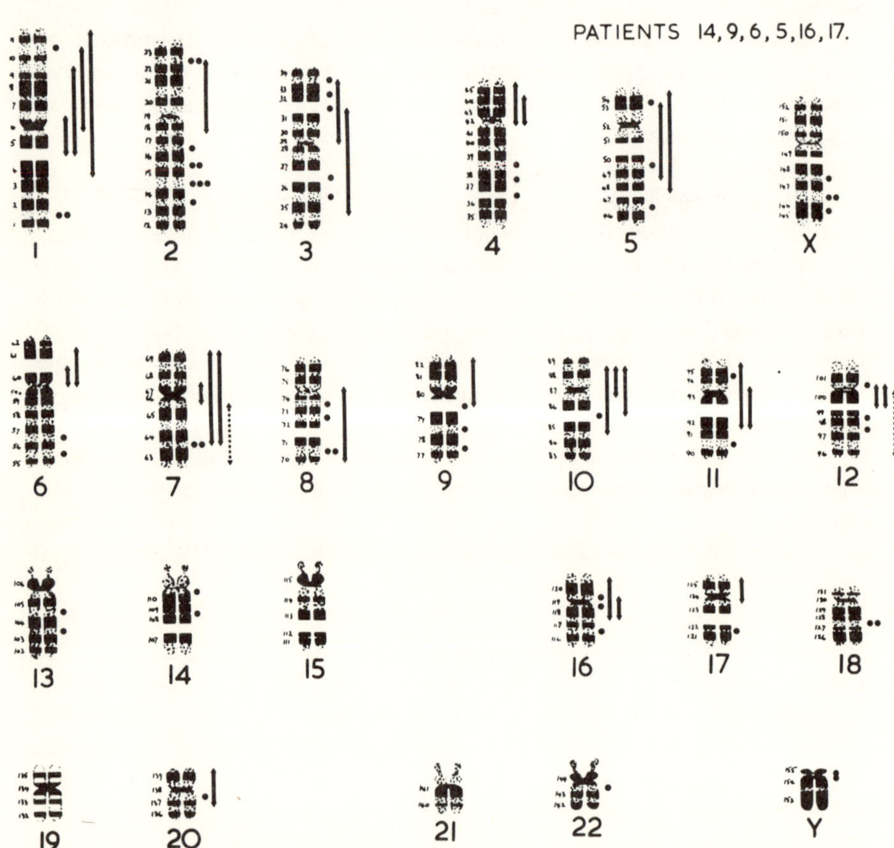

Figure 1. Frequency and distribution of break points of independent intrachange aberrations surviving in dermal fibroblasts following therapeutic irradiation. Solid arrows, pericentric inversions (29); dotted arrows, paracentric inversions (2); dots, terminal deletions (58: cell monosomic for the whole distal region beyond this point). Band numbering after Bigger *et al.* (1972).

not have sequential band differences to identify individual missing bands accurately (Savage 1977).

Pericentric inversions form the next major category, a total of 39 independent cases with nearly every chromosome represented (figure 1). The number observed relative to translocations is much less than might be expected from the corresponding ratios found for primary aberrations. There are, however, limitations in ease of detection for different chromosomes (reflected in figure 1) and smaller ones will certainly be missed. Their paucity also gives a clue to the mixed origin of many of the observed aberrations (see p.166).

Paracentric inversions, and their counterparts, interstitial deletions, were very rare; only two cases were recorded. In the main this is due to marked detection difficulties and the fact that the modal size is very small and will seldom result in pattern disruption.

Chromosome participation in aberrations and location of presumptive 'weak points'. The discontinuity in G-band pattern produced in the chromosome arm by each independent aberration ('break point') was assigned to one of the 364 (female) or 358 (male) light bands of the Harwell norm. The locations and frequencies for terminal deletions and inversions are shown in figure 1, and those for translocations in figure 2.

For purposes of analysis, the light (l) bands were either given equal weight or weighted on the basis of size and prominence. In the latter case, the l-bands of the norm were graded into four categories, and these were given integral weighting factors, 1 for the smallest bands and 4 for the largest. Individual chromosome lengths, or relative lengths for the Denver groups A-E (omitting X and Y) were derived from the weighted averages of the C and D columns of Paris Conference (1971). In the absence of evidence for time-related trends or sex differences, pooled data from all patients was used.

With regard to translocations, high numbers within cells make the data unsuitable for investigating the distribution of break points *between* chromosomes. We have however investigated the distribution *within* chromosomes and *within* chromosome arms.

The hypothesis of random distribution between l-bands (unweighted or weighted) was tested. This can only be an approximation, because the observed distribution is conditioned by the exchange distribution, which is unlikely to be a simple function of arm length. Table 3 shows individual χ^2 values for each chromosome. The results bear out the subjective impression gained from figure 2 that the distribution is not a simple function of either the number or size of the l-bands. For most chromosomes, distribution between opposite arms is in reasonable accord with the hypothesis (but overall is highly discrepant). In contrast, distributions within whole chromosomes are predominantly non-random, 15 out of 23 chromosomes deviating when bands are given equal weight, 13 when weighted. Overall χ^2 is reduced by $\sim$20% with weighting but is still highly significant.

For the autosomes, numbers of translocation break points make possible

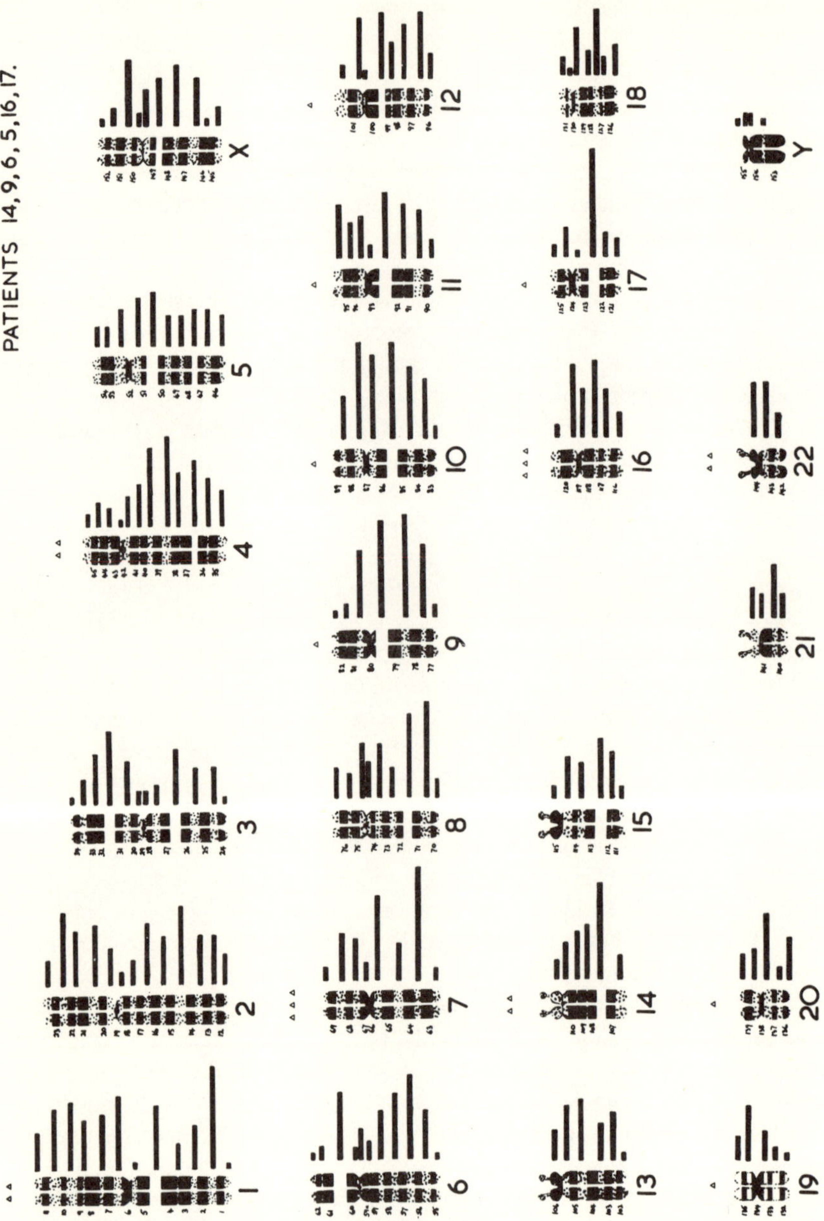

Figure 2. Frequency and distribution of break points of independent reciprocal translocations surviving in dermal fibroblasts following therapeutic radiation. Break points have been assigned to the light band where obvious pattern disruption occurs. A total of 1024 translocations was observed, but only about 600 were judged to be independent events. 1229 break points are mapped. Band numbering after Bigger *et al.* (1972).

Table 3. Tests for random distribution of translocation break points amongst light G-bands within chromosomes.

Chromosome number	Between opposite arms: unweighted χ^2(1df)	Between opposite arms: weighted χ^2(1df)	Within whole chromosomes: unweighted χ^2	df	Within whole chromosomes: weighted χ^2	df
1	2.67	5.47[1]	29.25[2]	11	27.33[2]	11
2	0.36	0.19	16.95	12	17.86	12
3	0.80	0.80	26.54[2]	11	14.74	11
4	17.36[3]	10.55[2]	30.29[2]	11	25.29[2]	11
5	1.96	1.40	5.79	9	5.69	9
6	4.19[1]	6.83[2]	35.67[3]	10	17.66	10
7	7.23[2]	0.96	33.82[3]	7	32.15[3]	7
8	2.04	3.07	20.90[2]	8	13.36	8
9	9.87[2]	2.96	32.98[3]	6	13.10[1]	6
10	1.91	1.15	27.42[3]	7	16.06[1]	7
11	0.15	0.03	9.73	7	15.52[1]	7
12	0.22	3.00	17.11[1]	7	14.24[1]	7
13	0.41	1.03	8.23	5	6.95	5
14	3.41	6.07[1]	14.26[1]	5	19.79[2]	5
15	0.91	0.04	9.06	5	10.83	5
16	0.27	2.04	11.85[1]	5	26.83[3]	5
17	2.18	1.41	35.91[3]	5	12.89[1]	5
18	5.04[1]	5.75[1]	14.00[1]	6	7.29	6
19	4.20[1]	5.92[1]	9.24	4	11.91[1]	4
20	0.97	0.52	8.07	4	9.62[1]	4
21	0.06	1.28	3.09	3	3.93	3
22	2.97	9.31[2]	10.36[1]	3	13.32[2]	2
X	1.39	1.39	24.17[2]	9	13.48	9
Y	not tested		not tested			
Total	70.57[3] (23df)	71.17[3] (23df)	434.68[3]	160	343.83[3]	160

[1] P=0.05−0.01. [2] P=0.01−0.001. [3] P<0.001.

a *within-arm* analysis. Table 4 shows results for tests of the overall distribution in (a) l-bands immediately proximal to the centromere, (b) the extreme terminal l-bands, and (c) the penultimate l-bands. Some of the smaller chromosomes do not have all three bands, in which case only (a) and (b) were tested.

Centromeric bands give excellent agreement, whilst terminal bands (but not penultimate bands) have a very significant deficiency. Consequently there is a tendency for excess break points to occur in the central arm regions. Part of this terminal deficiency will result from the inability to detect terminal l/l translocations, since they produce no pattern disruption. Given equal l-band weighting and truly random participation of arms amongst autosomes of the diploid karyotype, ~25.6% of all translocations involving a terminal band will be terminal/terminal. Even correcting for this 25% loss, the discrepancy is still very highly significant.

Table 4. Location of translocation break points in centromeric, penultimate and terminal l-bands of autosomes.

Position	Band weighting	Observed number	Expected number[1]	Significance of difference[2]		
				t	df	P
l-bands adjacent	–	285	292.2	0.296	151	0.77
to centromere	+	285	276.6	0.341	151	0.73
penultimate	–	297	261.2	1.52	151	0.13
l-bands	+	297	256.9	1.90	151	0.06
terminal	–	141	292.2	6.190	151	small
l-bands	+	141	256.0	5.460	151	small

[1] On the hypothesis of a random distribution of breaks within chromosomes.
[2] As compared with the overall difference between observed and expected numbers in all bands on the hypothesis of a random distribution of breaks within chromosomes.

Tests for the intrachromosomal aberrations (deletions and inversions) are simpler than those for interchanges. Using the seven Denver groups (autosomes only) distribution *between* chromosomes was investigated using Woolf's G-test. Neither pericentric inversions ($\chi^2_6=4.33$, P=0.63) nor terminal deletions ($\chi^2_6=10.05$, P=0.12) departed from expectation. Similar results were obtained using weighted and unweighted band frequencies.

There were too few break points for a within-chromosome or within-arm analysis.

Time interval between irradiation and biopsy. For the purposes of break-point analysis, the data from all patients were pooled. From table 1 it will be noted that biopsies were taken at various intervals after therapy up to 60 years. The material is too heterogeneous with respect to radiation dose, patient age and sex, and the data too sparse to attempt a detailed search for time-related trends in aberration types or break-point distributions. As a compromise, therefore, we have grouped patients into two blocks. Biopsies taken in under two years from therapy (patients 14, 9, and 6) and biopsies taken eight or more years from therapy (patients 5, 16, and 17) Using distributions of translocation break points within specific chromosomes from the two time blocks, a search was made for homogeneity using Fischer's exact test for a $2 \times n$ table. There was no evidence, for any chromosome tested, that time has influenced break-point distribution.

Selection against cells with higher damage levels might be another effect to show up with time. With 46 chromosomes, interchanges are expected to be distributed between cells according to the terms of a Poisson distribution, although heterogeneity in the population frequently leads to over-dispersion (Savage 1970). Using the total translocations observed (1024) uncorrected for clones, goodness-of fit to a Poisson distribution was tested using the dispersion index table 2 (Savage 1970). No individual patient distribution deviated significantly, but the early time block (and overall total) were significantly overdispersed. There is nothing, therefore, to suggest that cells with high frequencies of aberrations are at a disadvantage, or that time has a marked selective effect against such cells.

Evidence for *in vivo* division and migration: chromosomally marked clones. The facility to locate presumptive break points in relation to G-bands allows a much more accurate characterisation of any aberration. There are 364 l-bands in the Harwell ♀ norm, and, given random lesion induction and participation in exchange, the probability that the *same* two l-bands will be involved again *independently* in a second cell is very small ($\sim 5 \times 10^{-11}$). It follows, therefore, that the observation of two or more cells having an identical aberration constitutes strong evidence that they are members of a clone. The likelihood is increased if several identical changes are involved. The presence of such cells within the same culture flask may indicate no more than cell division *in vitro*, but when found in different flasks, we conclude that the progenitor cell must have divided (and its products migrated) *in vivo*. Three patients showed evidence of this nature, and in patient 17 one clone (having 5 translocations, 2 inversions, 1 deletion and an occasional pair of minute fragments) occurred in all four culture flasks. In one of these it was the dominant cell type, whereas in another only one or two cells were found.

Radiotherapeutic treatment is normally given in fractions, and the time interval between doses is sufficient to allow cell division to occur. Many clones should therefore show evidence of *in vivo* division by a second (and even third) order of damage superimposed on a basic set of changes. The majority of clones found were of this type, and the presence of examples within single flasks indicates that *in vivo* division is not necessarily accompanied by migration. Table 5 summarises the frequencies of different types of clones.

Time between biopsy and scoring. Usually a period of 14–18 d in

Table 5. Clone types and frequencies.

	Examples found	
	within same flask	in different flasks
I. All members with a basic set of identical changes	many	2
II. Some members with basic changes only, others with basic + various sets of additional changes.	—	—
III. All members with basic plus one order of various additional changes. No cells with basic only.	8	8
IV. Basic, basic + and basic ++, i.e. 2 or more orders of additional change.	—	1
V. Probable clones, but basic changes show only partial agreement.	1	1

culture elapsed before cytological preparations were made. During this time, cell division eliminates all unstable asymmetrical aberrations and acentric fragments that have not already been lost by division *in vivo*. Further samples of irradiated skin were obtained, and cytological preparations made at times earlier than 10 d. Table 6 shows that unambiguous primary dicentrics and centric rings can be found *in vitro*. Frequencies are low, as would be expected from the estimated loss rate of ~50% per cell generation, but high enough to rule out spontaneous origin in culture.

Table 6. Dicentric and centric-ring frequencies in early cell generations of short-term dermal fibroblast cultures.

Patient (sex)	Estimated X-ray dose (R)	Time of biopsy post-treatment (yr)	Origin of skin	Age at biopsy	Culture time (d)	No. of cells examined	No. of dicentrics and rings
18(M)	5500	0.25	chin	60	7	22	8
					10	60	0
20(F)	4320	1.5	cheek	33	5	88	15
28(M)	5540	5	nose	58	4	7	1
					5	19	0
					6	20	3
					7	77	8
27(F)	6000	19	knee	75	10	59	3
					10	20	0
					10	20	3
13(F)	?	49	lip	75	7	14	1

Participation of the X-chromosome. The tests so far have been confined primarily to autosomes in order to allow pooling of data from all patients. There is some evidence (Seabright 1973b) that for primary chromosome-type exchanges, the relative frequency of X/autosome translocations is substantially lower than expected. 39 independent translocations X/autosome were found in these skin cells, and these were partitioned amongst the Denver groups and observed and expected frequencies compared using Woolf's G-test. Two models for exchange formation were used, the first where order of arm participation is unimportant, the second where it is. The models gave probabilities of 0.35 and 0.33 respectively. Although this is a rather crude test, it provides fairly good evidence that there was no limitation of X-chromosome participation in these cells.

Discussion

Survival and Proliferation of Aberration-Bearing Cells

The data presented indicate that, following therapeutic radiation, dermal fibroblasts with large numbers of chromosomal changes (not all of which are balanced) can persist within the irradiated region for up to 60 years.

Such cells can proliferate and migrate several mm *in vivo* to accomplish tissue repair. The distributions of aberrations observed (table 2) provide evidence that in no case was there a massive influx of 'normal' cells from surrounding regions to effect repair, and no apparent bias against cells carrying large numbers of changes.

Aberration Types and Frequencies

In attempting to evaluate the aberration data, three things must be borne in mind: (a) The very high radiation doses used in these therapeutic treatments, considerably above anything that would be used for experimental purposes when aberration analysis is an end point. The levels of structural change recovered (table 2) give some indication of the enormous cytological damage originally produced in the tissue by the radiation, and at the same time present considerable problems in scoring.

(b) Evidence has been presented that some cells were stimulated to divide *in vivo* between dose fractions. It follows, therefore, that a proportion of the aberrations observed will have been derived secondarily from chromatid-type changes, including the possibility of mixed origin within single cells. The factors that determine relative frequencies of aberration kinds, and the participation of arms in exchange, vary throughout the cell cycle, so that great care is needed in making direct comparisons with published data for homogeneous primary aberration types.

(c) The cell samples available for analysis have undergone intense selection both *in vivo* and *in vitro*. Such selection will take a number of forms: (i) The elimination of cells with inviable chromosome combinations. This will include cells with major asymmetrical exchanges, for none were found in the main sample. Their presence when culture times were shortened to a few days shows that some of the irradiated fibroblasts had not divided *in vivo*. Since the time between irradiation and biopsy is known, a lower limit can be placed on the longevity of undivided fibroblasts. Patient 27 gives a certain lifetime of 19 years, and if the one aberration found in patient 13 is not spontaneous, this can be extended to 49 years. (ii) The favouring of faster growing cells and of those with advantageous genetic or karyotypic changes. The factors operating to produce this more positive selection will certainly differ between the *in vivo* and *in vitro* situations. The disparity between the frequencies of different clonal types is partly accountable for in this way. Surprisingly, quite a number of the cells (including some clones) were not cytologically balanced, but showed monosomy for often quite large chromosome regions. Such loss was not confined to particular chromosomes; in fact only three chromosomes (15, 19 and 21) did not show any deletions in the entire sample. Evidence for *untransformed* diploid hamster cells (Grote and Revell 1972) suggests that the loss of a fragment large enough to produce a micronucleus almost always leads to cell death or slow growth in culture. The difference between these findings may lie in the fact that nearly all the acentric fragments in Grote's experiments would be compound (i.e. double monosomy), being derived from major exchanges induced in G_1. Apart from cells in patient 6, who was constitutionally monosomic for part of Xq, only three examples of double monosomy were observed, all in patient 17.

Simple reciprocal translocations were by far the commonest aberration observed. In contrast to published data for primary changes, complex interchanges involving a common arm were rare.

The paucity of pericentric inversions relative to translocations is also noteworthy. The overall ratio $t/\text{PI} \simeq 21$ is much higher than published ratios for primary changes (~3, Seabright 1973a; ~5, Cooke, Seabright and Wheeler 1975; ~7, Buckton 1976). At first sight this might be taken as indicating selection against inversion, but it is more likely to reflect the chromatid origin of many of the exchanges, since ratios of interchanges/centric intrachanges are three to four times larger for chromatid primary changes than for chromosome ones. The large number of deletions is also a pointer to chromatid origin, since incompleteness of chromatid-type aberrations is much higher than chromosome-types (Savage 1976).

The rarity of paracentric inversions is not so surprising, as many of these are likely to be small and to escape detection. In addition there are marked observational differences between different chromosomes (Savage 1977).

Chromosome Participation in Aberrations

With regard to translocations in this experiment, the question 'do arms participate in exchange as a simple function of their measured length' cannot be tested. Valid tests require that exchanges be treated as entities, and require the partitioning of families of contingency tables, the dimensions of which depend upon the number of exchanges within a cell. Consequently very large amounts of data are needed even when numbers of exchanges per cell are low, which is certainly not the case here. The simpler tests frequently employed are compromises and are invalid when only particular classes of aberrations are considered.

For intrachanges (pericentric inversions and deletions), where interaction between chromosomes is not involved, the tests are simpler, and, in spite of the low total frequencies, the observed distributions between the Denver groups is in excellent agreement with that expected on the basis of measured length.

Distribution of Presumptive Break Points

In view of the inherent 3-band uncertainty for locating 'break points' in chromosome-type aberrations (Savage, Bigger and Watson 1976; Savage 1977) all of them were assigned to the l-band that would most readily produce the pattern disruption observed. There is evidence from primary aberrations for a preponderance of break points in l-bands (Seabright 1973a and b; Holmberg and Jonasson 1973; Savage, Watson and Bigger 1973b) or at l/d band junctions (Buckton 1976). If both l and d band regions can participate freely in exchange, then a reasonable frequency of l/d aberrations should be encountered. These will be detected by the presence of an additional band in one of the derivative chromosomes (Savage, Bigger and Watson 1976; Savage 1977). None were found in this work, which suggests that all exchanges occur (or appear to occur) exclusively between d/d or l/l bands, or that partial d-bands are unstable, and tend to merge with adjacent d-bands.

In the ordinary course of events radiation-induced lesions occur first, and a proportion of these interact to produce primary aberrations, *some* of which will be recognisable and their presumptive break points mapped. In the present data, a further level of selection (together with the conversion of primary chromatid aberrations to secondary chromosome ones) for viable types has occurred before mapping. Even if it were possible to record all surviving break points, there is no guarantee that their distribution would reflect any of the pre-selection ones. We cannot, however, record all break points, since many of the smaller surviving aberrations are missed. Moreover, intrinsic differences in probability of, and mode of, formation of different types, make pooling of break points illegitimate for simple tests. The difficulty lies in deciding what the expected distribution would be for any given hypothesis of arm participation. Disagreement with expectation on the basis of proportional arm-length is not therefore unexpected, and too much weight should not be given to it.

We are on surer ground when we consider within-chromosome distributions. The finding that there is a real terminal break point deficiency, which is not removed by correction for unobserved exchanges, is consistent with our primary chromatid data (Savage, Watson and Bigger 1973b) and with the data of Seabright (1973a) and Cooke, Seabright and Wheeler (1975), but not with that of Buckton and Jacobs (1973) and Buckton (1976). Again, in our case, this may be a reflection of the chromatid origin of many of the exchanges, and it is consistent with a large body of evidence suggesting a concentration of chromatid events towards the central regions of chromosome arms (Evans 1962, Savage 1975).

Participation of the X-Chromosome

In these data, there is no evidence for a restriction in X/autosome exchange. This is in contrast to the data of Seabright (1973b) (*contra* Buckton 1976) for primary chromosome changes. It may, however, be a further reflection of the mixed origin of the exchanges, as no restriction was found for primary chromatid exchanges (Savage, Watson and Bigger 1973b).

Summary

Samples of clinically normal human skin were removed from therapeutically X-irradiated areas at intervals of time ranging from one hour to 60 years after completion of radiation treatment.

Primary cultures of untransformed fibroblasts from these samples were analysed for surviving chromosomal structural changes using ASG banding techniques. Aberrations of four basic types, reciprocal translocations, terminal deletions, pericentric inversions and paracentric inversions (the last very rare) were found in all samples. Evidence indicates that many of these are secondary aberrations derived from primary chromatid types.

Presumptive break points for all aberrations were mapped, and various tests applied to investigate their within-chromosome distribution (the data are unsuitable for valid between-chromosome analysis). For translocations, the within-arm distributions are non-random, principally as the result of a very significant deficiency of break points in terminal segments. Tests for the intrachromosomal changes (pericentric inversions and deletions) are simpler, and in neither case were there significant departures from randomness.

Two lines of evidence are present in the data for division and migration of chromosomally abnormal cells *in vivo*: (a) the presence of identical aberrations in cells from different parts of the biopsy; (b) the presence of cells with sequential changes, indicating cell division between the dose fractions of the therapeutic regime.

Acknowledgements

We would like to thank Dr G. Wiernik of the Churchill Hospital, Oxford, for the skin samples, D. G. Papworth for statistical computions and Dr R. H. Mole for encouragement and advice.

References

Bigger, T.R.L., J.R.K.Savage & G.E.Watson (1972) A scheme for characterising ASG banding and an illustration of its use in identifying complex chromosomal rearrangements in irradiated human skin. *Chromosoma 39*, 297–309.

Buckton, K.E. (1976) Identification with G and R banding of the position of breakage points induced in human chromosomes by *in vitro* X-irradiation. *Int. J. Radiat. Biol. 29*, 475–88.

Buckton, K.E. & P.A.Jacobs (1973) The distribution of breakpoints in human reciprocal translocations, in *Chromosomes Today*, vol.4 (eds J.Wahrman & K.R. Lewis) pp.247–52. New York: Wiley.

Cooke, P., M.Seabright & M.Wheeler (1975) The differential distribution of X-ray-induced chromosome lesions in trypsin-banded preparations from human subjects. *Humangenetik 28*, 221–31.

Evans, H.J. (1962) Chromosome aberrations induced by ionizing radiations. *Int. Rev. Cytol. 13*, 221–321.

Grote, S.J. & S.H.Revell (1972) Correlation of chromosome damage and colony-forming ability in syrian hamster cells in culture irradiated in G_1. *Curr. Top. Radiat. Res. Q. 7*, 303–9.

Holmberg, M. & J.Jonasson (1973) Preferential location of X-ray-induced chromosome breakage in the R-bands of human chromosomes. *Hereditas 74*, 57–68.

Paris Conference (1971) *Standardization in Human Cytogenetics*. Birth Defects: Original Article Series vol. 8, no. 7. New York: National Foundation, March of Dimes.

Savage, J.R.K. (1970) Sites of radiation-induced chromosome exchanges. *Curr. Top. Radiat. Res. 6*, 129–94.

—— (1975) Radiation-induced chromosomal aberrations in the plant. *Tradescentia*: Dose-response curves. I. Preliminary considerations. *Radiat. Bot. 15*, 87–140.

——(1976) Classification and relationships of induced chromosomal structural changes. *J. Med. Genet. 13*, 103–22.

——(1977) Application of chromosome banding techniques to the study of primary chromosomal structural changes. *J. Med. Genet. 14* (in press).

Savage, J.R.K., T.R.L.Bigger & G.E.Watson (1976) Location of quinacrine mustard-induced chromatid exchanged points in relation to ASG bands in human chromosomes, in *Chromosomes Today*, vol.5 (eds P.L.Pearson & K.R.Lewis) pp.281–91. New York: Wiley.

Savage, J.R.K., G.E.Watson & T.R.L.Bigger (1973a) A brief comparison of the Paris Conference (1971) and Harwell schemes for G-band numbering in human chromosomes. *Chromosoma 41*, 123–7.

—— (1973b) The participation of human chromosome arms in radiation-induced chromatid exchange, in *Chromosomes Today*, vol.4 (eds J.Wahrman & K.R.Lewis) pp.267–76. New York: Wiley.

Seabright, M. (1973a) High resolution studies on the pattern of induced exchanges in the human karyotype. *Chromosoma 40*, 333–46.

——(1973b) Non-involvement of the human X chromosome in X-ray-induced exchanges. *Cytogenet. Cell Genet. 12*, 342–56.

Sumner, A.T., H.J.Evans & R.A.Buckland (1971) New technique for distinguishing between human chromosomes. *Nature New Biol. 232*, 31–2.

Visfeldt, J. (1966) *Radiation-induced chromosome aberrations in human cells.* Doctoral thesis published as Riso Report no.117, Danish Atomic Energy Commission.

M. SAGI, M. M. COHEN and T. SCHAAP

Non-Random Chromosome Damage in Human Lymphocytes induced by Streptonigrin and X-Rays

The apparent non-random distribution of chromosome breaks observed in human cells following exposure to certain chemicals, viruses and radiation has been widely investigated. Such studies were performed in the hope of better understanding chromosome architecture and the mechanism of action of particular clastogens. However, analysis of induced chromosome damage has been somewhat limited by available methods. With the development of the chromosome banding techniques, the detail with which breaks could be localised was significantly increased. Furthermore, the interpretation of banding patterns has led to the development of several models concerning the structure and function of various classes of chromatin. The known non-random clastogenic effects of streptonigrin and X-rays were investigated in an attempt to describe and correlate more precisely the induced chromosome damage with the differentiated staining pattern of the chromosome.

Materials and Methods

Microcultures of phytohaemagglutinin-stimulated peripheral leucocytes were established from four normal, healthy adults (two male and two female). 0.4 ml of heparinised blood were inoculated into 5 ml of Ham's Nutrient Medium F-10 containing 20% fetal calf serum, antibiotics, and glutamine. All reagents were obtained from Grand Island Biological Co. (GIBCO), New York. The cultures were maintained at 37°C for a total of 72 h with the clastogenic treatments being applied 24 h prior to harvest. Streptonigrin (SN), obtained from Chas. Pfizer & Co., was dissolved in sterile distilled water and added to cultures at a final concentration of 0.1 μg ml^{-1}. Other cultures were transferred to Petri dishes (60 mm) and exposed to a total of 200 rad delivered by a GE maximar machine at 250 kVp, 5 mA with a 1 mm Al filter, after calibration by a Victoreen dosimeter. Following treatment, these cells were replaced in test-tubes and returned to the incubator. Harvest of the cultures was according to standard techniques (Moorhead *et al.* 1960). Preparation of slides and chromosome banding were achieved by the 'hot-plate Giemsa 9' method of Patil, Merrick and Lubs (1971).

Chromosome preparations were systematically screened under low power

(200 ×). All metaphases that appeared complete and 'banded' were studied with an oil immersion objective (1250 ×) and included in the study. Cells with chromosome aberrations were photographed and the anomalies classified according to Cohen, Hirschhorn and Frosch (1967). Each simple break, (displacement of the distal fragment by more than the width of a chromatid) was categorised as either chromatid or isochromatid. In the quantitative evaluation, this group of aberrations was considered as single breaks. Structural rearrangements, e.g. interchanges, tri- and quadriradials (TR and QR), inversions, dicentrics and rings were all considered as the result of two breaks. Complex rearrangements, involving more than two breaks, were also scored on this basis. Gaps were noted but not included in the analysis. Each break was assigned to a specific chromosomal band (heterochromatin) or interband (euchromatin) according to the Paris Conference on Cytogenetics Nomenclature (1971).

Measurement of euchromatin and heterochromatin was achieved using the diagram of the Paris Conference (1971). Interband 2p23 was arbitrarily chosen as the 'reference band' and designated as equal to 'one unit of euchromatin'. By measuring the length of each interband in mm and converting this value to units of euchromatin, according to its relationship to band 2p23, the amount of euchromatin per interband as well as per chromosome was calculated. Conversely, the total chromosome length minus the amount of euchromatin yielded the heterochromatin content (bands) per chromosome. 'Pericentromeric interbands' were defined as that palely-staining region between the first proximal heterochromatic band in each chromosome arm visualised by G-banding. The 'terminal interband' was that euchromatin distal to the last heterochromatic band in either arm.

Chi-square techniques were applied to the observed distributions of clastogen-induced breaks on various levels of chromosome organisation by posing the following questions: (a) Are breaks randomly distributed between bands and interbands (i.e., heterochromatin and euchromatin)? (b) Are breaks distributed among the chromosomes according to their interband length (i.e. quantity of euchromatin)? (c) Are those breaks observed in the pericentromeric interbands randomly distributed among all chromosomes? (d) Is the number of breaks observed in a given interband related to its length?

Results

The distribution of the 819 chromosome breaks, induced by both clastogens among 1960 metaphases, is presented in table 1. A total of 426 SN breaks and 393 X-ray breaks was mapped to specific chromosome bands or interbands. It is of interest that the chromosome damage resulting from exposure to either 0.1 μg ml^{-1} SN or 200 rad X-radiation is quantitatively quite similar ($\bar{X}_{SN}=0.40$ breaks/cell; $\bar{X}_{X\text{-}ray}=0.44$ breaks/cell). No obvious differences were observed among the individuals tested, nor between the sexes with regard to either clastogen. Table 2 presents the numbers of cells

Table 1. Distribution of streptonigrin- and X-ray-induced chromosome damage among the four subjects studied.

		Streptonigrin (0.1 μg ml^{-1})			X-ray (200 rad)		
Subject	Sex	No. cells examined	No. breaks	Breaks/ cell	No. cells examined	No. breaks	Breaks/ cell
1	Female	232	99	0.43	301	118	0.39
2	Female	289	108	0.37	176	81	0.46
		521	207	0.40	477	199	0.42
3	Male	305	116	0.38	209	98	0.47
4	Male	250	103	0.41	198	96	0.48
		555	219	0.39	407	194	0.48
	Total	1076	426	0.40	884	393	0.44

Table 2. The number of breaks per cell induced by streptonigrin and X-rays in normal human lymphocyte cultures.

Streptonigrin (0.1 μg ml^{-1})			X-ray (200 rad)		
No. breaks	No. cells	%	No. breaks	No. cells	%
0	874	78.44	0	664	75.11
1	154	14.31	1	135	15.27
2	40	3.71	2	50	5.65
>3	38	3.53	>3	35	2.96
	1076	99.49		884	99.99

with 0, 1, 2 or more breaks. The similarity between the percentage of cells with a given number of breaks, irrespective of clastogenic treatment, is again striking, indicating a similar response of the cells to both agents.

Figures 1 and 2 illustrate the precision with which chromosome breaks and structural rearrangements can be assigned to individual bands and interbands. The distributions of SN- and X-ray-induced breaks to specific chromosomal bands or interbands are shown in figures 3 and 4. With the exception of 5 breaks, which fell into heterochromatic regions (bands), the remaining 814 'mapped' to interbands (euchromatin) or euchromatin-heterochromatin junction. With both clastogens a nonrandom distribution of breaks is readily apparent. However, since 99.4% of the breaks occurred in the interbands (figures 3 and 4), the randomness of this distribution within the euchromatin of individual chromosomes was tested. Table 3 indicates a highly non-random distribution of chromosome damage in the interbands induced by both SN and X-ray $P < 0.005$). It is obvious from figures 3 and 4

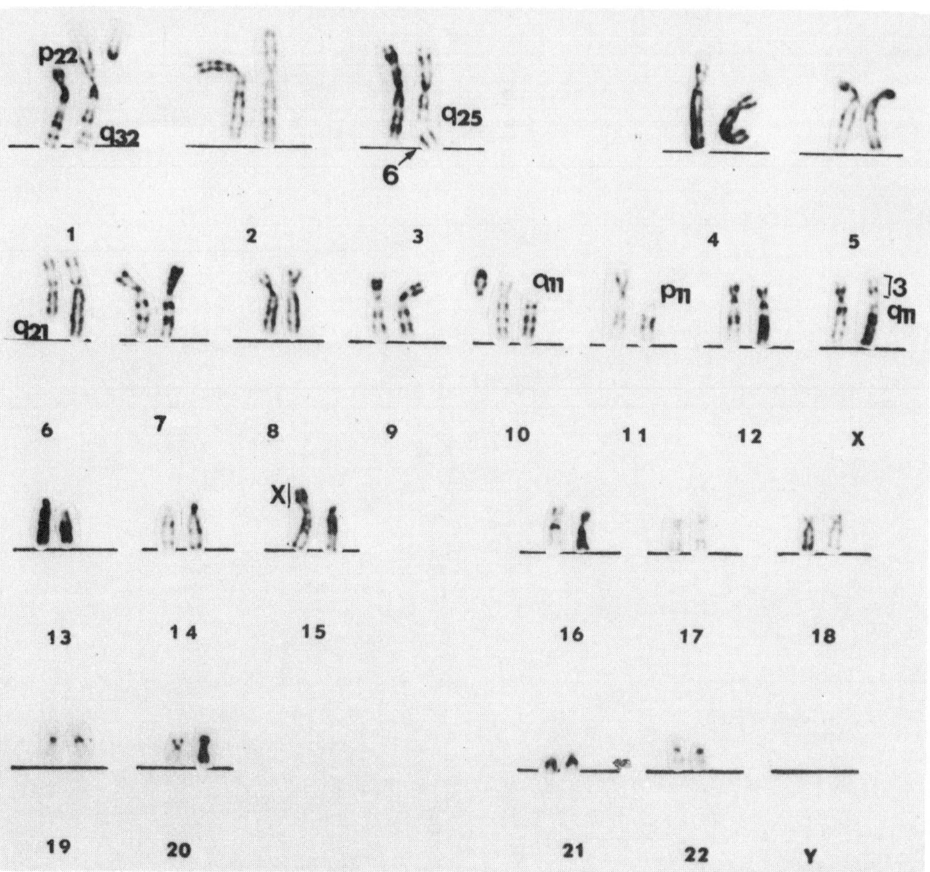

Figure 1. Karyotype of cell following 0.1 μg ml^{-1} SN treatment demonstrating multiple breaks and rearrangements. Chromosome 1: isochromatid break at band p22 in one homologue leading to an acrocentric fragment, and isochromatid break at q32 in second homologue. Chromosome 3: isochromatid break at q25 and interchange with the distal segment of 6q. Chromosome 6: isochromatid break at q21 and fragment translocated to 3q25. Chromosome 10: isochromatid break at q11. Chromosome 11: isochromatid break at p11. X chromosome: isochromatid break at q11 and distal segment of 3q25 → term translocated to short arm. Chromosome 15: translocation of short arm of X.

that a clustering of breaks occurs at the pericentromeric interbands of many chromosomes (46.5% of the SN and 40.4% of the X-ray breaks occurred in these regions). However, they were not randomly distributed among the chromosomes, and conspicuous concentrations appear in 1 and 2. Therefore, the data were re-analysed after eliminating the pericentromeric interbands and tested for randomness within the remainder of the euchromatin. Although the chi-square values were greatly reduced, the result remains statistically significant (table 4).

Figures 3 and 4 also indicate a general paucity of chromosome breaks

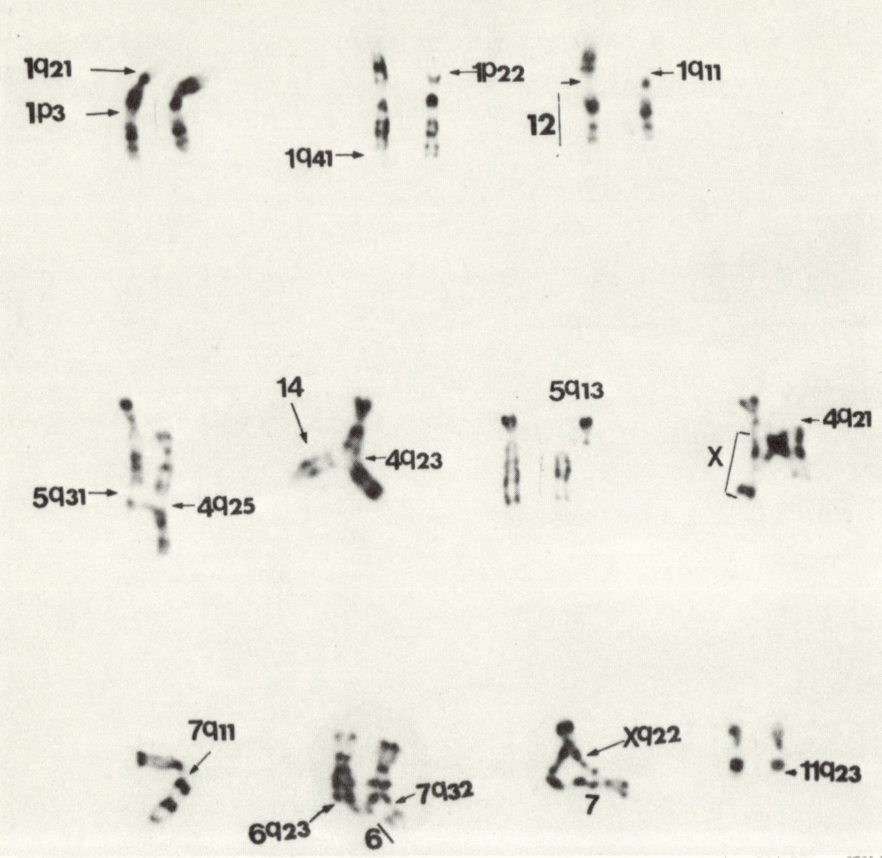

Figure 2. Composite of 'mapped' chromosomal abnormalities induced by X-rays and SN. Top, aberrations in chromosome 1; middle, aberrations in B group; bottom, aberrations in C group.

in the telomeric regions of the chromosomes. This may be due to technical bias in the inability of break detection in the terminal interbands by G-banding. To accurately ascertain such breaks, R or T banding techniques must be employed. Since breaks occurring beyond the terminal heterochromatic band of each chromosome arm are not readily apparent, the observed number of terminal breaks may be underestimated. To adjust for such bias, the terminal euchromatic interbands were eliminated and the data reanalysed. Upon removal of both the pericentromeric and terminal interbands, a distribution of breaks that is proportional to the total interstitial euchromatin per chromosome is observed (table 5).

The breaks in the pericentromeric interbands of the chromosomes were also tested for randomness (Table 6). Although pericentromeric interbands are generally more clastogen-sensitive than other interbands, the sensitivity of individual centromeric regions differs. Following SN treatment, the peri-

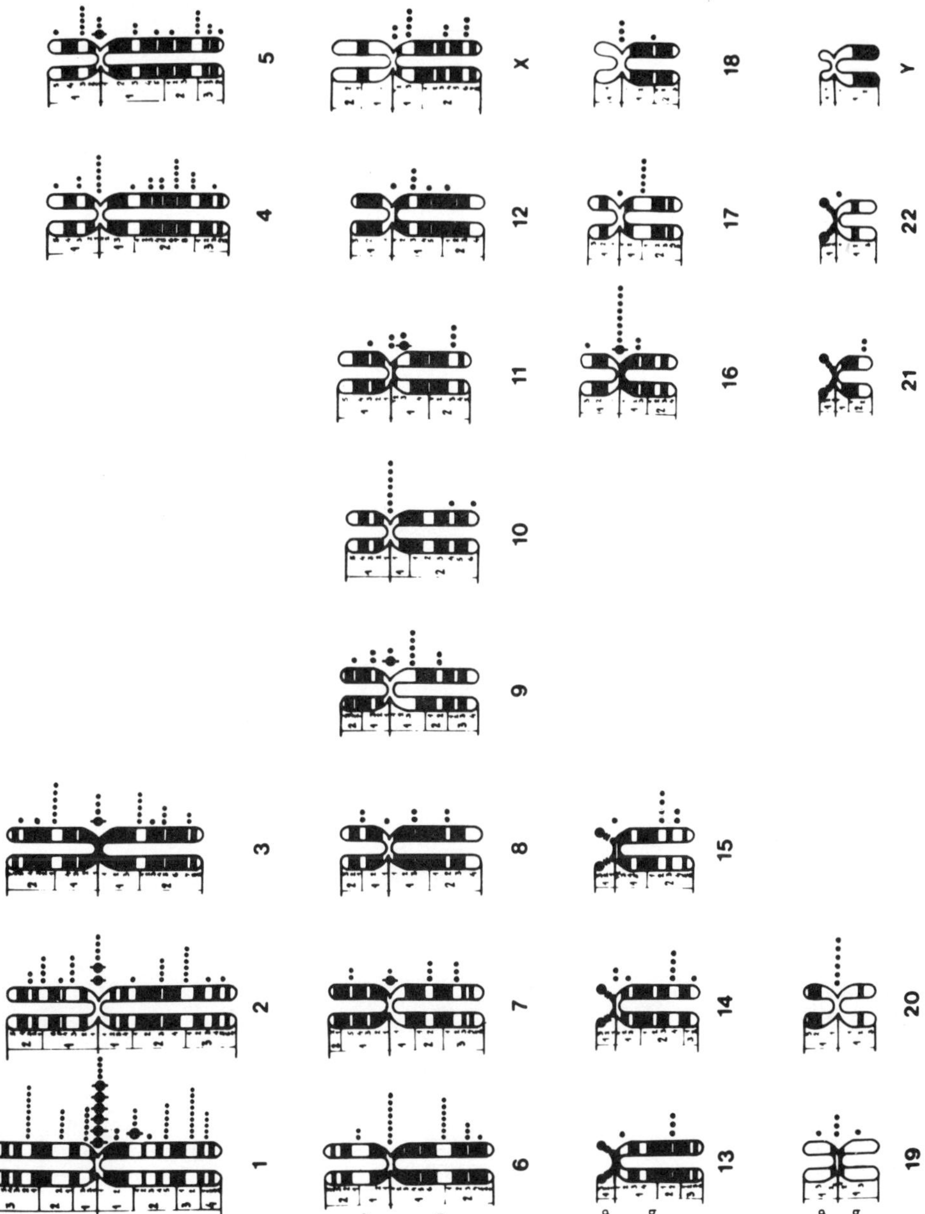

Figure 3. Distribution of chromosome damage induced by 0.1 μg ml^{-1} SN. Each small dot signifies 1 break; large dot signifies 10 breaks.

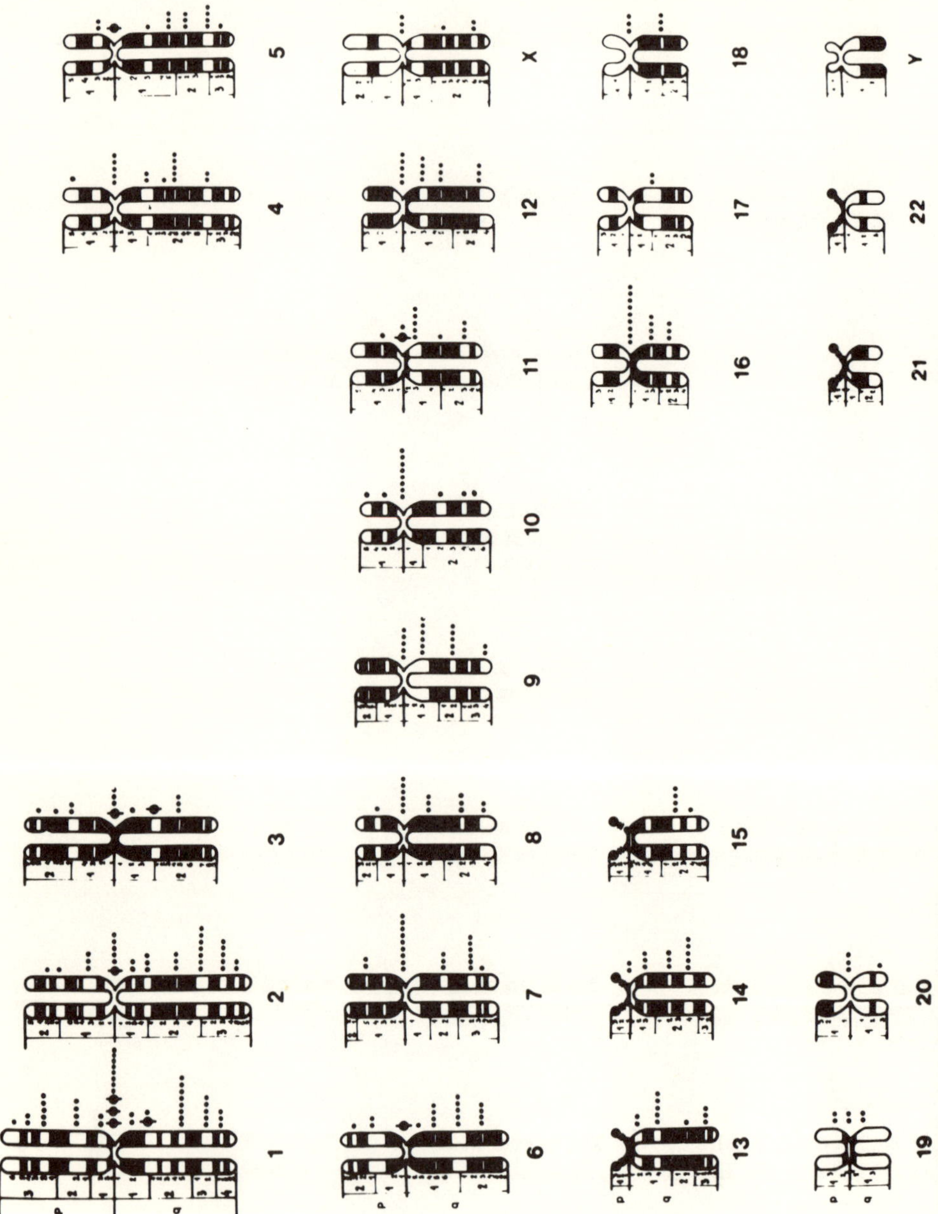

Figure 4. Distribution of chromosome damage induced by 200 rad X-rays. Each small dot signifies 1 break; large dot signifies 10 breaks.

Table 3. Test for random distribution of breaks based on length of euchromatin per chromosome (P < 0.005 for χ^2_{23}).

Chromosome No.	Streptonigrin breaks			X-ray breaks		
	observed	expected	χ^2	observed	expected	χ^2
1	119	39.2	162.4	81	36.2	55.4
2	48	34.9	4.9	42	32.2	3.0
3	32	14.9	19.6	31	13.8	21.4
4	23	23.0	—	16	21.2	1.3
5	29	20.4	3.6	24	18.9	1.4
6	21	20.4	—	27	18.9	3.5
7	19	18.7	—	21	17.3	0.8
8	7	17.9	6.6	15	16.5	—
9	21	20.4	—	18	18.9	—
10	9	18.7	5.0	13	17.3	1.1
11	17	17.9	—	21	16.6	1.2
12	6	14.5	5.0	15	13.5	0.2
13	4	8.5	2.4	11	7.9	1.2
14	8	11.1	0.9	13	10.3	0.7
15	7	9.8	0.8	5	9.0	1.8
16	21	11.9	6.9	16	11.1	2.2
17	6	16.6	6.8	2	15.3	11.6
18	4	15.3	8.3	6	14.1	4.7
19	5	20.0	11.3	6	18.5	8.4
20	6	16.6	6.8	4	15.3	8.3
21	2	6.4	3.0	0	5.9	5.9
22	1	10.2	8.3	0	9.4	9.4
X	11	27.6	10.0	6	25.5	14.9
Y	0	10.1	10.1	0	9.4	9.4
Total	426	426	282.7	393	393	167.8

centromeric interbands of chromosomes 1, 2 and 16 contain more breaks than expected (46% of all SN pericentromeric breaks were in 1 and 2), while 8, 12, 13, 15, 17, 21, 22 and Y appear 'more resistant'. With regard to the X-ray damage, the pericentromeric interbands of 1 and 2 are also more susceptible than expected (33% of all X-ray pericentromeric breaks), but not 16. Similar to the diferrential response to SN, the centromeres of 15, 17, 21, 22 and Y also appear resistant to X-ray induced damage.

In order to test for interstitial euchromatic 'hot spots', only those chromosomes with sufficient numbers of breaks for statistical analysis were studied, e.g. 1 and 2 for both SN and X-ray, and 4 for SN alone. In chromosome 1 a non-random distribution of breaks was observed following SN treatment (table 7) with interband lq42 containing more breaks than expected. Conversely, interbands lp22 and lq23 seem resistant. However, following X-ray exposure, interband lq25 appears to be a 'hot spot' while again interband lq23 is resistant. With regard to chromosome 2 a random distribution of breaks, proportional to the length of the interband, was obtained with

Table 4. Test for random distribution of breaks based on length of euchromatin per chromosome minus pericentromeric interband ($P < 0.005$ for χ^2_{22}).

Chromosome No.	Streptonigrin breaks			X-ray breaks		
	observed	expected	χ^2	observed	expected	χ^2
1	55	26.9	29.4	43	27.6	8.6
2	24	20.3	0.7	27	20.8	1.8
3	19	10.9	6.0	18	11.2	4.1
4	17	13.9	0.7	11	14.2	0.7
5	17	12.3	1.8	14	12.6	0.2
6	13	13.3	—	17	13.8	0.7
7	8	9.2	0.2	12	9.6	0.6
8	6	10.3	1.8	9	10.7	0.3
9	10	8.9	0.1	13	9.1	1.7
10	2	10.0	6.4	5	10.2	2.7
11	15	11.4	1.1	10	11.6	0.2
12	5	7.5	0.8	10	7.7	0.7
13	3	4.3	0.4	9	4.4	4.8
14	6	6.8	0.1	11	7.0	2.3
15	6	6.8	0.1	5	7.0	0.6
16	3	6.4	1.8	7	6.6	—
17	5	8.4	1.4	2	8.7	5.2
18	1	3.0	1.3	3	3.0	—
19	2	14.4	10.7	4	14.7	7.8
20	0	3.0	3.0	1	3.0	1.3
21	2	3.4	0.6	0	3.5	3.5
22	0	4.3	4.3	0	4.4	4.4
X	9	12.3	0.9	3	12.6	7.3
Y	—	—	—	—	—	—
Total	228	228	73.6	234	234	59.5

both clastogens ($P > 0.25$). The response of chromosome 4 to SN deviated significantly from random ($P < 0.05$, table 7) with the demonstration of a 'hot spot' in band 4q27 (figure 3).

Discussion

Streptonigrin is known to inhibit DNA synthesis, as well as cause the degradation of DNA (Rao and Cullen 1959, Radding 1963). On the other hand, it has been postulated that X-rays cause chromosome breaks directly by the action of ionising particles or indirectly through the formation of peroxides or free radicals (Evans 1962, Cerutti 1974). Since these clastogens act by differing mechanisms, different patterns of chromosomal response might be expected. Nonetheless, in the present studies, the damage induced by both clastogens is similar on several levels. That the concentration of SN and X-ray exposure tested were equivalent is evidenced by both the yield of chromosomal aberrations per cell (table 1) and the distribution of numbers of breaks in the cells, which were almost identical (table 2). Therefore, quanti-

Table 5. Test for random distribution of breaks based on length of euchromatin per chromosome minus both pericentromeric and terminal interbands (P > 0.10 for χ^2_{18}).

Chromosome No.	Streptonigrin breaks			X-ray breaks		
	observed	expected	χ^2	observed	expected	χ^2
1	54	37.1	7.7	42	32.4	2.8
2	24	27.4	0.4	27	26.8	—
3	19	15.0	1.1	18	14.7	0.7
4	16	15.0	0.1	10	14.7	1.5
5	15	13.6	0.1	14	13.4	—
6	13	18.4	1.6	17	18.1	—
7	8	10.1	0.4	12	10.0	0.4
8	6	8.3	0.6	7	8.1	0.1
9	10	6.8	1.5	5	6.6	0.4
10	1	5.6	3.8	5	8.2	1.2
11	15	10.9	1.5	10	10.7	—
12	5	4.8	—	7	4.7	1.1
13	3	4.8	0.7	9	4.7	3.9
14	5	7.0	0.6	10	7.9	0.6
15	6	8.7	0.8	5	8.5	0.1
16	3	3.5	—	7	3.4	3.8
17	5	7.5	0.8	2	7.2	3.6
18	1	2.6	1.0	3	2.6	—
19	—	—	—	—	—	—
20	—	—	—	—	—	—
21	—	—	—	—	—	—
22	—	—	—	—	—	—
X	9	10.9	0.3	3	10.3	5.0
Y	—	—	—	—	—	—
Total	218	218	23.0	213	213	25.2

tative comparisons between the two agents is valid. In both cases an almost identical non-random distribution of chromosome breaks was obtained (figures 3 and 4). Qualitatively, even though virtually all the breaks (99.4%) mapped to the euchromatic interbands (or the euchromatin-heterochromatin interface), they were not distributed at random in proportion to the amount of euchromatin per chromosome (table 3). There was a significant excess of breaks in the pericentromeric interbands with a concomitant paucity of breaks in the terminal interbands. Although the pericentromeric regions of most chromosomes were more sensitive to breakage, this was not true of all the chromosomes. Significant non-random distributions of pericentromeric breaks were obtained following exposure to both clastogens ($P < 0.005$, table 4). However, upon removal of both the pericentromeric and terminal interbands from the analysis, a significant correlation was observed between the length of remaining interstitial euchromatin and chromosome breaks induced by the clastogens (table 5). Such a correlation is not necessarily due

Table 6. Test for random distribution of breaks among the pericentromeric interbands of each chromosome ($P < 0.001$ for χ^2_{23}).

Chromosome No.	Streptonigrin breaks			X-ray breaks		
	observed	expected	χ^2	observed	expected	χ^2
1	64	8.25	365.7	38	6.625	148.6
2	24	8.25	30.1	15	6.625	10.6
3	3	8.25	3.3	13	6.625	6.1
4	6	8.25	0.6	5	6.625	0.4
5	12	8.25	1.7	10	6.625	1.7
6	8	8.25	—	10	6.625	1.7
7	11	8.25	0.9	9	6.625	0.9
8	1	8.25	6.4	6	6.625	0.1
9	11	8.25	0.9	5	6.625	0.4
10	7	8.25	0.2	8	6.625	0.3
11	2	8.25	4.7	11	6.625	2.9
12	1	8.25	6.4	5	6.625	0.4
13	1	8.25	6.4	2	6.625	3.2
14	2	8.25	4.7	2	6.625	3.2
15	1	8.25	6.4	0	6.625	6.6
16	17	8.25	9.3	9	6.625	0.9
17	1	8.25	6.4	0	6.625	6.6
18	3	8.25	3.3	3	6.625	2.0
19	3	8.25	3.3	2	6.625	3.2
20	6	8.25	0.6	3	6.625	2.0
21	0	8.25	8.2	0	6.625	6.6
22	1	8.25	6.4	0	6.625	6.6
X	2	8.25	4.7	3	6.625	2.0
Y	0	8.25	8.2	0	6.625	6.6
Total	198	198	488.8	159	159	223.6

to equal sensitivity of all interbands but indicates a random distribution of areas that are sensitive or resistant to breakage within individual chromosomes.

Chromosomes 1 and 2 appear to be extremely susceptible to breakage, as expected on the basis of their total length or euchromatic segments. Following SN and X-ray treatment 25% and 20% respectively of the total breaks observed were located in chromosome 1. For chromosome 2 the figures are somewhat lower, but significantly higher than expected; 11% for SN and 10% for X-ray. This may be due to the hypersensitivity of the pericentromeric regions of these two chromosomes. Increased susceptibility of 1 and 2 to SN action was observed by Cohen (1963) but this was not the case for X-ray exposure. Other investigations following X-ray exposure did not observe this increased sensitivity of these particular chromosomes (Caspersson, Lomahka and Zech 1971; Holmberg and Jonasson 1973; Jonasson and Holmberg 1973; San Roman and Bobrow 1973; Savage, Watson and Bigger 1973; Seabright 1973; Cooke, Seabright and Wheeler 1975).

Table 7. Test for interstitial 'hot spots' among interbands of chromosomes 1 and 4 ($P < 0.05$).

Inter-band	Streptonigrin breaks observed	Streptonigrin breaks expected	χ^2	X-ray breaks observed	X-ray breaks expected	χ^2
Chromosome 1						
p34	0	2.1	2.2	5[1]	3.2	1.0
p32	9	5.5	2.2	5	4.1	0.2
p22	5	11.1	3.3	4	8.3	2.2
p13	6	6.6	0.05	2	5.0	1.8
q21	12	8.8	1.2	11	6.6	2.9
q23	1	5.5	3.7	0	4.1	4.1
q25	6	4.4	0.6	8	3.3	6.7
q32	9	7.0	0.6	5	5.3	0
q42	5	2.2	3.6	—	—	—
Total	53	53.2	17.3	40	39.9	18.9
Chromosome 4						
p14	2	3.6	0.7			
q21	1	5.1	4.7			
q23, q25, q27, q33	10	3.6	11.3			
q31	3	3.6	0.1			
Total	16	15.9	16.8			

[1] Figure for (p34 + q42).

Several previous studies (Cohen 1963; Caspersson, Lomakha and Zech 1971; Cooke, Seabright and Wheeler 1975) have indicated increased breakage at the centromeric regions following exposure to various clastogens. This observation was particularly striking in the present study (figures 3 and 4). Of all the SN-induced breaks, 46.5% were mapped to the pericentromeric interbands while 40.4% of the X-ray breaks were similarly located. In chromosome 1, 54 and 47% of breaks were in the pericentromeric interband after SN and X-ray treatment, and 50 and 35% in chromosome 2.

This hypersensitivity of the pericentromeric regions of chromosomes 1 and 2 may be due to several reasons. The 'centromeres' of 1 (as well as 16) differ from the other chromosomes in their proximity to secondary constrictions that contain AT-rich satellite DNA (Corneo, Gianelli and Polli 1970; Jones and Corneo 1971). This specialised structure may constitute a 'weak point' at the junction or interface between the secondary constriction and the centromere itself. The centromeric region of chromosome 2 has evolutionary significance, since it seems to have resulted from a Robertsonian translocation between 13 and 17 of the chimpanzee (Warburton *et al.* 1973) as demonstrated by banding patterns (Turleau, de Grouchy and Klein 1972). Apparently, this translocation results from a fusion of two acrocentrics leading to a dicentric element whose stability was ensured by the inactivation of one centromere, which simulates a secondary constriction. This euchromatic-centromeric junction area also may constitute a 'weak point' and accumulate breaks more easily than other chromosomal regions.

Another interesting point is the apparent non-randomness of X-ray-induced breaks observed in this study. Non-banded studies initially indicated no differential sensitivity among or within chromosomes 1, 2 and 3 (Cohen 1963, Heddle 1965). Investigations performed with the aid of the chromosome banding techniques agree that a preponderance of X-ray-induced breaks occurs in euchromatin. However, their overall distribution among chromosome differs, Caspersson, Lomahka and Zech (1971) suggest that centromeres may be more sensitive, but not those of chromosomes 1 and 2. Although increased concentrations of breaks were localised to euchromatic regions by Savage, Watson and Bigger (1973), the centromeres were not included. San Roman and Bobrow (1973) found a concentration of breaks in 3 and the telomeres but not centromeres.

Holmberg and Jonasson (1973) found a direct relationship between the amount of 'R-banding material' in the chromosome and breakage rate, but no evidence of centromeric concentration. On the other hand, the X-ray experiments of Seabright (1973) indicate some similarities to our results. These authors found an increase of breaks at the centromere of chromosome 1 (Cooke, Seabright and Wheeler 1975) as well as a decrease of breaks in the telomeric regions, and a complete absence of breaks in the Y chromosome (Seabright 1973). They also point out the difficulty of localising breaks to the telomeres because of technical limitations of the G-banding procedure.

What possible explanation might reconcile these divergent results? All the experiments were performed on PHA-stimulated human lymphocytes and although the radiation doses differed slightly, this parameter would most likely influence the frequency of breaks but not their intrachromosomal distribution. The most variable parameter in these investigations was the time of X-ray exposure. Some workers irradiated the cells prior to culture (Holmberg and Jonasson 1973; San Roman and Bobrow 1973; Seabright 1973; Cooke, Seabright and Wheeler 1975), while others irradiated them at different times during the 72-h incubation period (Cohen 1963; Caspersson, Lomahka and Zech 1971; Savage, Watson and Bigger 1973). It may well be that various euchromatic interbands display different sensitivity to clastogenesis at different periods of the cell cycle. Support for this concept derives from the work of Beek and Obe (1977), who showed variable sensitivity to X-ray damage in human lymphocytes irradiated in G_0 and various stages of G_1 during the first division cycle following PHA stimulation.

Summary

The distribution of chromosome breaks induced in PHA-stimulated human lymphocyte cultures by two known clastogens, streptonigrin and X-rays, was investigated. Using G-banding techniques, more than 99% of the damage was mapped to interband regions (euchromatin) or the band-interband junctions. Response to both clastogens revealed a non-random distribution of breaks with respect to the amount of euchromatin per chromosome. A significant accumulation of breaks was observed in the pericentromeric interbands of certain chromosomes, particularly 1 and 2. The apparent lack of breaks in the telomeric interbands may be due to limitations of the G-banding technique. Interstitial 'hot-spots' were observed at different interbands in response to the two clastogens. It may well be that chromosome breaks at different interbands are related not only to the action of the specific clastogen but also to its time of action in relation to the cell cycle.

Acknowledgements

This study served as a partial requirement for an MSc degree (M.S.) at the Hebrew University of Jerusalem. The work was partly supported by grants from the National Foundation – March of Dimes, and the U.S. – Israel Binational Science Foundation. M.M.C. is an Established Investigator, Chief Scientist's Bureau, Israel Ministry of Health.

References

Beek, B. & G.Obe (1977) Differential chromosomal radiosensitivity within the first G-phase of the cell cycle of early-dividing human leukocytes *in vitro* after stimulation with PHA. *Hum. Genet. 35*, 209–18.

Caspersson, T., G.Lomahka & L.Zech (1971) 24 fluorescence patterns of human metaphase chromosome-distinguishing characters and variability. *Hereditas 67*, 89–102.

Cerutti, P.A. (1974) Effects of ionizing radiation on mammalian cells. *Naturwiss. 61*, 51–9.

Cohen, M.M. (1963) The specific effects of streptonigrin activity on human chromosomes in culture. *Cytogenet. 2*, 271–9.

Cohen, M.M., K.Hirschhorn & W.Frosch (1967) *In vivo* and *in vitro* chromosomal damage induced by LSD-25. *New Engl. J. Med. 277*, 1043–9.

Cooke, P., M.Seabright & M.Wheeler (1975) The differential distribution of X-ray induced chromosome lesions in trypsin-banded preparations from human subjects. *Humangenetik 28*, 221–31.

Corneo, G., E.Gianelli & E.Polli (1970) Repeated sequences in human DNA. *J. Mol. Biol. 48*, 319–27.

Evans, H.J. (1962) Chromosome aberrations induced by ionizing radiations. *Int. Rev. Cytol. 13*, 221–321.

Heddle, J. (1965) Randomness in the formation of radiation-induced chromosome aberrations. *Genetics 52*, 1329–34.

Holmberg, M. & J.Jonasson (1973) Preferential location of X-ray-induced chromosome breakage in the R-bands of human chromosomes. *Hereditas 74*, 57–68.

Jonasson, J. & M.Holmberg (1973) Evidence for an inverse relationship between X-ray induced chromatid and chromosome breakage in human chromosomes. *Hereditas 75*, 259–66.

Jones, K.W. & G.Corneo (1971) Location of satellite and homogeneous DNA sequences on human chromosomes. *Nature New Biol. 233*, 268—71.

Moorhead, P.S., P.C.Nowell, W.J.Mellman, D.M.Battips & D.A.Hungerford (1960) Chromosome preparations of leukocytes cultured from human peripheral blood. *Expt. Cell Res. 20*, 613–16.

Paris Conference (1971) ***Standardization in Human Cytogenetics***. Birth Defects: Original Article Series vol.8, no.7, 1972. New York: The National Foundation.

Patil, S.R., S.Merrick & H.A.Lubs (1971) Identification of each human chromosome with a modified Giemsa stain. *Science 173*, 821–2.

Radding, C.M. (1963) Incorporation of ^{3}H-thymidine by K12 (λ) induced by streptonigrin, in *Genetics Today*, vol.I (ed. S.J.Geerts) p.22. Oxford: Pergamon Press.

Rao, K.V. & W.P.Cullen (1959) Streptonigrin, an antitumor substance. *Antibiot. Annu.*, 950.

San Roman, C. & M.Bobrow (1973) The sites of radiation-induced breakage in human lymphocyte chromosomes, determined by quinacrine fluorescence. *Mutat. Res. 18*, 325–31.

Savage, J.R.K., G.E.Watson & T.R.L.Bigger (1973) The participation of human chromosome arms in radiation-induced chromosome exchange, in *Chromosomes Today*, vol. 4 (eds. J.Wahrman & K.R.Lewis) pp.267–76. New York: Wiley.

Seabright, M. (1973) High resolution studies on the pattern of induced exchanges in the human karyotype. *Chromosoma 40*, 333–46.

Turleau, C., J. de Grouchy & M.Klein (1972) Phylogénie chromosomique de l'homme et des primates hominiens (*Pan troglodytes*, *Gorilla gorilla* et *Pongo pygmaeus*) essai de reconstitution du caryotype de l'ancêtre commun. *Ann. Genet. 15*, 225–40.

Warburton, D., I.L.Firschein, D.A.Miller & F.E.Warburton (1973) Karyotype of the chimpanzee, *Pan troglodytes*, based on measurements and banding pattern: comparison to the human karyotype. *Cytogenet. Cell Genet. 12*, 453–61.

M. KUČEROVÁ and Z. POLIKOVÁ

In vitro Comparison of Normal and Trisomic Cell Sensitivity to Physical and Chemical Mutagens

The sensitivity of trisomic cells to ionising radiations has been studied for many years. In our laboratory these studies commenced ten years ago, when it was found that *in vitro* radiosensitivity was significantly higher in cells with trisomy 21 than in normal cells (Kučerová 1967). Sasaki, Tonomura and Matsubara (1970) observed that, regardless of the type of trisomy, all trisomic cells were more radiosensitive *in vitro* than cells with translocations, inversions and monosomy, all of which react to radiation in the same way as normal cells. The altered enzyme activity of trisomic cells is considered to influence the production of exchange aberrations and increase the aberration yield.

In the present paper this problem is considered from somewhat different aspects. As well as comparing the sensitivity of normal and trisomic cells to radiation a comparison is made of the *in vitro* sensitivity of these cells to a chemical mutagen–mitomycin C. In addition to gross chromosomal aberrations, changes in the frequency of sister chromatid exchanges (SCE) were also investigated.

Materials and Methods

The conventional cytogenetic micromethod described by Hungerford (1965) and the fluorescence plus Giemsa technique (FPG) introduced by Wolff and Perry (1974) were used.

Blood samples from four healthy males (A, aged 19; B, 35; C, 29; D, 38) and from four males with Down's syndrome (trisomy 21: E, 30; F, 55; G, 28; H, 32) were exposed to 100, 200, 300 and 400 rad of 200 kVp X-rays. The cells were irradiated with a TUR machine operated at 10 mA, with 0.5 mm Cu filter, h.v.l. 1 mm Cu. Other blood samples were exposed to 10^{-4}, 10^{-5} and 10^{-6} mg ml^{-1} concentrations of mitomycin C. The irradiated blood samples were cultured for 48 h, the others for 72 h, and for the final 24 hours they were exposed to mitomycin C. All cells were prepared for cytogenetic analysis by a conventional technique and the aberrations were scored according to WHO recommendations (Buckton and Evans 1973), scoring 100 cells for each dose and sample.

Further blood samples from each donor were treated with the FPG technique. The first sample was irradiated with 100 rad of X-rays, the second and third were exposed for 24 h to 10^{-5} and 10^{-6} mg ml^{-1} concentrations of mitomycin C. Two blood samples served as controls. For each dose and sample 50 cells were scored for SCEs, including centromeric SCEs. The latter, however, were also itemised separately.

Results

The comparison of aberration yields in trisomic and normal cells exposed to increasing X-ray doses showed that the percentage of aberrant cells is dose-dependent in both categories, with a significantly higher aberration level ($P < 0.001$) in the trisomics. This higher increase is even more pronounced when numbers of aberrations and numbers of dicentrics per cell are considered. Since the cells of trisomic and normal donors reacted similarly within each group, the results were pooled (figure 1a,b,c). The most common aberrations observed were dicentrics with fragments.

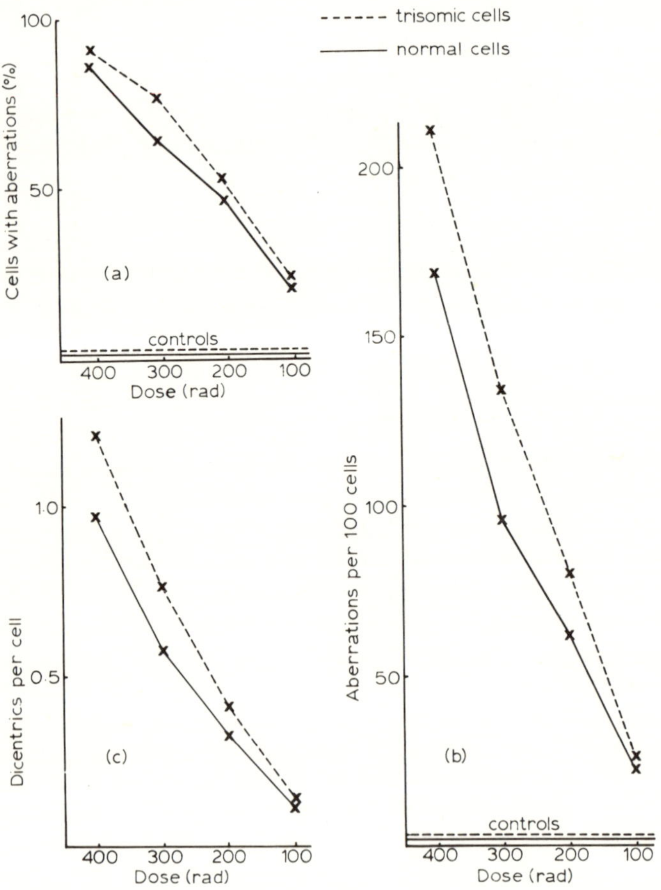

Figure 1. Sensitivity of normal and trisomic cells to radiation (pooled results).

The FPG technique showed no significant differences in the numbers of SCE in control cells and cells irradiated with 100 rad for either normal or trisomic material (figure 2).

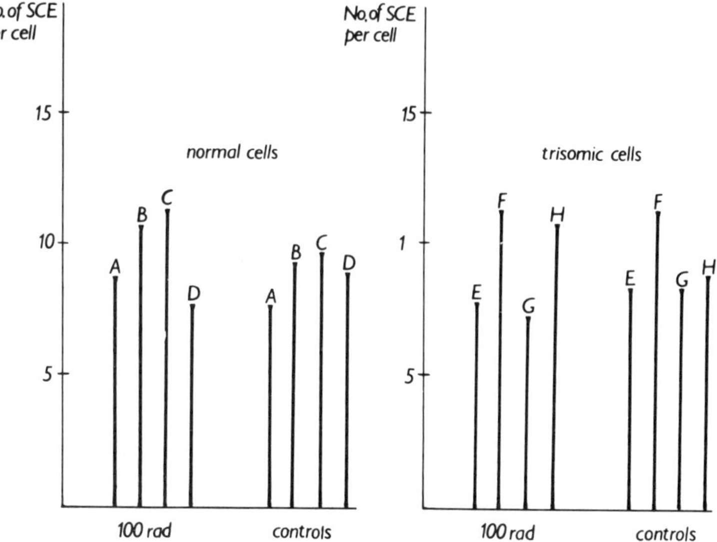

Figure 2. Sister chromatid exchanges in normal and trisomic cells after irradiation.

The sensitivity of donors to mitomycin C appeared to be less homogenous than their sensitivity to radiation. The numbers of aberrations were dependent on the drug concentration but the responses between donors within a group showed significant differences, so that pooling of results was not possible (figure 3). The most common aberrations found were chromatid and isochromatid breaks.

The numbers of SCEs in both cell types increased significantly ($P < 0.01$) with increasing concentrations of mitomycin C (figure 4).

Discussion

The higher radiosensitivity of cells with trisomy 21 as detected by analysis of chromosomal aberrations confirms previous results (Kučerová 1967; Sasaki, Tonomura and Matsubara 1970). A minimal increase of SCE yields after *in vitro* irradiation has been described by Perry and Evans (1975) and our results fully agree with this. A new observation is that irradiated trisomic cells do not react with an increased frequency of SCE, so that in this respect they behave as normal cells.

Individual differences in sensitivity to mitomycin C were observed both in normal and trisomic cells. The mechanism of mutagenic action of this cytostatic drug is not simple. Brøgger (1974) ascribes its action to a single alkylating effect with cross-linking of DNA strands. Individual differ-

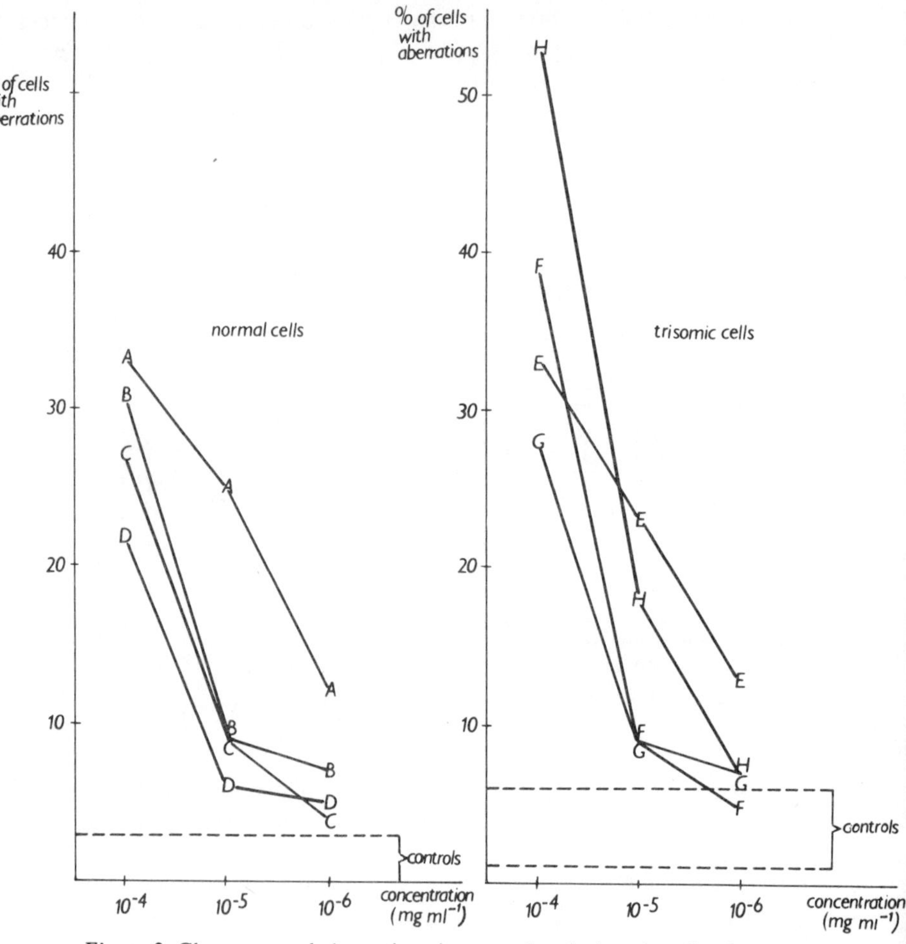

Figure 3. Chromosomal aberrations in normal and trisomic cells after exposure to mitomycin C.

ences in these and other steps of mutagenic action are possible and therefore the final effect could be variable.

In contrast to the results obtained with ionising radiation the sensitivity of trisomic cells to aberration induction by mitomycin C was not different to normal cells. The mechanism of radiation-induced damage of chromosomes and subsequent repair processes differ most probably from those after exposure to mitomycin C. It is possible that trisomic cells have an altered enzymatic status, which influences the post-radiation repair process but which is not important for the repair of the alkylated lesions induced by the cytostatic drug.

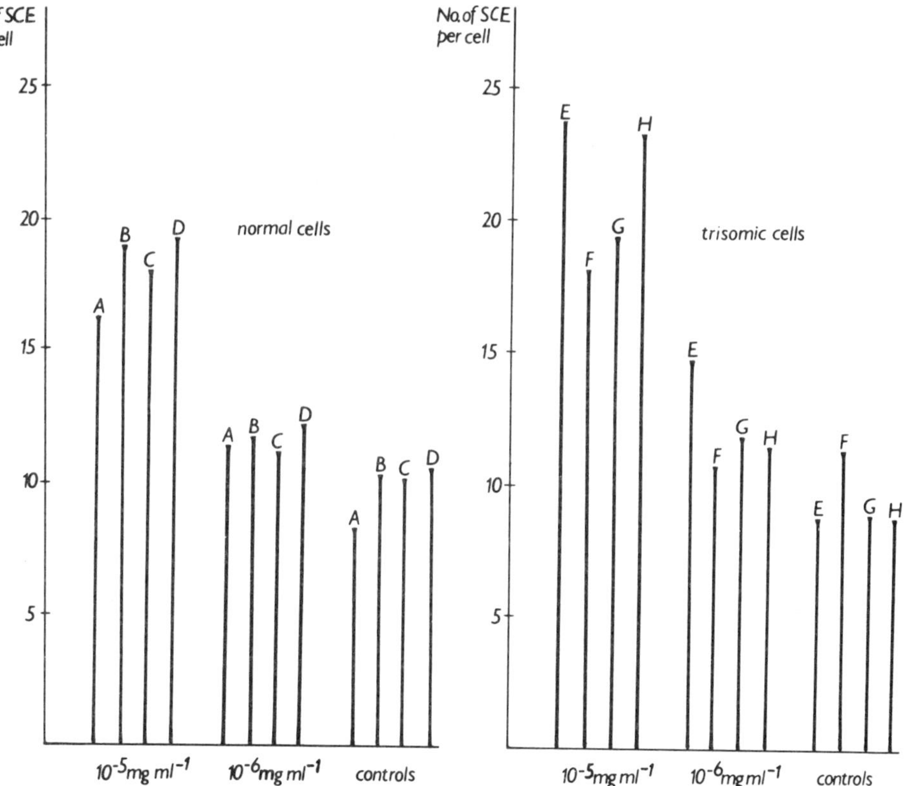

Figure 4. Sister chromatid exchanges in normal and trisomic cells after exposure to mitomycin C.

Summary

Lymphocytes from four healthy male donors and four males with Down's syndrome were irradiated with 100, 200, 300 or 400 rad, or exposed to different concentrations of mitomycin C. The numbers and types of chromosomal aberrations were analysed by the conventional technique and the yield of sister chromatid exchanges (SCE) was scored using the fluorescence plus Giemsa (FPG) technique.

As expected, the radiosensitivity of trisomic cells, measured by the number of chromosomal aberrations, was found to be higher than that of normal cells. The number of SCEs in both types of cells exposed to 100 rad was not significantly changed in comparison with controls.

The sensitivity of trisomic and normal cells to mitomycin C *in vitro* appears to be similar. The increased frequency of aberrations and SCEs in both types of cells was dependent on the drug concentration.

Acknowledgements

The authors are grateful to Dr V. Matoušek for his assistance with the mathematic evaluation of the results. This project was supported by International Atomic Energy Agency, Contract No. 1744/RB.

References

Brøgger, A. (1974) Different patterns of chromosome exchange induced by methylmethanesulphonate and mitomycin C in human cells. *Hereditas 77*, 617–33.

Buckton, K.E. & H.J.Evans (1973) *Methods for the Analysis of Human Chromosome Aberrations*. Geneva: WHO.

Hungerford, D.A. (1965) Leukocytes cultured from small inocula of whole blood and the preparation of metaphase chromosomes by treatment with hypotonic KCl. *Stain. Technol. 40*, 333–8.

Kučerová, M. (1967) Comparison of radiation effects *in vitro* upon chromosomes of human subjects. *Acta Radiol. 6*, 441–8.

Perry, P. & H.J.Evans (1975) Cytological detection of mutagen-carcinogen exposure by sister chromatid exchange. *Nature 28*, 121–5.

Sasaki, M.S., A.Tonomura & S.Matsubara (1970) Chromosome constitution and its bearing on the chromosomal radiosensitivity in man. *Mutat. Res. 10*, 617–33.

Wolff, S. & P.Perry (1974) Differential Giemsa staining of sister chromatid exchanges without autoradiography. *Chromosoma 48*, 314–53.

J. A. HEDDLE, R. D. BENZ and P. I. COUNTRYMAN

Measurement of Chromosomal Breakage in Cultured Cells by the Micronucleus Technique

Although measurements of chromosomal breakage in human cells have proven to be exceedingly valuable in many situations, the time and expertise required have often reduced the usefulness of this technique or the extent to which it could be exploited. We, and our colleagues, have been using one of the consequences of chromosomal breakage, the production of micronuclei, to make much easier and more rapid measurements of chromosomal damage. Recently, there have been a number of studies using micronuclei as a qualitative indicator of chromosomal breakage in the field of environmental mutagenesis (e.g. Schmid 1975, Heddle 1973, Matter and Schmid 1971). Although this is a novel assay, micronuclei have long been known to occur as a consequence of chromosomal loss or fragmentation. To our knowledge, however, only one attempt has been made previously to use micronuclei as a quantitative measure of chromosomal damage (Evans, Neary and Williamson 1959). At one time Fabergé (1940) suggested that chromosomal scoring could be made more rapid by simply counting the number of chromosomal bodies at metaphase and ignoring the types of aberrations Unfortunately it is the finding of metaphase cells and the counting of pieces that takes most of the time.

Micronuclei arise from chromosomes or chromosomal fragments that are not incorporated into one or other of the daughter nuclei during cell division. Usually they represent acentric fragments, which float passively in the cytoplasm because they lack the spindle attachment site and which are frequently excluded from the daughter nuclei in consequence. These micronuclei ordinarily resemble the main nuclei in all respects except that they are usually much smaller. Obviously the number of micronuclei present in cells will depend upon the number of acentric fragments present, i.e. on the extent of chromosomal breakage, but other factors, such as the proportion of the cell population that has divided, the probability that a fragment remains sufficiently far from the main nucleus to become a micronucleus, etc., are also important.

The main advantage in scoring micronuclei, rather than scoring the aberrations directly, is that the scoring is much more rapid and requires much less expertise. A secondary advantage is that the effect of sampling time, a major factor in aberration studies, is greatly reduced, as the micronuclei accumulate in the cell population and thus provide an approximate integration of the damage. The main disadvantage is that factors other than chromosomal breakage can affect the micronucleus frequency. Also, one cannot know what kind of aberration led to the micronucleus by just looking at it. In this paper we summarise a number of studies in which micronuclei have proved to be useful measures of chromosomal damage.

Materials and Methods

The culture technique used for most of the work has been described in detail by Countryman and Heddle (1977). In essence a miniature blood culture of the usual sort is set up in an Eppendorf 1.5 ml polypropylene microtube with 1 ml of medium and 0.01 ml of blood. The slides are made in the usual way after hypotonic treatment and fixation of the cells. Our standard sampling time has been 96 h after the culture is initiated. The microtubes permit the cells to be processed in their culture vessel and to be centrifuged down in 15 s, twelve tubes at a time. Typically, one slide is made per culture and 500 cells scored from it. Each point represents the mean of at least two cultures.

The criteria for scoring micronuclei are to be found in Countryman and Heddle (1976). The only deviation from these criteria that has evolved since that work was published is that we now score micronuclei even if three or more occur in the same cell, provided only that (a) there is a distinct main nucleus and (b) the main nucleus is more or less round rather than deeply lobed. This change was necessary because some of the cells of Bloom's syndrome, and of the other conditions studied, showed typical micronuclei in higher numbers than we had observed previously, and these occurred in cells that were clearly not polymorphs.

Results and Discussion

We began to study micronuclei in human lymphocytes for two reasons: first, the cells are initially synchronous (Bender and Prescott 1962) so that the potential difficulty of not knowing in what cell stage the chromosome breakage occurred is obviated and, second, we wanted to use this technique for studies on the human chromosome breakage syndromes. Since micronuclei arise primarily from acentric chromosomal fragments that are lost from the daughter nuclei at anaphase, it is of interest to know what fraction of fragments become micronuclei at each cell division. Presumably this fraction will vary depending upon the cell type because of interpolar distance, the extent of cytoplasmic flow, fragment size and possibly other factors. An indirect estimate of the rate of fragment loss was made by Carrano and Heddle (1973) using the data of Sasaki and Norman (1966). Their calculations suggested that only 20% of X-ray-induced (G_1) chromosome fragments would be lost per cell division in cultured human lymphocytes.

Somewhat similar figures (10–20%) have been obtained experimentally for Chinese hamster ovary cells containing X-ray-induced chromatid aberrations (Heddle, unpublished results). These results imply that fragment loss is surprisingly rare and that most fragments will be incorporated into a daughter nucleus. In human lymphocytes, those that are incorporated into a daughter nucleus will be replicated, so that each will be present in duplicate at the next mitosis (Sasaki and Norman 1966). Those cells that lose micronuclei will eventually be lost from the mitotic population, but may persist for at least one cycle (Carrano and Heddle 1973). As a consequence, the frequency of micronuclei begins to rise in a human lymphocyte culture at the onset of mitosis (about 44 h, Heddle, Evans and Scott 1967) and continues to rise for two or three cell cycles (figure 1). In our original experiments the peak always occurred at 96 h or later, but in more recent experiments it has occurred as early as 84 h. All of the lymphocyte data reported here represent 96-h cultures.

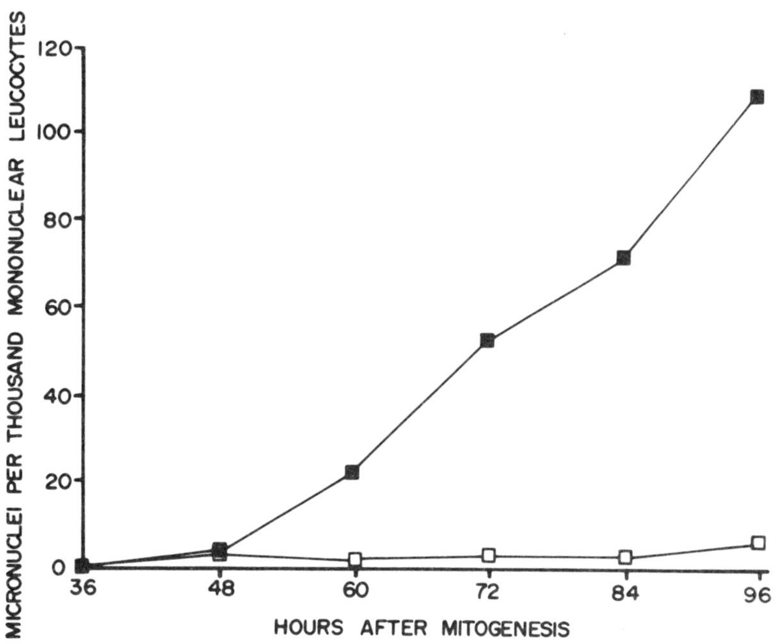

Figure 1. Increase in micronuclei with time in normal lymphocyte cultures, mean of five diploid donors. Solid symbols, cultures treated with 400 R 150 kVp X-rays; open symbols, unirradiated cultures. Data redrawn from Countryman and Heddle (1976).

The first experiments in lymphocytes were designed to test the assay to see if the results were consistent with what is known about chromosomal breakage in these cultures. The data in figure 2 show several aspects of the assay. First of all it is evident that the reproducibility is quite good from blood donor to blood donor. The normal diploid results were obtained with

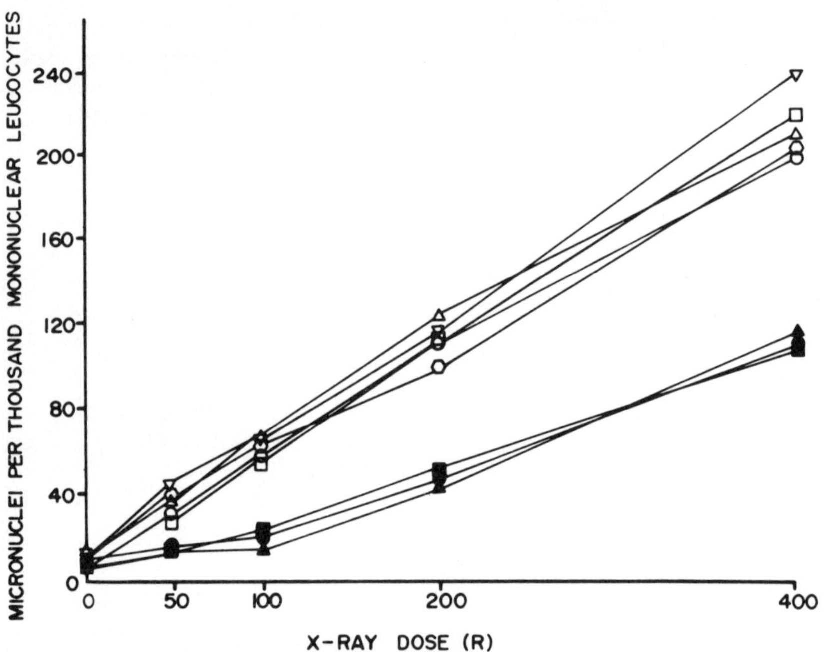

Figure 2. Dose response of micronuclei at 96 hours in cultures irradiated at the time of initiation for five Down's Syndrome (open symbols) and three normal diploid donors (solid symbols). Data redrawn from Countryman *et al.* (1977).

standard 5 ml cultures, and the microculture technique gives the same mean results at the expense of somewhat greater variability. Secondly, it is evident that the known two-fold difference in sensitivity between trisomy-21 and diploid lymphocytes is easily detected. The only discrepancy suggested by these results is that the dose-response curve is not of the form that one might expect. It is known that about 90% of X-ray-induced fragments are associated with dicentrics and rings in G_1 human lymphocytes and that these are two-hit aberrations with a dose-squared response (e.g. Sasaki and Tonomura 1969). The micronuclei, however, increase only as $D^{1.2}$. The reasons for this are, we believe, that fewer aberrations are expressed as micronuclei at higher doses primarily because fewer cell divisions occur. To test this possibility we fractionated the X-ray dose into two equal parts. In such an experiment, if the fractions are separated long enough for chromosome rejoining to occur before the second exposure, the frequency of two-hit aberrations will be halved (Wolff and Luippold 1955). If at the same time dose fractionation has no effect on interphase cell death or mitotic delay, the frequency of micronuclei should also be reduced to one half. As figure 3 shows, this is the case. It is of interest that in trisomy-21 cells the rejoining time measured by this technique is substantially less than in normal cells although the trisomic cells have more aberrations (Countryman, Heddle and Crawford 1977).

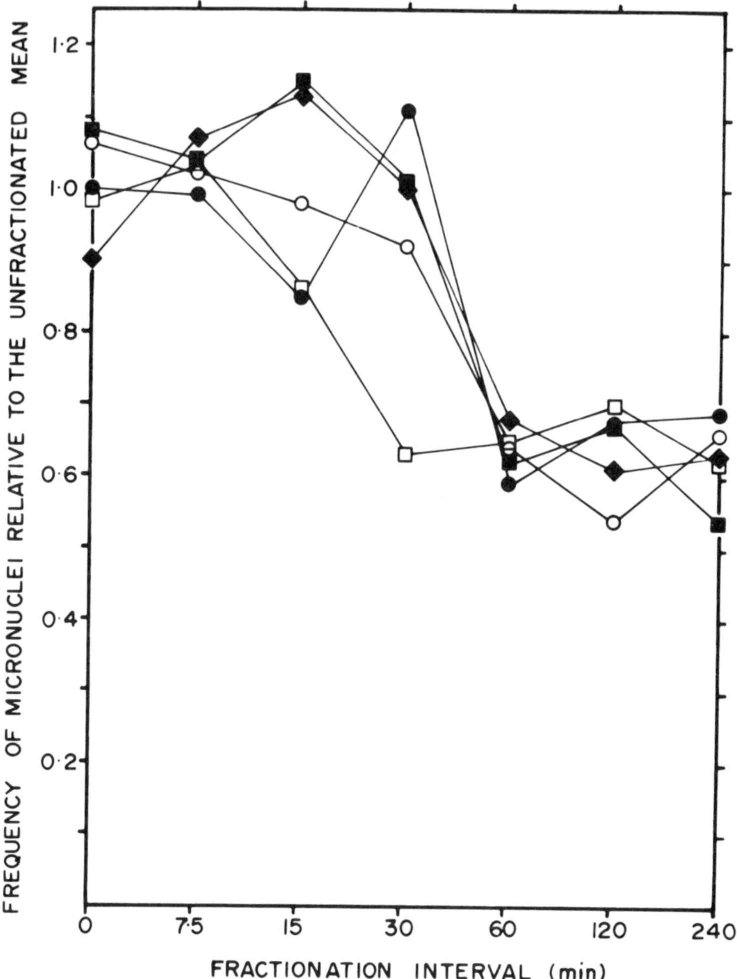

Figure 3. The result of giving 400 R in two 200 R fractions separated by various intervals of time is shown. Each symbol represents a different normal blood donor. In each case the data are expressed as a proportion of the mean response (of all five people) obtained with the unfractionated dose. Data redrawn from Countryman *et al.* (1977).

Having verified the micronucleus assay to our satisfaction, we then turned to the chromosome breakage syndromes (Bloom's syndrome, Fanconi's anaemia and ataxia telangiectasia). Figure 4 shows that the spontaneous frequency of micronuclei is elevated in these conditions. These data show that the frequency of spontaneous micronuclei in controls (presumably +/+) and in heterozygotes (bl/+ and fa/+) are similar, but that Fanconi's cells (fa/fa) and Bloom's cells (bl/bl) each have a more or less characteristic spontaneous frequency. The one heterozygote (bl/+) with an elevated frequency may represent a case of environmental or medical exposure to a

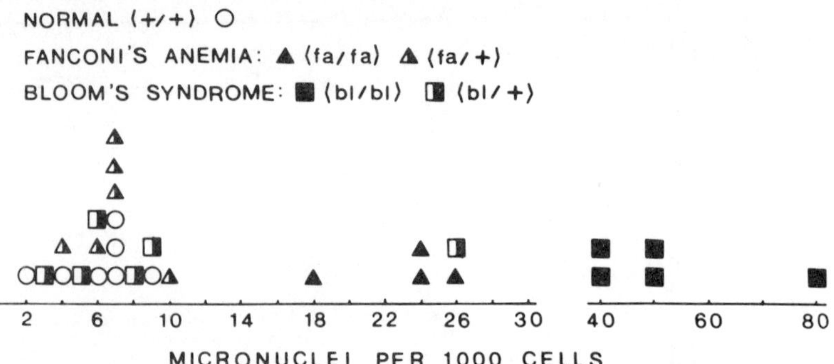

Figure 4. Histogram of the spontaneous micronucleus frequencies in lymphocytes from control (+/+), Fanconi's anaemia (fa/fa) and Bloom's syndrome (bl/bl) donors and their heterozygotes. Each symbol represents one person shown at the mean of all the untreated cultures of that person.

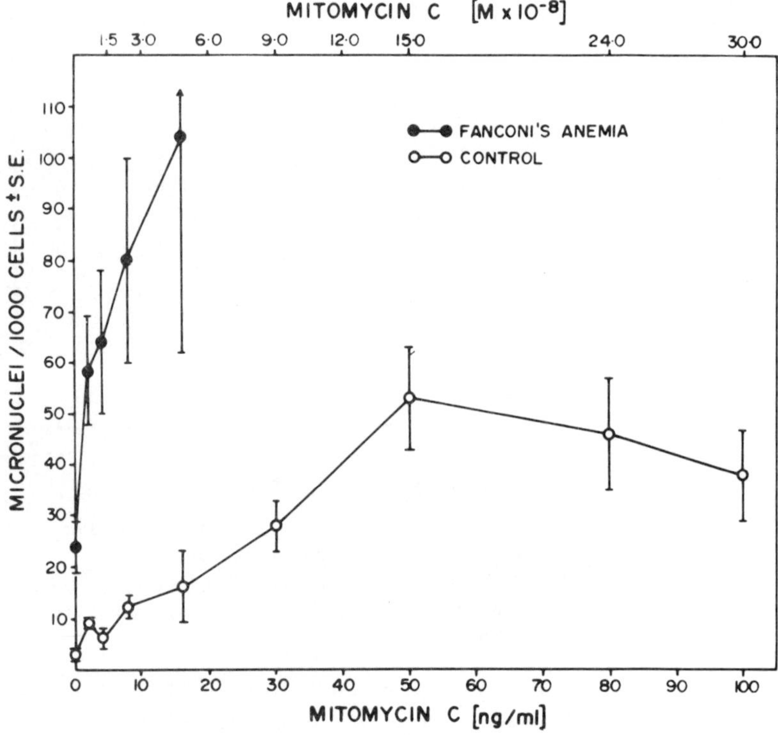

Figure 5. A comparison of the response of lymphocytes of four Fanconi's anaemia and four control donors to mitomycin C added to the culture upon initiation to give the concentrations shown. Not all donors are represented at all points, but no point with fewer than three is included. The large error bars on the Fanconi points are not the result of a large variation within an experiment but rather because each patient followed a different curve. It is not possible for us to know whether this is a difference between individuals or experiments at the moment, although preliminary data suggest that both factors are involved.

mutagen. Under the working hypothesis that these conditions, which show elevated risk of cancer as well as aberrations, are repair-deficient genotypes like xeroderma pigmentosum (Cleaver 1968), we expected that they would show an increased sensitivity to some mutagenic agent. This expectation has already been partly fulfilled. In cells of Fanconi's anaemia, Sasaki and Tonomura (1973) have shown that DNA crosslinking agents such as mitomycin C are much more effective at breaking chromosomes than in normal cells. We have confirmed this result with the micronucleus technique, as shown in figure 5. This result will, we hope, prove useful in completing the diagnosis of Fanconi's anaemia in any doubtful cases. Similarly, the increased sensitivity of ataxia telangiectasia fibroblast cells to X-rays (Taylor *et al.* 1975) is readily detectable (figure 6).

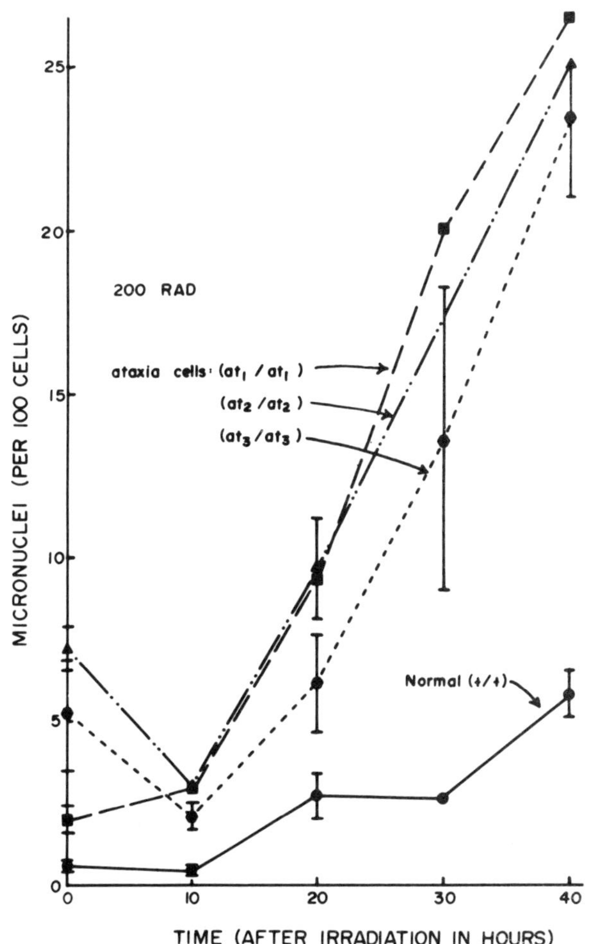

Figure 6. Increase in micronuclei with time in three ataxia and one control fibroblast line after anoxic γ-irradiation.

These results support our contention that, for a variety of studies in which the extent of chromosomal breakage, rather than the type, is of interest, the micronucleus assay provides a rapid and simple means by which to make the measurements. In situations in which chromosomal damage is responsible for much of the induced cell death, there should be a close relation between cell survival and micronuclei. In a recent study of the effects of psoralen and angelicin on Chinese hamster fibroblasts, it was found that the survival of cells was directly related to the micronucleus frequency, as shown in figure 7, although greatly different doses or treatments were used (Ashwood-Smith *et al.* 1977). This is to be expected if fragment loss (and thus micronucleus production) is the most common lethal event in mitotic cell populations in the 10–100% survival range.

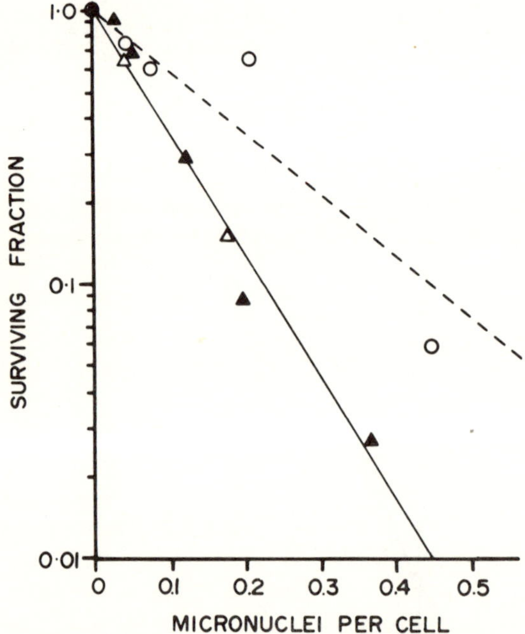

Figure 7. Relationship between survival (colony-forming ability) in CHO cells and the frequency of micronuclei induced by various mutagenic treatments. Open circles, X-irradiated cells; solid triangles, cells treated with psoralen + light; open triangles, cells treated with angelicin + light. Data redrawn from Ashwood-Smith *et al.* (1977).

Having indicated some of the successes of the micronucleus assay, we should also point out some of its faults. Two of these have already been mentioned, namely that one cannot tell from the appearance of a micronucleus what type of aberration gave rise to it, nor at what cell stage the damage was done. The great disadvantage, however, is that even if the aberration frequency is constant, the frequency of micronuclei will vary with the rate of cell division and the proportion of cells that divide. Thus no increase in the number of micronuclei could mean either that no chromosome breakage occurred or that the cells were killed and no cell division occurred. Any interpretation of frequencies of micronuclei must take both factors into account. One way to be sure that cell division has occurred (so that the absence of micronuclei shows that no aberrations were produced) is to irradiate the treated cultures with 100 R. This dose produces an easily detectable increase in the frequency of micronuclei if the cells are dividing normally. If this combined treatment should produce no micronuclei, the cells did not divide and the occurrence of aberrations is uncertain. In most cases we have found that the appearance of the interphase cells and the presence or absence of a few mitotic cells, even in the absence of colchicine, are reliable indicators of whether or not the cells are dividing.

In summary we have found this assay to be a useful measure of chromosomal breakage provided that its limitations are recognised. We intend to use it for further studies of the chromosomal breakage syndromes, and we hope that others will find it useful too.

Summary

The results of a series of experiments in which micronuclei were used to measure the extent of chromosomal damage are summarised. The data show that in most situations micronuclei accurately reflect chromosomal breakage and that they may be used for rapid and simple estimates of aberration frequency. The results of some studies on trisomy-21, Fanconi's anaemia, Bloom's syndrome and ataxia telangiectasia are included; the advantages and disadvantages of the technique are discussed.

Acknowledgements

We thank J. Burgess, B. Lue, J. Gingerich, and S. Noad for conducting many of the experiments; J. German, M. C. Paterson and F. Saunders for their cooperation and permission to use unpublished results on Bloom's, ataxia, and Fanconi's cells; and the many patients who have given blood samples. This work would not have been possible without the financial support of the Atkinson Charitable Foundation, the National Cancer Institute of Canada, and the National Research Council of Canada.

References

Ashwood-Smith, M.J., E.L.Grant, J.A.Heddle & G.B.Friedman (1977) Chromosome damage in Chinese hamster cells sensitized to near-ultraviolet light by psoralen and angelicin. *Mutat. Res. 43*, 377–86.

Bender, M.A. & D.M.Prescott (1962) DNA synthesis and mitosis in cultures of human peripheral leukocytes. *Exp. Cell Res. 27*, 221–9.

Carrano, A.V. & J.A.Heddle (1973) The fate of chromosome aberrations. *J. Theoret. Biol. 38*, 289–304.

Cleaver, J.E. (1968) Defective repair replication of DNA in xeroderma pigmentosum. *Nature 218*, 652–6.

Countryman, P.I. & J.A.Heddle (1976) The production of micronuclei from chromosome aberrations in irradiated cultures of human lymphocytes. *Mutat. Res. 41*, 321–32.

——(1977) A true microculture technique for human lymphocytes. *Human Genet. 35*, 197–200.

Countryman, P.I., J.A.Heddle & E.Crawford (1977) The repair of X-ray-induced chromosomal damage in trisomy 21 and normal diploid lymphocytes. *Cancer Res. 37*, 52–8.

Evans, H.J., G.J.Neary & F.S.Williamson (1959) The relative biological efficiency of single doses of fast neutrons and gamma rays on *Vicia faba* roots and the effect of oxygen. II. Chromosome damage: the production of micronuclei. *Int. J. Radiat. Biol. 1*, 216–29.

Fabergé, A.C. (1940) An experiment on chromosome fragmentation in *Tradescantia* by X-rays. *J. Genet. 39*, 229–48.

Heddle, J.A. (1973) A rapid *in vivo* test for chromosomal damage. *Mutat. Res. 18*, 187–90.

Heddle, J.A., H.J.Evans & D.Scott (1967) Sampling time and the complexity of the human leukocyte culture system, in *Human Radiation Cytogenetics* (eds H.J. Evans, W.M.Court Brown and A.S.McLean) pp.6–19. Amsterdam: North-Holland.

Matter, B. & W.Schmid (1971) Trenimon-induced chromosomal damage in bone marrow of six mammalian species, evaluated by the micronucleus test. *Mutat. Res. 12*, 417-25.

Sasaki, M.S. & A.Norman (1966) Proliferation of human lymphocytes in culture. *Nature 210*, 913–14.

Sasaki, M.S. & A.Tonomura (1969) Chromosomal radiosensitivity in Down's syndrome. *Jpn. J. Human Genet. 14*, 81–92.

——(1973) A high susceptibility of Fanconi's anemia to chromosome breakage by DNA cross-linking agents. *Cancer Res. 33*, 1829–36.

Schmid, W. (1975) The micronucleus test. *Mutat. Res. 31*, 9–15.

Taylor, A.M.R., D.G.Harnden, C.F.Arlett, S.A.Harcourt, A.R.Lehmann, S.Stevens & B.A.Bridges (1975) Ataxia telangiectasia: a human mutation with abnormal radiation sensitivity. *Nature 258*, 427–9.

Wolff, S. & H.E.Luippold (1955) Metabolism and chromosome-break rejoining. *Science 122*, 231–2.

P.E.PERRY, M.JAGER and H.J.EVANS

Mutagen-Induced Sister Chromatid Exchanges in Xeroderma Pigmentosum and Normal Lymphocytes

Using the recent methods for studying sister chromatid exchanges (SCES) based on incorporation of BUdR into chromosomal DNA (Latt 1973, Perry and Wolff 1974), much is already known about the effect of various chemical agents upon the frequency of these events in cultured mammalian cells (Perry and Evans 1975) and in cells exposed *in vivo* (Allan and Latt 1976). Moreover, it has now been realised that SCEs also give an important insight into the mechanism of DNA repair in the mammalian cell following chemical damage, and some recent work has examined the induction of SCEs by chemicals (Latt 1975; Wolff, Rodin and Cleaver 1977) and by UV irradiation (de Weerd-Kastelein, Keijer and Rainaldi 1976; Schönwald and Passarge 1977) in cells of patients suffering from syndromes associated with defective DNA repair capacity. One such syndrome is xeroderma pigmentosum (XP), which is characterised by sensitivity of areas of skin exposed to sunlight leading to pigmentation and susceptibility to skin cancer. The disease is usually characterised by an abnormally low level of unscheduled DNA synthesis (UDS) (Cleaver 1968), and an abnormally high level of chromosome aberrations (Parrington, Delhanty and Baden 1971), in cells exposed to 254 nm UV irradiation. And it has been shown that the cells are defective in the initial step of excision repair, which may involve a UV-specific endonuclease (Setlow *et al.* 1969).

Recent experiments by Schönwald and Passarge (1977) and de Weerd-Kastelein, Keijer and Rainaldi (1976) have demonstrated that the SCE response in typical XP cells differ from that in normal cells following exposure to 254 nm UV irradiation, and the observed increased frequency of SCE correlates with the ability of thymine dimers to give rise to SCE (Kato 1977) and the well-known inability of XP cells to repair such dimers.

Studies on the induction of DNA damage and repair by chemical mutagens have shown that chemical agents can be loosely classified into two types on the basis of their repair response, being either 'UV-like' or 'X-ray-like' (Cleaver 1971; Regan and Setlow 1974; Stich, San and Kawazoe 1973). Damage caused by UV-like chemicals, such as 4-nitroquinoline-1-oxide (4NQO), is not repaired as efficiently in XP cells as in normal cells (Regan and Setlow 1974, Sasaki 1973), and this is associated with a decreased level of UDS (Stich and San 1971) and an increased frequency of chromosome aberrations with this mutagen (Stich, Stich and San 1973). Damage due to X-ray-like chemicals, however, such as ethyl methanesulphonate (EMS) or nitrogen mustard (HN2), can be handled by the XP excision repair system (Cleaver 1971) resulting in a similar level of UDS and of chromosome aberrations to normal cells (Sasaki 1973).

Recent work by Wolff, Rodin and Cleaver (1977) has shown that transformed XP fibroblasts give abnormally high SCE yields following exposure to a wide range of chemical mutagens, including both UV- and X-ray-like agents, therefore substantiating the already considerable evidence that the mechanism of formation of SCE cannot be directly equated with the processes involved in UDS or in the development of chromosome aberrations.

This report deals with the induction of SCE in XP cells by UV irradiation and by chemical agents that induce a range of types of damage to DNA. The chemicals used were 4NQO, which elicits increased chromosome aberration production and decreased UDS in XP cells; EMS, a monofunctional alkylating agent that produces monoadducts; and HN2, a bifunctional alkylating agent that causes inter- and intra-strand cross links as well as monoadducts. The cells used were cultured peripheral blood lymphocytes, and, to gain some additional insight into the nature of the repair defect in XP cells, cells were treated at either G_0 (before stimulation with PHA) or at 26–29 h post-stimulation, at which time the first wave of stimulated lymphocytes were in their first S phase in culture.

Materials and Methods

Clinical Status of Patient

The patient was a 7-yr-old Arabian girl, who was noticed at the age of 18 months to be developing spots on her cheeks. These gradually increased and the skin colour has become darker than that of her relatives. At the age of two, spots were noticed on the hands and face, and more recently there has been involvement of the lips and trunk and the eyes do not tolerate light well. Her intelligence is normal, she is rather short in stature, light in weight, shows multiple freckles on her face, with rather less on her arms and trunk, and her skin shows prolonged erythema on exposure to UV light. Her 2-yr-old brother is also showing symptoms of the disease, and there is some evidence for consanguinity between the parents, who are believed to be second cousins. The nature of the repair deficiency has not yet been further characterised in this patient.

As a control, blood was obtained from a normal 22-yr-old female.

Cell Culture

The cells were cultured in McCartney bottles containing 5 ml RPMI 1640 medium with 15% fetal calf serum, 0.8% PHA, penicillin and streptomycin (100 IU ml^{-1}) and 25 μM BUdR, at 37°C in a light-proof box (to prevent photolysis of BUdR containing DNA) for 72 or 96 h. Colcemid (final concentration 5×10^{-7}M) was added to the culture medium 2.5 h prior to harvest to arrest the cells in mitosis.

UV Exposure

A Hanovia low-pressure Hg vapour ultraviolet source that emits UV predominantly at 254 nm was used. Because of the low penetrance of UV light, the cells were exposed in a thin layer: 0.4 ml aliquots of whole blood (G_0 stage), or 0.4 ml aliquots of whole blood cultured for 26 h (S phase), spun down and made up to 0.4 ml with colourless phosphate buffered saline, were spread in a thin layer approximately 7 cm in diameter over the centre of 8.5 cm diameter petri dishes, using a rubber policeman. The exposures used ranged from 1 s at 45 cm distance to 3, 10, 30 and 100 s at 30 cm distance.

Chemical Exposure

As with UV exposure, cells were exposed to chemical mutagens either at the time of stimulation and addition of BUdR (G_0 exposure), or after 26 h in BUdR-containing medium (mainly S exposure). The chemicals were diluted and dispensed using sterile automatic pipettes. 4NQO was dissolved initially in ethanol and further dilutions were made in water. HN2 was dissolved in water and EMS is a liquid miscible with water. Further dilutions of these compounds were also made in water. The concentrations of the chemicals in the dilution series were arranged such that a 1 in 100 dilution in medium gave the required final molarity. The final concentrations of 4NQO and HN2 were 10^{-5}, 10^{-6}, 10^{-7} and 10^{-8}M; those of EMS were 10^{-3}, 10^{-4}, 10^{-5} and 10^{-6}M.

Slide Preparation and Staining

The cells were treated with 0.075 M KCl for 7 min, fixed in 3:1 methanol:acetic acid, followed by two changes of fixative, and chromosome preparations made by placing drops of cells suspended in fixative onto clean microslides.

To stain the sister chromatids differentially, the FPG technique of Perry and Wolff (1974) was used. The slides were stained in 0.5 μg ml^{-1} Hoechst 33258 fluorescent stain for 15 min, rinsed, dried and mounted in distilled water. The coverslips were ringed with 'Holdtite' rubber cement to prevent evaporation and the slides exposed to the fluorescent light from an X-ray viewing box for 5–6 h, after which time the coverslips were removed and the slides incubated in 2× SSC at 60°C for 30 min. The slides were then rinsed and stained in 4% Giemsa in pH 6.8 Sörensens buffer, rinsed, and mounted in DPX. The slides were coded, randomised and scored for SCE by a single observer. Where possible 20 cells were scored for each culture.

Results and Discussion

The results of the UV experiments are summarised in table 1. It can be seen that the control SCE levels in XP cells are somewhat higher than in the normal cells (see also figure 1 control values) and a similar result is evident in some published data (Wolff, Rodin and Cleaver 1977), but not in others (Schönwald and Passarge 1977, Wolff *et al.* 1975) in which no differences were found between XP, XP variant and normal cells. However, our finding of a consistent small increase in SCE in XP vs normal cells was evident in all our experiments and in all our repeat samples studied. The presence of BUdR itself

Table 1. Induction of SCE in normal (N) and XP cells following UV irradiation at G_0 or S phases of the cell cycle.

Duration of UV exposure (s)	Distance of cells from source (cm)	G_0 irradiation		S irradiation	
		N	XP	N	XP
0	—	9.85	—	10.25	16.9
1	45	11.5	—	11.85	19.4
3	30	9.5	34.15	12.7	35.2
10	30	15.0	—	15.45	27.8
30	30	9.7	—	—	21.95
100	30	—	—	—	27.4

results in the formation of SCE, and if our findings are substantiated then they may reflect the fact that XP cells may be less efficient than normal cells at dealing with the type of DNA damage caused by BUdR substitution.

UV-irradiation of control (normal) cells gives a slightly higher yield of SCE if conducted during S than during G_0, and it is tempting to speculate that the long G_1 stage of these cells offers sufficient time for a proportion, at least, of the UV-induced lesions, which would otherwise give rise to SCE during S, to be repaired by the normal excision repair mechanism. Regan and Setlow (1974) have found that a substantial proportion of extensive damage caused by UV irradiation is repaired in 18–20 h in normal cells.

No SCE results were obtained for most of the 100 s UV exposures because few metaphases were seen, and none of these were in the second division following stimulation, showing that the UV treatment had induced considerable delay in cell progression. Also, although many metaphases were to be seen in the G_0-irradiated XP cells, these cells all exhibited staining characteristic of cells that had passed through only one DNA S phase in the presence of BUdR, indicating also that some delay in the cell cycle had occurred. Presumably the G_0-irradiated XP cells were less able to excise the UV-induced dimers and this unexcised damage caused a delay during the first DNA S phase. Irradiation of XP cells during the S phase, however, may have resulted in many of the dimers forming in post-replication DNA, which did not prevent progression through the first S phase. The same level of SCE is induced in XP cells by a 3 s exposure in G_0 and S cells (34.15 and 35.2 respectively), implying equivalence in the amounts of residual damage and indicating that the dimers induced in G_0 were still present at the following S phase, at which time SCE formation occurs (Wolff, Bodycote and Painter 1974).

Figure 1 shows the induction of SCE in XP and normal cells by various doses of chemicals applied 26 h after initiation of the cultures. Only at the highest concentration of each chemical was there any increase in SCE frequency in the normal cells, whereas the XP cells showed an increased yield for all chemicals at all concentrations, except the lowest dose of EMS. The greater response of XP cells relative to normal cells to 4NQO treatment is to be ex-

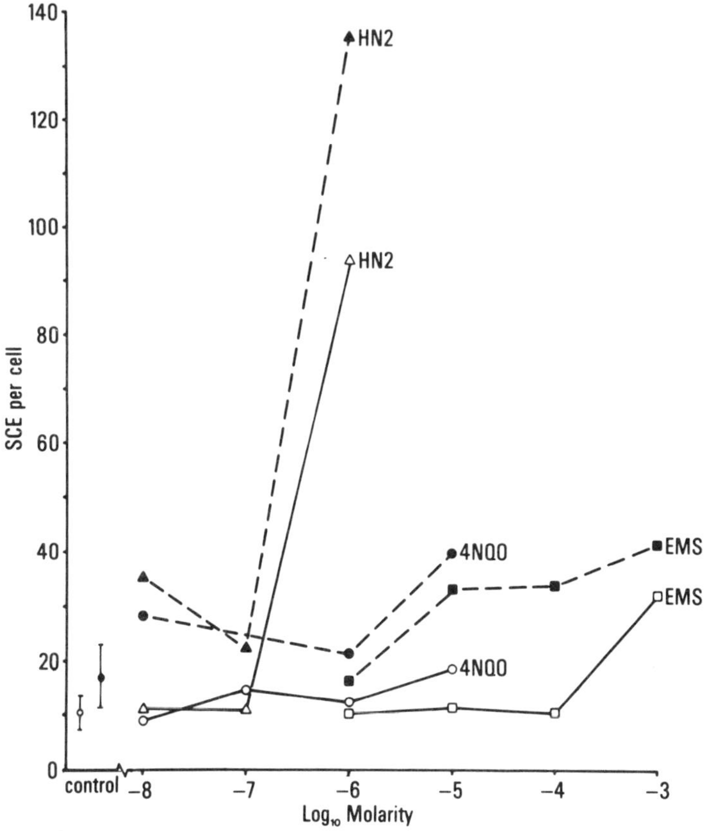

Figure 1. Sister chromatid exchange induction in normal (solid lines) and xeroderma pigmentosum (dashed lines) peripheral blood lymphocytes exposed to nitrogen mustard (HN2), 4 nitroquinoline-1-oxide (4NQO) and ethylmethanesulphonate (EMS) from 26 h after initiation of culture until harvest at 72 h.

pected, as this chemical has been classified as UV-like in that the induced damage is repaired by the same mechanism as UV-induced damage. On the other hand, previous work has shown that damage due to HN2 (inter- and intra-strand cross links and monoadducts) and EMS (monoadducts) can be handled normally by XP cells when the end points measured are levels of UDS and frequency of chromosome aberrations. However, the data presented in figure 1 substantiate the recent work of Wolff, Rodin and Cleaver (1977) on SCE induction, which indicates that XP cells cannot handle damage by these chemicals as efficiently as normal cells. There is no direct evidence that the type of lesions induced by HN2 or EMS can be handled by the UV dimer excision repair system, and both these agents elicit normal responses to aberration induction and UDS in XP cells (see above). It therefore seems that this XP patient and others (Wolff, Rodin and Cleaver 1977), may be deficient not only in UV-like excision repair, but also in possible repair processes that enable the cell to remove or bypass other types of single-strand damage and interstrand cross links in DNA.

Clearly the DNA repair deficiency in this XP patient, which is not simply defective repair of UV-induced dimers but also of chemically-induced lesions, suggests that SCE induction provides a highly sensitive technique capable of resolving the presence of a variety of types of damage in DNA.

Summary

The induction of sister chromatid exchanges (SCE), by ultra-violet irradiation and by three chemical mutagens that differ in the type of repair response that they elicit, has been compared in lymphocytes from a control and from an individual suffering from the DNA excision repair deficiency syndrome, xeroderma pigmentosum (XP). The XP lymphocytes were found to be more sensitive in terms of SCE response, not only to UV irradiation, but also to all of the chemicals studied. The results indicate that the abnormality of DNA repair in this XP patient is expressed not only in the defective excision of thymine dimers, or other UV photoproducts, but also in a reduced ability to repair other types of DNA lesion.

Acknowledgement

We wish to thank Dr J. A. Savin for access to the patient and for providing clinical details.

References

Allan, J.W. & S.A.Latt (1976) Analysis of sister chromatid exchange formation *in vivo* in mouse spermatogonia as a new test system for environmental mutagens. *Nature 260*, 499–51.

Cleaver, J.E. (1968) Defective repair replication of DNA in Xeroderma pigmentosum. *Nature 218*, 652–6.

——(1971) Repair of alkylation damage in ultraviolet sensitive (Xeroderma pigmentosum) human cells. *Mutat. Res. 12*, 453–62.

de Weerd-Kastelein, E.A., W.Keijer & P.Rainaldi (1976) Induction of sister chromatid exchanges in Xeroderma pigmentosum cells following UV exposure. *2nd Int. Workshop on Repair Mechanism in Mammalian Cells*, Noordwijkerhaut, Holland, May 2–6.

Kato, H. (1977) Mechanisms for SCEs and their relation to the production of chromosomal aberrations. *Chromosoma 59*, 179–92.

Latt, S.A. (1973) Microfluorometric detection of deoxyribonucleic acid replication in human metaphase chromosomes. *Proc. Natl. Acad. Sci. U.S.A. 70*, 3395–9.

——(1975) Induction by alkylating agents of sister chromatid exchanges and chromatid breaks in Fanconi's Anemia. *Proc. Natl. Acad. Sci. U.S.A. 72*, 4066–70.

Parrington, J.M., J.D.A.Delhanty & H.P.Baden (1971) Unscheduled DNA synthesis, UV induced chromosome aberrations and $S.V._{40}$ transformation in cultured cells from Xeroderma pigmentosum. *Ann. Hum. Genet. Lond. 35*, 149–60.

Perry, P. & H.J.Evans (1975) Cytological detection of mutagen-carcinogen exposure by sister chromatid exchange. *Nature 258*, 121–5.

Perry, P. & S.Wolff (1974) New Giemsa method for the differential staining of sister chromatids. *Nature 251*, 156–8.

Regan, J.D. & R.B.Setlow (1974) Two forms of repair in the DNA of human cells damaged by chemical carcinogens and mutagens. *Cancer Res. 34*, 3318–25.

Sasaki, M.S. (1973) DNA repair capacity and susceptibility to chromosome breakage in Xeroderma pigmentosum cells, *Mutat. Res. 20*, 291–3.

Schönwald, A.D. & E.Passarge (1977) U.V. light induced sister chromatid exchange in Xeroderma pigmentosum lymphocytes. *Hum. Genet. 36*, 213–18.

Setlow, R.B., J.D.Regan, J.German & W.L.Carrier (1969) Evidence that Xeroderma pigmentosum cells do not perform the first steps in the repair of UV damage to their DNA. *Proc. Natl. Acad. Sci. U.S.A. 64*, 1035–41.

Sitch, H.F. & R.H.C.San (1971) Reduced DNA repair synthesis in Xeroderma pigmentosum cells exposed to the oncogenic 4-nitroquinoline 1-oxide and 4-hydroxy-aminoquinoline 1-oxide. *Mutat. Res. 13*, 279–82.

Stich, H.F., W.Stich & R.H.C.San (1973) Elevated frequency of chromosome aberrations in repair deficient human cells exposed to the carcinogen and mutagen 4-nitroquinoline 1-oxide. *Proc. Soc. Exp. Biol. Med. 142*, 1141–4.

Stich, H.F., R.H.C.San & Y.Kawazoe (1973) Increased sensitivity of Xeroderma pigmentosum cells to some chemical carcinogens and mutagens. *Mutat. Res. 17*, 127–37.

Wolff, S., J.Bodycote & R.B.Painter (1974) Sister chromatid exchanges induced in Chinese hamster cells by UV irradiation of different stages of the cell cycles: The necessity of cells to pass through S. *Mutat. Res. 25*, 73–81.

Wolff, S., J.Bodycote, G.H.Thomas & J.E.Cleaver (1975) Sister chromatid exchanges in Xeroderma pigmentosum cells that are defective in DNA excision repair or post replication repair. *Genetics 81*, 349–55.

Wolff, S., B.Rodin & J.E.Cleaver (1977) Sister chromatid exchanges induced by mutagenic carcinogens in normal and Xeroderma pigmentosum cells. *Nature 265*, 347–9.

Chromosomal Effects of Mutagenic Carcinogens and the Nature of the Lesions leading to Sister Chromatid Exchange

The first of the known chemical mutagens, nitrogen mustard, is an alkylating agent capable of forming lesions that can lead to mutations several cell generations after treatment (Auerbach 1976). Other mutagenic carcinogens, as well as ultraviolet light, share this property, which can be noted not only for mutations, but also for chromosome aberrations (Evans and Scott 1969, Ikushima and Wolff 1974). These chemicals also induce large numbers of sister chromatid exchanges (SCEs) at doses far lower than those required to produce measurable increases in ordinary chromosome aberrations (Latt 1974; Kato and Shimada 1975; Solomon and Bobrow 1975; Perry and Evans 1975; Stetka and Wolff 1976; Wolff, Rodin, and Cleaver 1977). The ease of scoring exchanges and the sensitivity of the SCE system have led to its adoption in many laboratories as the most sensitive mammalian system for the detection of a potentially mutagenic, and thus potentially carcinogenic, effect of a chemical. Numerous papers have now been published in this area (see Wolff 1977 for review).

Although the chemical lesions that lead to SCE formation are still unknown, several experiments have been carried out that give some clues to their nature.

Long-Lived Lesions

That the lesions leading to UV-induced SCEs are long-lived was shown in studies by Wolff, Bodycote and Painter (1974) in which Chinese hamster cells were irradiated at different times in the cell cycle. The results showed that the cells had to pass through S before an SCE could be formed. If the cells were irradiated at G_2 no induced SCEs were found at the first subsequent metaphase, but if the cells were allowed to progress further and pass through an S phase, an increased number of SCEs could be found at the second metaphase after irradiation.

X-rays, in contrast, do not induce SCEs very efficiently and do not seem to form this type of long-lived lesion in DNA (Wolff, Bodycote and Painter 1974; Perry and Evans 1975). On the other hand, chemical mutagens do. Table 1 presents the results of an experiment in which we treated Chinese

Table 1. Sister chromatid exchanges in Chinese hamster ovary cells: SCEs per large acrocentric chromosome in first and second cell cycles after treatment with chemicals.

Treatment 30 min in G_1		1st cycle (twins) No. of SCE/ chromosome pairs	SCE per pair	2nd cycle (singles) No. of SCE/ chromosomes	SCE per chromosome	Ratio of singles to twins in 4*n* cells
Control		69/283	0.24	132/566	0.23	1.9
EMS	10^{-4} M	109/281	0.39	192/562	0.34	1.8
MMS	10^{-5} M	92/284	0.32	183/578	0.32	2.0
4NQO	10^{-8} M	126/288	0.44	184/576	0.32	1.5
Control		29/ 98	0.30	52/196	0.27	1.8
AAAF	10^{-8} M	26/ 97	0.27	69/194	0.36	2.7
AAAF	5×10^{-8} M	41/ 95	0.43	74/190	0.39	1.8
AAAF	10^{-7} M	55/ 96	0.57	108/192	0.56	2.0

hamster ovary (CHO) cells in culture with low concentrations of ethanemethane sulphonate (EMS), methylmethane sulphonate (MMS), 4-nitroquinoline-1-oxide (4NQO), or acetoxyacetylaminofluorene (AAAF). Synchronised cells in G_1 were treated with the chemicals for 30 min and the cells were washed and recultured in McCoy's 5A medium supplemented with 15% fetal calf serum and 10 μM bromodeoxyuridine (BUdR). The medium also contained 10^{-7} M colcemid to inhibit cell division. 24–26 h later, when the cells had replicated twice in the presence of BUdR, they were collected by centrifugation, exposed to hypotonic KCl solution, and fixed in 3:1 methyl alcohol: acetic acid as reported in previous publications (Perry and Wolff 1974; Wolff and Perry 1974, 1975). Air-dried preparations were made and the slides stained by the FPG or harlequin chromosome technique of Wolff and Perry (1975). The technique was slightly modified, however, in that after staining in Hoechst 33258 the slides had cover-glasses mounted with M/15 Sorensen's buffer, pH 6.8, and were exposed for approximately 45 s to light from a 100 W high-pressure mercury burner at 625 J m^{-2} s^{-1}. The slides were then placed in 10 × SSC at 62°C for 20 min and washed and stained in 3% Giemsa.

Those cells that were made tetraploid by the colcemid were scored for twin and single SCEs. If an exchange is induced at the first round of replication after treatment, then, when the cell becomes tetraploid and a chromosome containing the SCE is replicated, the same SCE is in both of the pair of daughter chromosomes and is recorded as a twin exchange (Taylor 1958). If, however, the exchange is produced in the second cell cycle, after the daughter chromosomes are formed, the exchange occurs in only one of the two and will be recorded as a single. In this way one can determine the number of exchanges induced in the first cell cycle and the second cell cycle after treatment. Ordinarily, because there are twice as many chromosomes in the tetraploid cells, there are twice as many singles as there are twins (Taylor 1958).

The results presented in table 1 were obtained from the large acrocentric chromosome in CHO cells. The original CHO cells, which have 21 chromosomes, are monosomic for this chromosome. Therefore in tetraploid cells there are only two such acrocentric chromosomes, which precludes the scoring of false twins that can arise from an incorrect pairing of the four similar chromosomes that would ordinarily be present had such monosomy not occurred. The results show that a 30 min treatment of G_1 cells with either EMS, MMS, 4NQO or AAAF increases the numbers of SCEs produced in the first cell cycle and also increases the number of SCEs produced in the second cell cycle. The ratio of single to twin exchanges remains close to 2 except after treatment with 4NQO, indicating that with EMS, MMS and AAAF no repair of DNA damage occurred between the two cell cycles. With 4NQO the ratio dropped below 2 but remained higher than 1, indicating that partial repair occurred. Even in this case, however, the data show that the lesions induced in G_1 of one cell cycle can persist and give rise to SCEs in the next cell cycle.

If the cells are allowed three rounds of replication in the presence of BUdR, then one-half of the genome will be harlequinised (Wolff and Perry 1974, Dutrillaux *et al.* 1974). Any SCEs that occur in harlequinised regions can be observed, and will represent SCEs that occur at the third cell cycle (Tice, Chaillet and Schneider 1975). Table 2 shows the yields of SCEs found in the third cell cycle. In this case, all the chromosomes could be scored unambiguously, and the number of SCEs per cell were recorded. The numbers have been multiplied by a factor of 2 to normalise them to the SCEs found in the first cell cycle where the total genome can be scored. AAAF, which has a half-life in water of only 7 min (Miller 1970) and gives rise to equal numbers of SCEs in the first and second cell cycle, appears to produce lesions that are repaired by the third cell cycle, since the yield there is equal to the control

Table 2. Sister chromatid exchanges in Chinese hamster ovary cells: SCEs per chromosome in third cell cycle after treatment for 30 min in G_1 with AAAF

Treatment	No. of SCE/ chromosomes	SCE per chromosome[1]
Control	457/2048	0.45
AAAF 10^{-8} M	431/2066	0.42
AAAF 5×10^{-8} M	487/2075	0.47
AAAF 10^{-7} M	512/2053	0.50

[1] Ratio obtained by multiplying quotient from preceding column by two, since only one-half of the genome contains differentially stained chromatids in third cycle cells.

yield. It thus appears that the lesions that give rise to SCEs after AAAF treatment are long-lived lesions that last for two cell cycles but not three.

Nature of the Lesions

Although there is ample evidence that the lesions that lead to the formation of SCEs are long-lived, their exact nature is unknown. When an SCE occurs, both polynucleotide strands of the DNA of one chromatid exchange with the two of its sister (Taylor 1958, Wolff and Perry 1974, Wolff *et al.* 1975). This can be deduced from the ratio of the single to twin exchanges, as well as from the fact that single polynucleotide strand exchanges, which would result in harlequinisation after one round of replication in the presence of BUdR, do not occur (Wolff and Perry 1975, Kihlman and Kronborg 1975). This is not to say, however, that the initial lesions could not reside on but one of the polynucleotide strands, and could then be converted to single-strand breaks that could lead to double-strand breaks (Kato 1977). Nevertheless the involvement of single-strand breaks has not been shown directly, and the fact that X-rays do not induce SCEs very efficiently would indicate that single-strand breaks, or, for that matter, double-strand breaks themselves, are not likely candidates for being in the major pathway to the production of SCEs.

In addition, although thymine dimers are long-lived lesions, and although UV light can cause an increase in SCEs, thymine dimers do not seem to be lesions that can lead to the exchanges. This was first shown by Wolff, Bodycote and Painter (1974) in cultured Chinese hamster cells that had different excision repair capacities. The amount of increase induced by UV light was independent of the repair capacity of the various cell lines. This finding has been dramatically confirmed in experiments with xeroderma pigmentosum (XP) cells by de Weerd-Kastelein *et al.* (1977). They UV-irradiated XP cells that varied in their capacity to carry out unscheduled DNA synthesis. The cells, which came from all five complementation groups, ranged from less than 5% repair to about 50% repair ability. They also used a variant XP cell that was normal in regard to unscheduled DNA synthesis. The results showed that there was no relation between the amount of repair and the sensitivity as manifested by an increased yield of SCEs. One cell line, XP21RO from complementation group C, had a repair capacity 10–15% of normal and was extremely sensitive, whereas another cell line, XP1TE, from the same complementation group with the same repair capacity was only one-third as sensitive. A similar lack of correspondence existed throughout the data.

Although the induction of SCEs seems to be independent of unscheduled DNA synthesis and repair replication, which are both forms of excision repair, it has been postulated that repair might still be involved in their production and, in particular, it has been suggested that post-replication repair phenomena could be involved. The data on this point, however, are quite equivocal. For instance, treatment with caffeine (which is an inhibitor of post-replication repair) has been reported to increase and to decrease (Kato 1977) the induction of SCEs in Chinese hamster cells after fluorescent light irradiation. Also,

XP variant cells, which are defective in post-replication repair, react to UV irradiation as though they were normal cells. Furthermore, in normal cells treatment with caffeine after chemicals can increase the yield of chromosome aberrations but has no affect on SCEs (Kihlman 1975).

As implied by this last result with caffeine, the lesions that result in the formation of SCEs are different from those that lead to the production of ordinary chromosome aberrations. This had previously been shown in experiments with normal and XP cells that were treated with both UV-like and X-ray-like mutagenic carcinogens (Wolff, Rodin and Cleaver 1977). XP cells from the line XP12RO, complementation group A, are defective in the level of excision repair that they can carry out after UV irradiation and after certain UV-like mutagens such as 4NQO. They are, however, able to carry out a normal level of excision repair after treatment with X-rays or various alkylating agents. With these two classes of chemical agents the cells' ability to carry out excision repair is directly correlated with the number of chromosome aberrations induced and the number of cells killed. For instance, after treatment with the UV-like mutagen 4NQO the number of chromosome aberrations (Sasaki 1973; Stich, Stich and San 1973) and the cell killing (Stich, San and Kawazoe 1973) are higher in XP cells than in normal cells, whereas after treatment with an X-ray-like chemical like MNNG the aberrations and killing are as in normal cells. When SCEs were studied, however, it was found that the yields of SCEs were higher in XP cells after treatment with both types of chemicals (Wolff, Rodin and Cleaver 1977).

If the XP12RO cells were defective only in excision repair, the increased yield of SCE found in these cells after treatment with X-ray-like alkylating agents would have to be the result of unexcisable damage in DNA and would thus indicate that a minor unexcised fraction of the total damage could lead to SCEs. This fraction would have to be too small to be resolved by the usual measurements of repair replication or unscheduled synthesis and also too small to give rise to higher than normal amounts of chromosome aberrations or cell killing. Because SCEs are very sensitive indicators of unexcised damage in DNA, however, the chemicals presumably would be able to produce an increase in them.

Recent experiments have raised the question whether or not there is a causal relation between the increased sister chromatid exchanges found in XP cells and unexcised O^6-alkylguanine (Goth-Goldstein 1977). After treatment with many alkylating agents, N^7-alkylguanine is a major alkylation product. For instance, after treatment with methylnitrosourea there is ten times as much N^7-methylguanine formed as O^6-methylguanine in both XP and normal cells. The N^7-alkylguanine is readily excised from both types of cells. Ethylnitrosourea, however, leads to a different ratio of the two forms of alkylated guanine. 40% of the alkylations are at the O^6 position and 60% at the N^7 position, making it possible to study the relative loss of the two types of alkylated products in XP and normal cells. In the normal cells, O^6-alkylguanine was lost from DNA as fast or perhaps even faster than N^7-alkylguanine. In XP cells, however, N^7-alkylguanine was lost much faster than O^6-alkylguanine. This difference in the excision of the two types of alkylated guanine in XP cells raises the question whether there is a causal relation between the unexcised O^6-alkylguanine and the increase in SCE noted in the same cells.

There are additional observations indicating that the lesions resulting in chromosome aberrations differ from those that lead to SCEs: SCE are induced at very high frequencies by concentrations of chemicals that induce few aberrations (Perry and Evans 1975); SCEs are not markedly increased by low doses of ionising radiations, whereas aberrations are (Wolff, Bodycote and Painter 1974; Perry and Evans 1975); SCEs saturate with increasing doses of ionising radiation (Marin and Prescott 1964) and tritiated thymidine (Gibson and Prescott 1972) whereas aberrations do not; SCEs react differently to treatments with caffeine than do chromosome aberrations (Kato 1973, Kihlman *et al.* 1977); and SCE are not correlated in any consistent fashion with the increased aberrations seen in the human diseases Bloom's syndrome, ataxia telangiectasia and Fanconi's anaemia, in that SCEs are high in Bloom's syndrome (Chaganti, Schonberg and German 1974), normal in ataxia telangiectasia (Galloway and Evans 1975), and low in Fanconi's anaemia (Latt *et al.* 1975).

Summary

Sister chromatid experiments to date indicate that sister chromatid exchanges (SCEs) are induced by long-lived lesions in the DNA, which then can lead to SCE formation when the cells pass through S. When SCEs are formed, both strands of the DNA double helix exchange with those in the sister chromatid. After UV treatment, the long-lived lesions do not seem to be thymine dimers, since the production of SCEs is independent of excision repair as measured by unscheduled synthesis or repair replication. Treatment of repair-defective xeroderma pigmentosum cells have indicated that the lesions responsible for SCE formation might be a minor unexcisable fraction of the damage. One possible candidate for this would be alkylation at the O^6 position of guanine, which is ordinarily a minor alkylation product.

Acknowledgement

This work was performed under the auspices of the U.S. Energy Research and Development Administration.

References

Auerbach, C. (1976) *Mutation Research.* London: Chapman and Hall.

Chaganti, R.S.K., S.Schonberg & J.German (1974) A manyfold increase in sister chromatid exchanges in Bloom's syndrome lymphocytes. *Proc. Natl. Acad. Sci. U.S.A. 71*, 4508–12.

de Weerd-Kastelein, E.A., W.Keijzer, G.Rainaldi & D.Bootsma (1977) Induction of sister chromatid exchanges in xeroderma pigmentosum cells following exposure to ultraviolet light. *Mutat. Res.* (in press).

Dutrillaux, B., A.M.Fosse, M.Prieur & J.Lejeune (1974) Analyse des échanges de chromatides dans les cellules somatiques humaines. *Chromosoma 48*, 327–40.

Evans, H.J. & D.Scott (1969) The induction of chromosome aberrations by nitrogen mustard and its dependence on DNA synthesis. *Proc. Roy. Soc. B. 173*, 491–512.

Galloway, S.M. & H.J.Evans (1975) Sister chromatid exchange in human chromosomes from normal individuals and patients with ataxia telangiectasia. *Cytogenet. Cell Genet. 15*, 17–29.

Gibson, D.A. & M.Prescott (1972) Induction of sister chromatid exchanges in chromosomes of rat kangaroo cells by tritium incorporated into DNA. *Exp. Cell Res. 74*, 397–402.

Goth-Goldstein, R. (1977) Reduced excision of O^6-alkylguanine from the DNA of xeroderma pigmentosum-derived fibroblasts after treatment with alkylating carcinogens. *Nature 267*, 81–2.

Ikushima, T. & S.Wolff (1974) UV-induced chromatid aberrations in cultured Chinese hamster cells after one, two, or three rounds of DNA replication. *Mutat. Res. 22*, 193–201.

Kato, H. (1973) Induction of sister chromatid exchanges by UV light and its inhibition by caffeine. *Exp. Cell Res. 82*, 383–90.

—— (1977) Mechanisms for sister chromatid exchanges and their relation to the production of chromosomal aberrations. *Chromosoma 59*, 179–91.

Kato, H. & H.Shimada (1975) Sister chromatid exchanges induced by mitomycin C: A new method of detecting DNA damage at chromosomal level. *Mutat. Res. 28*, 459–64.

Kihlman, B.A. (1975) Sister chromatid exchanges in *Vicia faba*. II. Effects of thiotepa, caffeine and 8-ethoxycaffeine on the frequency of SCEs. *Chromosoma 51*, 11–18.

Kihlman, B.A. & D.Kronberg (1975) Sister chromatid exchanges in *Vicia faba*. I. Demonstration by a modified fluorescent plus Giemsa (FPG) technique. *Chromosoma 51*, 1–10.

Kihlman, B.A., S.Sturelid, F.Palitti & A.Becchetti (1977) Effects of caffeine, an inhibitor of post-replication repair in mammalian cells, on the frequencies of chromosomal aberrations and sister chromatid exchanges induced by mutagenic agents. *Mutat. Res. 46*, 130–1.

Latt, S.A. (1974) Sister chromatid exchanges, indices of human chromosome damage and repair: Detection by fluorescence and induction by mitomycin-C. *Proc. Natl. Acad. Sci. U.S.A. 71*, 3162–6.

Latt, S.A., G.Stetten, L.A.Juergens, G.R.Buchanan & P.S.Gerald (1975) Induction by alkylating agents of sister chromatid exchanges and chromatid breaks in Fanconi's anemia. *Proc. Natl. Acad. Sci. U.S.A. 72*, 4066–70.

Marin, G. & D.M.Prescott (1964) The frequency of sister chromatid exchanges following exposure to varying doses of ^{3}H-thymidine or X-rays *J. Cell Biol. 21*, 159–67.

Miller, J.A. (1970) Carcinogenesis by chemicals: an overview (G.H.A.Clowes Memorial Lecture). *Cancer Res. 30*, 559–76.

Perry, P. & H.J.Evans (1975) Cytological detection of mutagen-carcinogen exposure by sister chromatid exchange. *Nature 258*, 121–5.

Perry, P. & S.Wolff (1974) New Giemsa method for the differential staining of sister chromatids. *Nature 251*, 156–8.

Sasaki, M.S. (1973) DNA repair capacity and susceptibility to chromosome breakage in xeroderma pigmentosum cells. *Mutat. Res. 20*, 291–3.

Solomon, E. & M.Bobrow (1975) Sister chromatid exchanges—a sensitive assay of agents damaging human chromosomes. *Mutat. Res. 30*, 273–8.

Stetka, D.G. & S.Wolff (1976) Sister chromatid exchange as an assay for genetic damage induced by mutagen-carcinogens. II. *In vitro* test for compounds requiring metabolic activation. *Mutat. Res. 41*, 343–50.

Stich, H.G., R.H.C.San & Y.Kawazoe (1973). Increased sensitivity of xeroderma pigmentosum cells to some chemical carcinogens and mutagens. *Mutat. Res. 17*, 127–37.

Stich, H.F., W.Stich & R.H.C.San (1973) Chromosome aberrations in xeroderma pigmentosum cells exposed to the carcinogens, 4-nitroquinoline-1-oxide, and N-methyl-N′-nitro-nitrosoguanidine (37194). *Proc. Soc. Exp. Biol. Med. 142*, 1141–4.

Taylor, J.H. (1958) Sister chromatid exchanges in tritium labeled chromosomes. *Genet. 43*, 515–29.

Tice, R., J.Chaillet & E.L.Schneider (1975) Evidence derived from sister chromatid exchanges of restricted rejoining of chromatid subunits. *Nature 256*, 642–4.

Wolff, S. (1977) Sister chromatid exchange, in *Annual Review of Genetics*, Annual Reviews, vol.11, California: Palo Alto (in press).

Wolff, S., J.Bodycote & R.B.Painter (1974) Sister chromatid exchanges induced in Chinese hamster cells by UV irradiation of different stages of the cell cycle: the necessity for cells to pass through S. *Mutat. Res. 25*, 73–81.

Wolff, S., J.Bodycote, G.H.Thomas & J.E.Cleaver (1975) Sister chromatid exchange in xeroderma pigmentosum cells that are defective in DNA excision-repair or post-replication repair. *Genetics 81*, 349–55.

Wolff, S. & P.Perry (1974) Differential Giemsa staining of sister chromatids and the study of sister chromatid exchanges without autoradiography. *Chromosoma 48*, 341–53.

—— (1975) Insights on chromosome structure from sister chromatid exchanges and the lack of both isolabelling and hetero-labelling as determined by the FPG technique. *Exp. Cell Res. 93*, 23–30.

Wolff, S., B.Rodin & J.E.Cleaver (1977) Sister chromatid exchanges induced by mutagenic carcinogens in normal and xeroderma pigmentosum cells. *Nature 265*, 347–9.

Sister Chromatid Exchanges in Lymphocytes treated with 8-Methoxypsoralen and exposed to Long-Wave Ultraviolet Light

Photochemotherapy, using 8-methoxypsoralen, taken orally, and whole-body irradiation with long wave UV, is an effective and clean method for the treatment of psoriasis (Parrish *et al.* 1974), a distressing but not malignant condition affecting 1–2 per cent of the population. This therapy is already being used in several centres throughout the world and its use is likely to become much more widespread. Thus it seems probable that a considerable number of people are likely to receive the treatment, and it is therefore particularly important that we are aware of any genetic hazard that may exist.

The condition is due to the excessive proliferation of epidermal cells, and the therapy appears to work by killing or impairing the reproductive potential of these cells. In the presence of irradiation in the region of 360 nm, 8-methoxypsoralen (8-MOP) forms adducts with pyrimidines (Musajo, Rodighiero and Dall'Aqua 1965). Monoadducts or DNA cross-links may be formed in a ratio of about 3:1 (Cole 1971). Failure to remove these alkylation products, particularly the crosslinks, clearly inhibits the further multiplication of the cells. When removal does occur, errors in repair may result in mutation.

We therefore decided to look at sister chromatid exchange in human lymphocytes with three main objectives. First, to examine the effect of 8-MOP/UVA in relation to SCE induction in cultured cells when this treatment was administered prior to differential labelling of the chromosomes with BUdR; second, to examine the possibility that *in vitro* irradiation of blood of patients who had taken 8-MOP orally would permit the demonstration of the presence of 8-MOP in the peripheral circulation; and third, to discover whether the SCE level in patients undergoing photochemotherapy was higher than that in those who were not.

Material and Methods

For the *in vitro* study blood samples were obtained from 13 healthy adults (Mourelatos, Faed and Johnson 1977). We also studied 8 patients with psoriasis prior to any treatment involving 8-MOP, and 11 patients who were receiving photochemotherapy (Mourelatos *et al.* 1977). Samples were obtained from patients on treatment two hours after oral ingestion of 8-MOP but before UVA irradiation and also 30 min later, following irradiation. Samples were collected into light-shielded, heparinised tubes. Cultures were established from 0.4 ml of whole blood added to 4 ml of culture medium (Eagle's medium with 15% calf serum, 100 units ml^{-1} penicillin and 0.1 $\mu g\ ml^{-1}$ streptomycin) in universal containers. These containers were placed in a light-tight box, injected, through the lid of the box, with 0.1 ml phytohaemagglutinin and 0.1 ml BUdR solution to give a final concentration of 12 μM, and incubated at 37°C for 72 h. Colchicine (0.002% v/w) was added for the final three hours before harvest. Chromosome preparations were made and stained with 0.7 μg ml Hoechst 33258 and Giemsa (Perry and Wolff 1974). Slides were coded before being scored. Only well-spread mitoses with good differential staining of chromatids and a count of 46 chromosomes were selected for scoring for SCEs. A minimum of 30 cells was scored from each culture.

For the experiments with blood from normal subjects, 8-MOP was added to the cultures and, when appropriate, irradiation by UVA followed 30 min later. The cultures were irradiated in 80 cm^2 NUNC tissue culture flasks under parallel UV tubes (Atlas non-filter 20 W fluorescent UV) with an emission peak at about 350 nm (irradiance 0.41 mW cm^{-2}) and immediately after transferred to universal containers, placed in the dark and then injected with phytohaemagglutinin and BUdR. Blood from psoriatic patients was irradiated in the same manner but an exposure time of 30 min was consistently used.

Results and Discussion

The results of the experiments on blood from normal subjects (tables 1 and 2) showed that, up to a concentration of 4 $\mu g\ ml^{-1}$, 8-MOP alone has no measurable effect on SCE, but in the presence of 8-MOP and UVA there is an increase in SCE level related to the dose of 8-MOP. The activating effect of the UVA on the 8-MOP appears to be complete in about 15 min, and although there is a continuing slow rise in SCE with increasing exposure this is no greater than that due to the effect of the UVA alone.

The results for psoriatic patients are shown in table 3. Patients who have not had 8-MOP/UVA therapy have SCE levels that are not significantly different from normal subjects. Patients on treatment with photochemotherapy do

Table 1. SCE levels in lymphocytes from normal subjects after *in vitro* exposure to 0.74 J cm^{-2} UVA.

8-MOP conc. ($\mu g\ ml^{-1}$)	No. of cells	Mean SCE rate	S.D.
0	150	8.6	2.12
0.004	90	12.2	3.82
0.04	33	13.0	3.88
0.4	74	18.8	5.18
4.0	149	21.0	6.15

Table 2. Effect of varying the exposure time to UVA (irradiance 0.41 mW cm^{-2}) on mean SCE rate in the presence and absence of 8–MOP. The number of cells contributing is shown in brackets.

8-MOP conc. (μg ml^{-1})	UVA exposure time (min) 0	7.5	15	30	60
0	7.4 (702)	7.5 (60)	8 (120)	8.6 (150)	10.9 (120)
4	7.9 (60)	15.4 (109)	19.7 (120)	21.0 (149)	22.9 (103)

Table 3. Mean SCE rates in blood of controls and psoriatic patients before and after *in vitro* UVA (0.74 J cm^{-2}). Standard deviation shown in brackets.

Patient group	No UVA	*in vitro* UVA
Controls	7.4 (2.59)	8.2 (2.97)
Psoriatics without photochemotherapy	8.0 (2.95)	7.9 (2.61)
Psoriatics with photochemotherapy	9.0 (2.93)	14.2 (4.30)

have a marginally higher mean SCE rate but the difference is not significant. When the blood is irradiated *in vitro* the mean SCE rate rises to a level that we estimate, by reference to our control curves, to indicate a circulating 8-MOP concentration of about 0.8 μg ml^{-1} (Mourelatos *et al.* 1977). Irrespective of the length of time for which these patients have been on therapy, the level found in their blood before and after *in vivo* irradiation is not distinguishable (figure 1).

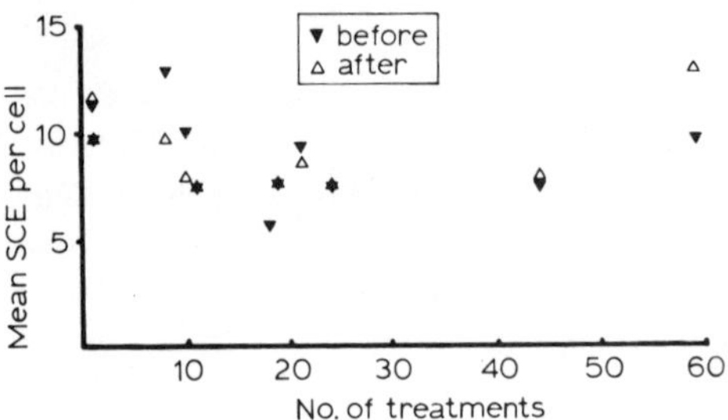

Figure 1. Mean sister chromatid exchanges in lymphocytes of patients on photochemotherapy before and after *in vivo* UVA irradiation, shown against the number of treatments given.

Wolff-Schreiner *et al.* (personal communication) have recently undertaken similar studies and have obtained essentially similar results. They have taken the further step of subdividing the psoriatic patients into different clinical groups, and they come to the interesting conclusion that the SCE counts of patients with arthropathy are higher than those with conventional psoriasis.

We would stress that the values associated with our experiments on lymphocytes are provisional. Of particular concern is the evidence that the SCE distribution becomes bimodal as 8-MOP concentration rises, a significant proportion of the population of cells showing little or no increase in SCE rate (Mourelatos, Faed and Johnson 1977). Whether this implies uneven distribution of exposure to the drug or radiation, or the presence of cells of different sensitivity, is still being investigated. This may have particular relevance for psoriasis in which the proportion of B and T cells has been reported to be disturbed (Guilhou *et al.* 1976). Thus the levels found after *in vitro* irradiation of the cells of psoriatic patients may not be strictly comparable with those of normal subjects.

We have also used the lymphocyte as a conveniently accessible cell to examine the effect of this treatment on SCE level as a possible indicator of mutagenic potential in man. It is clear that SCEs are induced by the combination of 8-MOP and UVA, but the therapy does not appear to result in a significant increase in lymphocyte SCE levels, presumably because these cells are largely shielded from UVA. It has been shown, however, that 8-MOP, in the dark, is a frame shift mutagen in bacteria (Clarke and Wade 1975, Bridges and Mottershead 1977). This effect becomes apparent at concentrations greater than 10 μg ml^{-1}. Although we have failed to find any effect of the drug in the dark on SCE level, the highest concentration we have so far adequately tested is 8 μg ml^{-1}, which is about ten times higher than the concentration we believe to be present in the circulation two hours after ingestion of the drug. Studies are continuing on the effect of higher concentrations in the dark.

Summary

Human lymphocytes have been used to examine the effect of 8-methoxypsoralen and UVA on chromosome damage as indicated by sister chromatid exchange rates. An *in vitro* study on the blood of normal subjects shows an effect on SCE that is related to the dose of the drug for a fixed UVA exposure, and has provided a dose-response curve that has been used to give an indication of the concentration of the drug in the blood of psoriatic patients two hours after its oral ingestion. 8-MOP/UVA therapy appears to be without effect on the SCE levels found in the lymphocytes of psoriatic patients.

References

Bridges, B.A. & R.P.Mottershead (1977) Frameshift mutagenesis in bacteria by 8-methoxypsoralen (methoxalen) in the dark. *Mutat. Res.* (in press).

Clarke, C.H. & M.T.Wade (1975) Evidence that caffeine, 8-methoxypsoralen and steroidal diamines are frame shift mutagens for *E. coli* K-12. *Mutat. Res. 28*, 123–5.

Cole, R.S. (1971) Psoralen monoadducts and interstrand cross links in DNA. *Biochim. Biophys. Acta 254*, 30–9.

Guilhou, J.J., J.Meynadier, J.Clot, E.Charmasson, M.Dardenne & J.Brochier (1976)

Immunological aspects of psoriasis. II. Dissociated impairment of thymus dependent lymphocytes. *Br. J. Dermatol. 95*, 295–301.

Mourelatos, D., M.J.W.Faed & B.E.Johnson (1977). Sister chromatid exchanges in human lymphocytes exposed to 8-methoxypsoralen and long wave UV radiation prior to incorporation of bromodeoxyuridine. *Experientia* (in press).

Mourelatos, D., M.J.W.Faed, P.Gould, B.E.Johnson & W.E.Frain-Bell (1977) Sister chromatid exchanges in lymphocytes of psoriatics after treatment with 8-methoxypsoralen and long wave ultraviolet radiation. *Br. J. Dermatol.* (in press).

Musajo, L., G.Rodighiero & F.C.Dall'Aqua (1965) Evidence of a photoreaction of the photosensitising fuorocoumarins with DNA and with pyrimidine nucleosides and nucleotides. *Experientia 21*, 24–6.

Parrish, J.A., T.B.Fitzpatrick, L.Tannenbaum & M.A.Pathak (1974) Photochemotherapy of psoriasis with oral methoxalen and long wave ultraviolet light. *New Eng. J. Med. 291*, 1207–11.

Perry, P. & S.Wolff (1974) New Giemsa method for the differential staining of sister chromatids. *Nature 251*, 156–8.

A. BRØGGER, H. WAKSVIK and P. THUNE

No Evidence for Chromosome Damage in Psoriasis Patients treated with Psoralen and Long-Wave Ultraviolet Light

When irradiated with long-wave ultraviolet light (UVA from 320–370 nm) psoralen molecules will react with thymine residues in the DNA to form monoadducts or crosslinks (Cole 1970). This is probably the reason why psoralen/UVA (PUVA) is a clastogen (Sasaki and Tonomura 1973; Swanbeck *et al.* 1975; Waksvik, Brøgger and Stene 1977) and increases the frequency of sister chromatid exchange (Carter, Wolff and Schnedl 1976; Waksvik, Brøgger and Stene 1977). Both the chromosome damage and the SCE frequency is enhanced in a synergistic way by caffeine (Waksvik, Brøgger and Stene 1977).

During the last few years PUVA has been introduced in the treatment of psoriasis, and this seems to be a very promising cure (Wolff *et al.* 1976). Swanbeck *et al.* (1975) found an increase in the chromosome aberration frequency in some of their psoriasis patients treated with PUVA. In order to examine further the question whether chromosome damage occurs in the lymphocytes of psoriasis patients as a result of PUVA treatment, we have studied five patients very closely during the treatment period. One advantage with such a prospective study is that the problem of adequate controls may be circumvented by using the patients as their own controls. We have also extended our previous *in vitro* studies to doses of psoralen comparable with the plasma doses of the patients at the time of the UVA irradiation.

Material and Methods

Five patients, aged 23 to 64 and suffering from chronic, stable psoriasis, covering at least 50% of their body surface, were investigated. The patients had not previously received any treatment that might induce chromosome damage, such as methotrexate (Ryan, Boddington and Spriggs 1965) or dithranol (Swanbeck *et al.* 1975). 8-methoxypsoralen (8-MOP) of a new micronised formulation (supplied by Nyegaard et Co. A/S, Oslo) was given orally 1 h before UVA irradiation in a dose of about 0.5 mg kg^{-1} body weight. The plasma concentrations of 8-MOP in the patients were then 10–320 ng ml^{-1}. The patients were irradiated in an apparatus with Philips Black Light lamps for 30 min. Treatment was given four to five times a week with doses starting at 1.2 J cm^{-2} and gradually increasing to 3.6 J cm^{-2}. The number of treatments varied from 15 to 26, after which four of the patients were 100% clear and the fifth 80% clear.

Blood samples for chromosomes studies were taken at the first day of treatment prior to the administration of psoralen; 2 h after the UVA irradiation at the 1st, 5th, 10th and 20th treatments; and finally at a maintenance treatment 6 months after the start of therapy (doses of the last treatment are given in table 2). Whole blood cultures were prepared and harvested at 68 h as previously described (Brøgger 1967). Preparations for SCE studies were made with the FPG technique using 1 μg ml^{-1} of BUdR. The slides were coded and analysed blindly according to the aberration criteria previously defined (Brøgger 1971).

The *in vitro* experiment was made with lymphocytes from the same donor and conditions as previously described (Waksvik, Brøgger and Stene 1977), except that whole blood cultures were used.

Results

There was an increase in the number of constrictions and gaps after 10 treatments, particularly in one patient, but after 20 treatments the aberration level was not abnormally high (table 1). Break events were only in one case higher after treatment (patient A.L. at the 5th). The incidence of exchanges,

Table 1. Chromosome studies of five psoriasis patients treated with PUVA. The data are average per cent constrictions and gaps (CG) and break events (BE). 100 cells were examined from each patient at each sampling time.

	Before treatment	After treatments				6 months
		1	5	10	20	
CG	6.3	6.3	4.2	9.2	4.0	2.2
BE	1.7	0.6	1.2	0.8	1.0	0.0

Table 2. Sister chromatid exchange in four psoriasis patients before and after receiving a maintenance dose of PUVA at 6 months after start of therapy. Counts from 50 cells.

Patient	Dose (J cm^{-2})	SCE per cell	
		before	after
R.M.	3.0	3.74 ± 2.28	4.18 ± 1.92
P.H.	2.3	3.68 ± 1.61	4.24 ± 2.31
A.L.	4.5	4.22 ± 1.99	3.54 ± 2.13
A.O.	1.5	3.14 ± 1.61	3.14 ± 1.45

0.0008 per cell, was not higher than the level of exchanges found in our laboratory among 1800 cells from a healthy blood donor, 0.001 per cell.

The SCE frequency was determined before and after PUVA treatment, when the patients received a maintenance dose, and was found not to be increased (table 2).

The *in vitro* experiment showed that nanogram doses of 8-MOP and 24 mJ cm^{-2} UVA did not induce chromosome aberrations and did not increase the SCE frequency (table 3).

Table 3. Chromosome aberrations (100 cells examined) and sister chromatid exchange (50 cells examined), in human lymphocytes treated *in vitro* with various low doses of 8-methoxypsoralen for 2 h and then irradiated with 24 mJ cm^{-2} UVA.

8-MOP ($ng\ ml^{-1}$)	Per cent CG	Per cent BE	SCE per cell
0	1.0	0	3.90 ± 2.06
50	1.9[1]	0	3.62 ± 1.23
125	2.0	0	3.88 ± 1.69
250	3.0	0	3.78 ± 1.18
500			3.90 ± 1.28

[1] 52 cells.

Discussion

We did not find any chromosome damage in the psoriasis patients treated with PUVA. The following possible reasons may be considered:

(1) The psoralen or UVA doses were too small.

The *in vitro* experiments, where we used doses similar to the *in vivo* plasma doses, support this interpretation. On the other hand, Swanbeck *et al.* (1975) found that chromosome damage was induced by *in vitro* UVA irradiation of lymphocytes from patients who had taken 60–80 mg 8-MOP orally 2 h before the blood was drawn. This indicates that the drug gets into the lymphocytes *in vivo* and could mean that the reason why our results were negative is that the UVA irradiation was too low. We do not know what doses the lymphocytes received *in vivo*. The UVA irradiation is absorbed by melanin in the skin and the porphyrin groups of the haemoglobin, but a certain amount of energy might reach the peripherally circulating lymphocytes.

In our *in vitro* experiment, the UVA dose was 24 mJ cm^{-2}, which was sufficient to induce damage with mg doses of psoralen *in vitro* (Waksvik, Brøgger and Stene 1977). Thus, our data point to the low psoralen dose as the limiting factor.

(2) A sufficient number of exposed cells were not present in the blood sample.

We do not know the size of the population of lymphocytes exposed during the 30 min the patient was receiving the UV. The proportion of the exposed cell population being sampled by venipuncture 2 h after exposure is also unknown.

(3) The PUVA-induced damage was eliminated by repair or cell death.

The lymphocytes had received different exposures: the acute exposure immediately before sampling, and the various doses received at previous treatments. We have no information on the persistence of PUVA damage except that we know from our *in vitro* experiments that treatment given 24 h before harvest will lead to measurable amounts of chromosome aberrations and increased SCE (Waksvik, Brøgger and Stene 1977).

(4) The damage was lost because the cells were harvested at 68 h.

There is no general agreement between laboratories whether to harvest lymphocyte cultures at 48 or 72 h. The radiation experiments of Heddle, Evans and Scott (1967) and others, show clearly that the aberration yield decreases from 48 to 72 h because of the difference in frequency of first and second metaphases. On the other hand, with chemical mutagens, Kilian and Picciano (1976) claim that their group has found no essential difference between 48 and 72 h.

Most laboratories seem to report a peak of mitotic activity at 72 h, so that the yield of analysable cells is highest at this time of harvest. This is one of the reasons why we have chosen the 68 h culturing time. Using the same slides as for SCE counting, we have determined the proporions of first, second and third (or later) divisions at 68 h to be 23.75, 54.25, and 22.0% respectively. We do not know whether exposed cells follow this pattern, but in any case it seems improbable that the sampling time alone can account for our entirely negative results.

Our results are not in complete agreement with the observations of Swanbeck *et al.* (1975), who found a significant increase in chromosome damage in one case and a slight increase in six other patients, whereas only one case had a normal, low incidence of damage. The SCE method is much more sensitive than the conventional study of chromosome aberrations (Perry and Evans 1975). However, with the nanogram doses of psoralen not even the SCE method was able to reveal any effect of PUVA.

We have altogether analysed 34 blood samples for chromosome aberrations. Our resources might have been used to study 17 patients only twice. We decided to perform this intensive study in the search for an eventual gradual build-up of chromosome damage. Taking both *in vivo* and *in vitro* results into account, we do not consider them to warrant any warning against a limited use of PUVA in the treatment of psoriasis. But we recommend that patients should be watched for possible future development of malignant disease. Psoralen has been used in the treatment of vitiligo for 20 years (Parrish *et al.* 1974) without any report of systemic toxicity, and it has not been linked with any increase in skin cancer or other neoplasia. But a careful monitoring of each patient for 15–20 years has not been reported.

Summary

Five psoriasis patients treated with 8-methoxypsoralen and UVA (PUVA) were studied by lymphocyte cultures at the 1st, 5th, 10th and 20th treatment and at a maintenance treatment 6 months later. Abnormal amounts of chromosome aberrations were not found, and the frequency of sister chromatid exchange (examined at the last treatment) was not increased. *In vitro* experiments with nanogram doses of psoralen (similar to plasma levels in patients) showed no increase in chromosome aberration or SCE frequency.

The results may indicate that therapeutic doses of PUVA have no clastogenic effect, but other explanations are possible: a sufficient number of exposed cells were not present in the blood sample; the PUVA damage was eliminated by repair or cell death; or the damage was lost because the cells were harvested at 68 hours.

Acknowledgements

The technical help provided by Nyegaard & Co. A/S, Oslo, is greatly appreciated. We are grateful to Bitten Lunde and Reidun Bjørge for skilful technical assistance with the chromosome studies.

References

Brøgger, A. (1967) Translocation in human chromosomes with special reference to mental retardation and congenital malformations. Oslo: Universitetsforlaget.

—— (1971) Apparently spontaneous chromosome damage in human leukocytes and the nature of chromatid gaps. *Hum. Genet. 13*, 1–14.

Carter, D.M., K.Wolff & W.Schnedl (1976) 8-methoxypsoralen and UVA promote sister-chromatid exchanges. *J. Invest. Dermatol. 67*, 548–51.

Cole, R.S. (1970) Light-induced cross-linking of DNA in the presence of a furocoumarin (psoralen). Studies with phage λ, *Escherichia coli* and mouse leukemia cells. *Biochim. Biophys. Acta 217*, 30–9.

Heddle, J.A., H.J.Evans & D.Scott (1967) Sampling time and the complexity of the human leukocyte culture system, in *Human Radiation Cytogenetics* (eds. H.J. Evans, W.M.Court-Brown & A.S.McLean) pp.6–19. Amsterdam: North Holland.

Kilian, D.J. & D.Picciano (1976) Cytogenetic surveillance of industrial populations, in *Chemical Mutagens. Principles and Methods for Their Detection*, vol.4 (ed. A.Hollaender) pp.321–39. New York & London: Plenum Press.

Parrish, J.A., T.B.Fitzpatrick, L.Tanenbaum & M.A.Pathak (1974) Photochemotherapy of psoriasis with oral methoxsalen and longwave ultraviolet light. *New Engl. J. Med. 291*, 1207–11.

Perry, P. & H.J.Evans (1975) Cytological detection of mutagen-carcinogen exposure by sister chromatid exchanges. *Nature 258*, 121–5.

Ryan, T.J., M.M.Boddington & A.J.Spriggs (1965) Chromosomal abnormalities produced by folic acid antagonists. *Br. J. Dermatol. 77*, 541–55.

Sasaki, M.S. & A.Tonomura (1973) A high susceptibility of Fanconi's anemia to chromosome breakage by DNA cross-linking agents. *Cancer Res. 33*, 1829–36.

Swanbeck, G., M.Thyresson-Hök, A.Bredberg & B.Lambert (1975) Treatment of psoriasis with oral psoralens and longwave ultraviolet light. *Acta Dermatol.-Venereol. (Stockh.) 55*, 367–76.

Waksvik, H., A.Brøgger & J.Stene (1977) Psoralen/UVA treatment and chromosomes. I. Aberrations and sister chromatid exchange in human lymphocytes *in vitro* and synergism with caffeine. *Hum. Genet.* (in press).

Wolff, K., T.B.Fitzpatrick, J.A.Parrish, F.Gschnait, B.Gilchrest, H. Hönigsmann, M.A.Pathak & L.Tanenbaum (1976) Photochemotherapy for psoriasis with orally administered methoxalen. *Arch. Dermatol. 112*, 943–50.

Effects of Twelve Folate Analogues on Human Lymphocytes in vitro

For the last four years the writer has been investigating the effects on lymphocytes of compounds that may be termed folate analogues, in that they exert their inhibitory effects on metabolism by reason of their structural similarities to folates. These substances are extensively used as anti-bacterial, anti-parasitic and anti-neoplastic agents. They may be grouped roughly into the 2,4-diaminopyrimidines (or pteridines), typified by the anti-malarial pyrimethamine (2,4-diamino-5(p-chlorophenyl)6-ethylpyrimidine), and those derived from, or similar in structure to, aminopterin (4-amino-4-deoxyfolate), typified by amethopterin (methotrexate, MTX).

These compounds have proved very useful in investigation of folate metabolism in unicellular organisms that do or do not synthesise their own folates. It is mainly from studies of such unicellular organisms that our ideas of the modes of action of folate antagonists are derived.

It is generally considered that the most important primary effect of all these compounds is that they bind with dihydrofolate reductase and interfere with the generation of folate in tetrahydrofolate form (THF) from dihydrofolate (DHF):

$$\mathrm{DHF + NADPH + H^{+} \rightleftharpoons NADP^{+} + THF}$$

Folic acid and folates are similarly reduced at a much lower rate

$$\mathrm{folate + 2NADPH + 2H^{+} \rightleftharpoons 2NADP^{+} + THF}$$

Every organism has this reductase in some form, and the basis of therapy is that the compound used should bind much more closely with the reductase of the organism than that of the host. Thus trimethoprim binds 50000 times more strongly with *E. coli* than with human reductase (Burchall and Hitchins 1965). It has never been demonstrated that the reductases of neoplastic cells are different from those of the host, but presumably the drugs are useful in part at least because rapidly dividing tissues have a very big turnover of folates.

The only source of thymidine and its phosphorylated ribonucleotides is by methylation of deoxyuridine or its monophosphate. The methyl precursor has to come from 5,10-methylene THF in the so-called thymidylate synthetase reaction

$$\mathrm{dUMP + 5,10\text{-}methylene\ THF \rightarrow dTMP + DHF}$$

so that there is a continual production of DHF during DNA synthesis and the depleted stocks of folate in THF form have to be replaced by the DHF reductase reaction.

The Analogues Investigated

The folate analogues with which I have mainly worked are MTX, aminopterin, methyldichlorophen, pyrimethamine, trimethoprim, and seven diaminopyrimidines (unidentified when tested) kindly supplied by Dr J.J. Burchall, Department of Mircobiology, Wellcome Foundation Institute, North Carolina (specified in table 1). These twelve compounds, which, by binding with DHF reductase, secondarily interfere with the methylation of uridine to thymine, showed the following effects on lymphocytes in culture:

(a) All except trimethoprim produced chromatid-type damage in lymphocytes that were exposed in culture to appropriate concentrations during DNA synthesis.

(b) All, even at lower concentrations than those producing significant levels of chromatid damage in 100 metaphases, produced giant lymphocytes. At higher concentrations, where few (if any) cells reached metaphase, these giant cells were no longer produced.

(c) Some of these substances appeared to be able to enter lymphocytes during G_0/G_1 and subsequently induce chromatid damage and giant lymphocytes.

(d) All of these effects, from all the compounds tested, could be modified or prevented entirely by the presence of folates during S.

Table 1. Seven compounds screened 'blind'.

1. 2,4-diamino-1-(4′-butylphenyl)-6,6-dimethyl-1,6-dihydro-1,3,5-triazine hydrochloride
2. 5,6-trimethylene-2,4-diamino quinazoline
3. 2,4-diamino-5-methyl-6-sec.butylpyrido (2,3d) pyrimidine
4. 2,4-diamino-5-methyl-6-sec.amylpyrido (2,3d) pyrimidine
5. 2,4-diamino-6-butylpyrido (2,3-d) pyrimidine hydrochloride
6. 1-(p-n-butoxyphenyl)-4,6-diamino-1,2-dihydro-2,2-dimethyl-5-triazine hydrochloride
7. 4,6-diamino-1,2-dihydro-2,2-dimethyl-1-(3′,4′,5′-trimethoxyphenyl)-s-triazine hydrochloride

Typical Findings as Exemplified by Methotrexate

MTX is the substance most carefully investigated, and a summary of the findings from a large number of experiments will serve to illustrate the phenomena that can be studied.

Findings in Patients Taking MTX

100 metaphases were examined from each of 24 patients that had taken the drug for psoriasis for periods between three months and three years. The total accumulated dose varied between 400 and 4000 mg. In the 2400 meta-

Table 2. Chromatid-type damage in two experiments where lymphocytes were exposed to varying concentrations of methotrexate over 72 h. Exp.1 at MTX concentrations 5 and 15 $\mu g\ ml^{-1}$: no metaphases. Exp.2 at concentration 4 $\mu g\ ml^{-1}$: no metaphases.

Types of chromosome damage found	MTX 0.5 $\mu g\ ml^{-1}$ Exp.1 Donor A	MTX 2.0 $\mu g\ ml^{-1}$ Exp.1 Donor A	Control Exp.1 Donor A	MTX 0.25 $\mu g\ ml^{-1}$ Exp.2 Donor B	MTX 0.5 $\mu g\ ml^{-1}$ Exp.2 Donor B	MTX 1.0 $\mu g\ ml^{-1}$ Exp.2 Donor B	MTX 2.0 $\mu g\ ml^{-1}$ Exp.2 Donor B	Control Exp.2 Donor B
Chromatid gaps	2	3	4	6	6	21	35	1
Chromatid breaks	2	15	1	4	6	30	33	0
Iso-chromatid gaps	3	9	2	5	9	11	31	1
Iso-chromatid breaks	13	24	1	6	8	12	32	0
Quadriradials	1	1	0	2	5	4	13	0
Triradials	0	0	0	1	1	2	5	0
Simple figures	1	2	0	2	4	8	13	0
Complex figures	0	0	0	1	2	3	9	0
Dicentric + fragment	0	0	0	0	0	1	0	0
Total metaphases with damage	17	33	2	14	21	47	56	0
Total metaphases scored	100	100	200	100	100	100	100	100

phases only 10 chromatid breaks (0.4%) and 14 chromosome breaks (0.6%) were found, figures not significantly different from control levels in this laboratory (0.65% chromatid breaks and 0.44% chromosome breaks in 24 106 metaphases). These findings agree with those of Voorhees *et al.* (1969), Kaong and Swartzenruber (1969) and Valenti, Bird and Bird (1967), although Ryan, Boddington and Spriggs (1965) and Locher and Franz (1967) considered their results to indicate some damage. All these authors were reporting much smaller numbers of patients and metaphases than this author. Further, it is inherently unlikely that damage would be found in lymphocytes sampled from the peripheral circulation; these divide so seldom that they would be exposed to MTX very infrequently while in the S phase of a cycle *in vivo.*

It would be more reasonable to expect damage in marrow cells a few hours subsequent to single large injections of the drug. The writer has not had an opportunity to test this. Only Jensen (1967) has examined marrow shortly after an injection; his findings were negative, although he reports damage in the cells from one patient several months after a single ingestion of 25 mg MTX.

Findings from in vitro Experiments with MTX

When lymphocytes are cultured with appropriate concentrations of MTX, chromatid, but not chromosome, type damage occurs and the duration of the cell cycle is increased. Individual lymphocytes are slowed down or finally blocked in synthesis, and increase greatly in size. The extent of all these induced anomalies vary greatly for the same drug concentration between different individuals. In all the experiments here reported, Lederle intravenous injection ampoules were used as the source of MTX. This preparation was more stable than preparations made up from powder for each experiment. MTX powder is rather unstable, and sensitive to humidity, light and pH in particular.

Chromatid damage. Exposure of cells in culture to concentrations between 0.05 and 4.0 μg ml^{-1} over 48 or 72 h produces detectable chromatid damage (chromatid breaks, isochromatid breaks and exchange figures) beginning at concentrations between 0.05 and 0.25 μg ml^{-1}, depending on the donor, and increasing in frequency up to a level where 40–60% of metaphases show some damage. At yet higher levels few, if any, cells reach metaphase. Findings in two typical screening experiments are shown in table 2.

Exposure of the cells for the first 22 h of culture only, the drug then being removed by repeated centrifuging and addition of fresh medium, does not result in any damage when the cultures are subsequently harvested at 48 to 72 h. This suggests that the drug cannot enter the cell *in vitro* during G_0/G_1. Similarly, when the drug is only present for the last 2.5 h of culture no damage is found, suggesting resistance to damage during G_2 + prophase.

Prevention of chromatid damage by folates and thymine. It has been shown in a large series of experiments that damage from MTX in cultures can be prevented by adequate concentration of stable tetrahydrofolates and

Table 3. Evidence for protection of lymphocytes by various folates against methotrexate in 72-h cultures. (DHF = dihydrofolic acid, THF = tetrahydrofolic acid, 5- & 10-FTHF = 5- and 10-formyltetrahydrofolic acid, 5-MTHF = 5-methyltetrahydrofolic acid.)

Types of chromosome damage found	MTXI 0.5 μg ml^{-1} + DHF 100 μg ml^{-1} Exp.E Donor D	MTXI 0.5 μg ml^{-1} + THF 100 μg ml^{-1} Exp.E Donor D	MTXI 0.5 μg ml^{-1} + 5-FTHF 100 μg ml^{-1} Exp.E Donor D	MTXI 0.5 μg ml^{-1} + 5-MTHF 100 μg ml Exp.E Donor D	Control MTXI 0.5 μg ml^{-1} no folate Exp.E Donor D	MTXI 0.25 μg ml^{-1} + folic acid 100 μg ml^{-1} Exp.F Donor E	MTXI 0.25 μg ml^{-1} + 10-FTHF 100 μg ml^{-1} Exp.F Donor E	Control MTXI 0.25 μg ml^{-1} Exp.F Donor E
Chromatid gaps	3	16	0	1	38	2	0	9
Chromatid breaks	2	38	0	1	60	4	0	9
Iso-chromatid gaps	2	18	2	0	25	2	0	5
Iso-chromatid breaks	3	22	0	0	50	3	0	9
Quadriradials	0	1	0	0	9	0	0	0
Triradials	0	1	0	0	2	0	0	0
Simple figures	0	7	0	0	16	0	0	1
Complex figures	0	0	0	0	2	0	0	0
Total metaphases with damage	5	48	0	1	98	7	0	17
Total metaphases scored	100	100	100	100	200	100	100	100

modified by folic acid or dihydrofolic acid. The folates tested were 5-formyl (and 10-formyl) tetrahydrofolic acid, 5-methyl tetrahydrofolic acid and tetrahydrofolic acid. The latter is very unstable, which is presumably why it afforded no protection (table 3). The others gave protection against concentrations of MTX that induced some damage in 50% of metaphases. However, these tetrahydrofolates only afforded protection if present during S, and were apparently not able to enter cells and build up a protective concentration during G_0/G_1. If one of them was present only for the first 22 h of culture and then removed and the culture carried on with added MTX, no protection was afforded.

It is therefore of some interest that it has been possible to show that the resistance to induced chromatid damage *in vitro* by MTX can be greatly increased by giving folic acid to volunteers. The washed lymphocytes of these volunteers were each exposed to 0.1, 0.25 and 0.5 μg ml^{-1} MTX throughout 72-h cultures. The donors then each took 10 mg of folic acid daily for a week, and their washed lymphocytes were tested again against the same MTX concentrations. Over the period their red cell folate levels had risen by factors between 5 and 7, and the chromosome damage was dramatically reduced in cultures from all donors. This suggests that the intracellular THF levels in lymphocytes had risen markedly *and* afforded protection against MTX over one or two S phases *in vitro*. In view of the high turnover of THF this result is perhaps surprising.

It is noteworthy, however, that in experiments with cultures of washed lymphocytes cultured in Eagle's MEM medium made up without folate, and which has no purine or pyrimidine bases, lymphocytes will pass through at least two cycles *in vitro*, although the cycle is prolonged and giant lymphocytes appear not in cultures harvested at 48 h but in those harvested at 72 h.

Perhaps surprisingly, folic acid and dihydrofolic acid in concentrations of 100 μg ml^{-1} proved to be partially protective against MTX in culture. However, protection fell off rapidly with both at lower concentrations. A possible explanation for this apparently paradoxical finding is that these substances may bind more strongly with DHF reductase when in competition with MTX, and may even be capable of displacing MTX if in high concentrations. It has to be remembered that small amounts of DHF reductase are continually being synthesised anew. Protective effects of folates are illustrated in table 3.

Complete protection against chromatid damage from MTX in culture is also given by thymine in concentrations above about 30 μg ml^{-1}. This short-circuiting of the thymidylate synthetase reaction by exogenous thymine recalls the protection against MTX by thymine in bacteria that require exogenous folate, such as *L. caseii* and *S. faecalis*. In experiments none of the other end-products of purine or pyrimidine metabolism tested (adenine, guanine, cytidine, glycine, histidine, methionine, and serine) afforded any protection. Nor was protection afforded by minimal concentrations of thymine augmented by adenine, guanine or cytosine.

Large lymphocytes in cultures treated with MTX. It is noteworthy that, even at concentrations of MTX so low that significant increase of chromatid damage is not detectable in 100 metaphases, very large lymphocytes begin to appear. In processing, the cultures were given hypotonic treatment (0.56% KCl for 5 min at 37°C) so that only nuclei could be measured. As a convenience in measuring the diameters of these nuclei with an eyepiece micrometer, only lymphocytes of 10 μm or more in diameter were measured; this eliminated all non-transformed lymphocytes, but also some in early stages of transformation.

In untreated cultures the majority of lymphocytes reach prophase just before the nuclear diameter reaches 20 μm, the proportion exceeding 20 μm varying in different individuals from about 5 to 15%, being greater on average in women. Lymphocytes with diameters over 25 μm are uncommon, occurring only in frequencies of up to 0.5%.

In MTX-treated cultures, even at concentrations where no detectable chromatid damage can be detected, the proportion of cells over 20 μm in diameter may rise to 25% and include cells of up to 35 μm in diameter. These effects only occur if the MTX has been present during S, and they can be prevented by appropriate folates and thymine precisely as can induced chromatid damage. As concentrations of MTX are further increased the proportion of giant lymphocytes falls; eventually no nuclei reach a diameter even of 20 μm and no cells reach metaphase.

The explanation of these phenomena appears to be that, as far as giant lymphocytes are concerned, they are produced by lack of thymine, due to failure of methylation of uridine, inhibiting (or blocking) DNA synthesis, although RNA and protein synthesis continue. Certainly giant cells continue to take up ^{3}H-lysine and [5]-^{3}H-uridine. Why, at still higher concentrations, the blocked lymphocytes never reach a large size is less obvious. It could be that the lack of THF interferes at the two stages in the chain of purine synthesis where it is necessary for the transfer first of a methyl and then of a formyl group to the precursors of inosinic acid. However, I could not in the presence of high MTX concentrations alter the situation by adding inosinic acid and/or adenine and/or guanine.

It is perhaps significant, however, that when cultures are treated with 5-fluorodeoxyuridine (5-FUdR), which acts as an analogue of uridine, giant cells are produced, preventable by thymine but not by folates, *but* lymphocyte size continues to increase even when blockage is such that no cells reach metaphase. It is also of interest that 5-bromodeoxyuridine, which appears to be inserted into DNA instead of thymine, does not produce giant lymphocytes.

The whole subject of giant cells is an extremely interesting one. The analogy with megakaryocytes in the bone marrow in megalocytic anaemias is obvious, where folates are not available either from failure of absorption or secondarily because they cannot be utilised due to absence of cyanocobalamin. In these megakaryocytes the RNA:DNA ratio is much higher than in normal marrow cells.

Slowing of the cell cycle. It is relatively easy to demonstrate by autoradiography and liquid scintillography that, *pari passu* with the induction of chromatid damage and of giant blocked lymphocytes, the cell cycle is slowed and that the slowing is entirely determined by slowing of DNA synthesis. Even with concentrations that do not prevent any cells reaching metaphase there appears to be no prolongation of G_0/G_1 prior to the onset of DNA synthesis in culture.

Effects of Other Folate Analogues

The pattern of response to all the folate analogues tested is essentially the same, although the spectrum of concentrations at which the phenomena appear vary from compound to compound. Considerably lower concentrations of aminopterin and methyldichlorophen cause similar degrees of damage as MTX, whereas pyrimethamine and the other substances are less 'toxic'. Methyldichlorophen, pyrimethiamine and numbers 1, 2 and 3 of the unknown compounds (table 1) appear to gain entry to lymphocytes *in vitro* in G_0/G_1, or at least cannot be washed from the cells and so cause damage in the subsequent S. It appears that all these compounds are lipid soluble.

It is of interest to note that in the case of trimethoprim, and probably two of the seven compounds unknown when tested, giant lymphocytes are readily produced but, up to the high concentrations where no lymphocytes reach metaphase and DNA synthesis is virtually blocked, no chromatid damage occurs. This is interesting, for it suggests non-identical mechanisms for induction of chromatid damage and DNA synthesis blocking, although they both appear to have non-availability of thymine in common.

Typical findings from the screening experiments of one of the unknown compounds are shown in table 4. Similar data in greater detail are available

Table 4. Typical findings from an experiment with an unknown 2,4-diamino-pyrimidine.

Concentration of drug (μg ml^{-1})	Duration of exposure of lymphocytes	Additives	No. per 1000 transformed lymphocytes 20+μm dia.	No. per 200 metaphases with chromatid damage
0.000 (control)			71	2
0.0625	0-72 h	none	99	2
0.125	,,	,,	184	14
0.25	,,	,,	224	21
0.5	,,	,,	285	19
1.0	,,	,,	289	17
2.0	,,	,,	335	6/129
4.0	,,	,,	109	no metaphases
8.0	,,	,,	86	,,
16.0	,,	,,	3	,,
4.0	0-22 h	,,	170	8
4.0	0-72 h	leukovorin 100 μg ml^{-1}	96	3
4.0	,,	thymine 100 μg ml^{-1}	111	0

for all the 12 compounds studied. As can be seen, with increasing concentrations of the compound the proportions of lymphocytes over 20 μm in diameter rose from 99/1000 at 0.0625 μg ml^{-1} to 335 per 1000 at 2 μg ml^{-1} and then fell rapidly to about zero for 4 to 16 μg ml^{-1}. The proportion of metaphases with chromatid damage followed a similar pattern, rising to a peak at lower concentrations than the nuclear diameter peak, and then at concentrations over 2 μg ml^{-1} no cells reached metaphase. The protection afforded by 5-FTHF (in the form of leukovarin) and thymine against the high concentration of 4 μg ml^{-1} is obvious in terms of nuclear diameters, cells reaching metaphase and chromatid damage.

Taking the findings at face value it seems likely that some of the drug was intracellular or bound to the cell surface, even after washing the cells at 22 h. This drug was subsequently identified as lipid soluble.

Giant Lymphocytes

As already noted, giant lymphocytes may plausibly be attributed to a slowing down and eventual complete blockage of DNA synthesis due to thymine lack, while RNA and protein synthesis continue. Loh (1960) noted that, after six days of culture with MTX in the medium, HeLa cell proliferation slowed down almost completely and there was marked loss of cells. Also the cells and their nuclei were greatly enlarged, and nucleoli were enlarged and basophil. In these enlarged cells protein, and particularly RNA, continued to increase relative to DNA. Further, these cells would support growth of polio virus, which is RNA-dependent, but not vaccinia or herpes viruses that are DNA-dependent. It is interesting to see that, in the experimental system used, enlargement of the lymphocytes in S is detectable at lower concentrations of these folate analogues than any other induced damage in the cells.

Of all the compounds tested, apart from the folate analogues, only five have been found to cause giant cells. 5-fluorodeoxyuridine induces such 'megalymphocytes', and these still seem to be occurring when the concentration is so high that no cells reach metaphase. As already noted, these effects of 5-FUdR can be prevented by thymine but *not* by folates. Daunorubomycin induces giant cells larger than those from any other drug, up to 50 μm in nuclear diameter. As the antibiotic can enter the lymphocytes during G_0/G_1 *in vitro*, these large cells can be seen in the first S phase in culture even if the lymphocytes have been thoroughly washed at 22 h and the culture carried on in fresh medium. Ethoglucid (Epodyl ICI) is an alkylating agent and also an epoxide. This substance behaves like daunorubomycin in entering lymphocytes during G_0/G_1 *in vitro* and causing enlargement of cells in the subsequent S. The antibiotic causes chromosome damage attributable to effects in G_0/G_1 as well as chromatid damage induced during S.

Mycophenolic acid is an antibiotic product of *Penicillium brevi-compactum*, isolated by Raistrick as far back as 1932 and tested as a cytotoxic agent in therapy of neoplasms. Although it induced giant lymphocytes up to concentrations at which blockage is such that no cells reach metaphase, no chromatid damage was found.

Finally, giant cells are consistently produced by cytosine arabinoside in culture. With this agent chromosome- and chromatid-type damage and giant cell induction do not appear to be prevented by cytidine, but *are* prevented by dCMP.

It is perhaps worth while, in view of this first publication dealing with the phenomenon of giant lymphocytes, to list the many other compounds that the writer has tested against lymphocytes in culture in recent years, where, whether or not chromosomal damage occurred and S was prolonged, the phenomenon of giant cells was *not* observed. These are therefore set out in Table 5.

Table 5. Substances tested that do not induce giant lymphocytes in culture.

Base analogues
6-azauracil, 6-azauridine, 5-bromo-2-deoxyuridine, 6-azathymine, primidone (anti-convulsant) pyrimidine analogue, 5-bromodeoxycytidine, 8-bromoadenosine, cordycephin (9-cordyposidadenine; 3-deoxyadinosine), thioguanine, proguanil, allopurinol, oxypurinol, azathioprine, 6-mercaptopurine.
Alkylating agents
Chlorambucil, melphalan, nitrogen mustard, quinacrine mustard, busulphan, cyclophosphamide (after incubation with liver).
Antibiotics
Erythromycin, streptomycin, ligomycin, cyclohexamide, dactinomycin, chloramphenicol, penicillamine, penicillamide.
Anti-inflammatory agents
Phenylbutazone, oxyphenbutazone, naproxen, indomethacin, sodium aurothiomalate.
Diuretics
Bentrofluorazide, chlorothiazide.
Miscellaneous
Nalidixic acid[1], practolol, propranolol, vincristine, vinblastine, H_{11} (tumour-inhibiting agent, Hosa Labs), sulphamethaxozole, amylobarbitone, phenobarbitone, pentobarbitone, phenotoin sodium.

[1] Nalidixic acid is an effective urinary antiseptic. According to the manufacturer's literature (Negrim Winthrop) it probably acts by causing thymineless death in *E. Coli* and other organisms. *In vitro* it produced budding of bacteria (*E. Coli* and *L. Caseii*) like folate analogues. However, in *in vitro* culture with human lymphocytes the metaphase index begins to fall at a concentration of about 75 μg ml^{-1}. At higher concentrations dead lymphocytes are seen, and no metaphases are seen after about 250 μg ml^{-1}. At no concentration are enlarged lymphocytes induced.

Discussion and Conclusions

The mechanism of induction of chromatid damage by the folate analogues appears to be clear-cut; thymine lack for DNA synthesis. The slowing of DNA synthesis and accompanying enlargement of lymphocytes can also be explained on the same basis—thymine lack—as suggested by the prevention of both kinds of phenomena by tetrahydrofolates or thymine.

It is not possible to reproduce all the data in a review such as this, but in respect of all the compounds full data, as in table 3, are available, and for methotrexate, methyldichlorophen, trimethoprim and pyrimethamine a great variety of other experimental findings are as yet unpublished. It is thus not possible to discuss here the relative severities of the various effects of these substances.

It is notable that non-availability of the pyrimidine bases appears to cause the specific phenomena of chromatid damage and lymphocyte enlargement, whereas none of the purine analogues listed in table 5, which have all been investigated by the author, caused chromatid damage. Probably there are alternative sources of purine intracellularly and in the culture medium, whereas the pyrimidines must be derived from uridine *via* a single metabolic pathway.

It would certainly appear that the prerequisite for the development of giant lymphocytes is that the agent attacks only DNA, having minimal effects on other metabolic processes. The mode of action of ethoglucid is not known. Daunorubomycin probably attacks predominantly DNA polymerase. However, this explanation does not account for the giant cells produced by cytosine arabinoside.

Summary

An account is given of findings of *in vitro* experiments where lymphocytes in culture were exposed to twelve folate analogues. Chromatid damage was induced by most of these compounds, probably all except trimethoprim. However, all caused slowing of DNA synthesis and consequent prolongation of the cell cycle. The lymphocytes slowed down became much larger than normal, and this enlargement occurred at concentrations of the analogues lower than those needed to induce a detectable amount of chromatid damage. All these effects could be prevented by adequate concentrations of stable tetrahydrofolates and modified by folic acid or dihydrofolic acid. They are also preventable by thymine, but not by any other of the end productions of purine or pyrimidine metabolism.

Above certain concentrations lymphocytes seem to be completely blocked early in DNA synthesis and never reach even normal sizes.

Possible explanations of these phenomena are discussed.

Acknowledgements

The author is in receipt of a grant from the Medical Research Council. Dr Leo Bernstein of the Wellcome Foundation, Beckenham, and Dr J.J. Burchall of the Wellcome Foundation, North Carolina, have been most helpful in discussions and in arranging for supplies of certain chemicals, and the latter in supplying the seven 'unknown' folate analogues. Mr C. Patel, Mr Robert Wright and Mrs Morag Whitehead have given invaluable technical help, and Mrs Anne Naylor has made the data assembly and other clerical work seem easy.

References

Blakely, R.L. (1969) *The biochemistry of folic acid and related pteridines.* Amsterdam: North-Holland.

Borsa, J. & G.F.Whitmore (1969) Cell killing studies on the mode of action of methotrexate on L-cells *in vitro. Cancer Res. 29*, 737–44.

Burchall, J.J. & G.H.Hitchings (1965) Inhibitor binding analysis of dihydrofolate reductases from various species. *Mol. Pharmacol. 1*, 126–9.

Jensen, M.K. (1967) Chromosome studies in patients treated with azathioprine and methotrexate. *Acta Med. Scand. 182*, 445–55.

Kaong, D.T. & A.A.Swartzenruber (1969) Effect of chemotherapeutic agents on chromosomes of patients with lung cancer. *Dis. Chest 55*, 98–100.

Locher, H. & J.Franz (1967) Chromosomen eranderongen bei methotrexate behandlungen. *Med. Welt. 34*, 1965–6.

Loh, P.C. (1960) Effects of amethopterin on cell growth and viral synthesis. *Proc. Soc. Exp. Biol. Med. 105*, 296–300.

McBurney, M.W. & G.F.Whitmore (1975) Mechanism of growth inhibition by methotrexate. *Cancer Res. 35*, 586–90.

Ryan, T.J., M.M.Boddington & A.I.Spriggs (1965) Chromosomal abnormalities produced by folic acid antagonists. *Br. J. Dermatol. 77*, 541–55.

Valenti, C., L.Bird & L.Frank (1970) Chromosomal studies of patients with psoriasis treated with amethopterin. *Acta Dermatol.* (*Stockholm*) *50*, 393–6.

Voorhees, J.J., M.K.Janzen, E.R.Harrell and S.G.Chakrabarti (1969) Cytogenetic evaluation of methotrexate-treated psoriatic patients. *Arch. Dermatol. 100*, 269–74.

Williams, W. (1971) Effects of amethopterin treatment on thymidylate synthetase in human leukocytes and bone marrow cells. *Ann. N.Y. Acad. Sci. 186*, 365–77.

S.D.LAWLER and P.A.M.WALDEN

Chromosome Studies in Patients treated with Chemotherapy for Trophoblastic Tumours

An increased risk of malignant disease developing after radiotherapy is well established and the question has arisen whether cytotoxic chemotherapy incurs a similar hazard. Sixty-four malignancies following cyclophosphamide therapy have been assembled from the literature by Puri and Campbell (1977). Leukaemia is known to occur with increased frequency in patients with multiple myeloma and there is evidence suggesting that melphalan treatment is associated with the development of acute myeloid leukaemias, which are unusual in that all show gross chromosomal abnormalities (Hossfeld *et al.* 1975).

The increased risk of malignancy following radiotherapy has been shown to increase linearly with dose (Court Brown and Doll 1957) and is also related to the site and volume of tissue irradiated. Radiation dose can be monitored by studying the chromosome aberrations produced in lymphocytes, and the emergence of radiation-induced clones in the haemopoietic system can be detected by studying bone marrow cells. Chromosome damage following cytotoxic therapy has not been extensively reported.

In looking for other evidence of carcinogenic effects of chemotherapy, it is necessary to accept that the time interval between exposure and the appearance of a second malignancy may be many years. It is also desirable to study patients with a malignancy that does not involve haemopoietic tissue. Patients that have been treated by chemotherapy for trophoblastic tumours, that is invasive mole or choriocarcinoma, fulfil this criterion and provide a group of patients with long-term survivors.

Materials and Methods

The Patients

The patients were females aged between 19 and 41, treated in the Department of Medical Oncology at the Charing Cross Hospital, London. All but one of them had a trophoblastic tumour and the therapy was monitored by measuring levels of human choriongonadotrophin (HCG). The exceptional patient had a dysgerminoma of the ovary that was monitored in the same way because HCG was secreted by the tumour. The patients with trophoblastic neoplasms in this series were categorised into three prognostic groups prior to therapy, according to various criteria (Bagshawe 1976). For the chromosome study, patients were selected from each of the groups. Group A comprised seven patients in the low risk, good prognosis group. Patients in this category received courses of methotrexate (50 mg) on alternate days × 4 with folinic acid 30 h after each injection of methotrexate. A treatment course of 8 days was followed by a rest period of 7 days. Some patients received in addition 6-azouridine 0.5 g 6-hourly × 8 before each course of methotrexate. The number of courses was determined by the estimated time to achieve complete tumour eradication.

Group B comprised ten patients whose tumours were judged to be likely to show resistance to methotrexate alone. They received five or more drugs in a series of courses each of 3–8 days duration, but usually not more than two drugs were given in any one course. These drugs included methotrexate, 6-azouridine, hydroxyurea, actinomycin D, cytosine arabinoside, vincristine, cyclophosphamide, adriamycin, melphalan, chlorambucil and bleomycin. Not all patients received all these drugs, and rest periods between drug courses were of 7–10 days duration.

Patients in group C were judged at the outset to have highly unfavourable prognostic features, and they were treated with a regimen that included seven or eight of the drugs listed for group B patients during a course of 6–9 days. Toxicity from these courses was marked, and the interval between successive courses was 10–20 days. Some of these patients also received intrathecal methotrexate and radiotherapy for brain metastases.

Cytogenetic Technique

Lymphocyte cultures were prepared from heparinised whole blood, 0.4 ml in 10 ml of medium composed of TC-199 and 20% AB serum + phytohaemagglutinin (Wellcome PHA reagent grade reconstituted 0.1 ml per 10 ml culture). The cultures were incubated at 37°C and terminated at 48 h except on occasions when, for convenience, they were terminated at 72 h.

Bone marrow was processed by a modification of the technique of Tjio and Whang (1962). The chromosomes were stained conventionally by Giemsa. The aim was to examine the chromosomes in 100 lymphocytes before, during and after treatment and in 100 bone marrow cells after treatment. The aberrations noted were gaps, breaks, fragments, abnormal chromosomes, polyploid cells and abnormal mitotic figures.

Results

All the patients had a normal 46,XX constitutional karyotype. The findings in the patients in group A are given in table 1. Three of the patients (397, 410 and 411) listed in table 1a had received a single dose of methotrexate before the first blood sample was taken. Patient 395, who contributed 1 gap, 1 fragment, 1 exchange figure and 1 endoreduplication to the total number of aberrations, had been given progesterone (Primulot Depot) in Lusaka before evacuation of a molar pregnancy. The aberration frequency

Table 1. Chromosome studies in lymphocytes of 7 patients in group A. (Ga = gaps, Br = breaks, Fr = fragments, Di = dicentrics, Ex = exchanges, En = endoreduplication.)

Patient number	Treatment status	Chromosome damage Ga	Fr	Ex	En	Br	Di	No. of cells
(a) *Before chemotherapy*								
394		1	0	0	0			100
395		1	1	1	1			100
397		1	0	0	0			100
400		2	1	0	0			100
402		3	0	0	0			100
410		0	0	0	0			100
411		1	1	0	0			100
Total		9	3	1	1			700
(b) *During chemotherapy*								
394	after 1 course	6				0		100
	after 2 courses	0				0		100
395	after 2 courses	0				0		100
397	after 3 courses	1				0		100
400	after 3 courses	1				0		100
	after 5 courses	0				0		100
	after 7 courses	0				0		32
402	after 3 courses	3				0		100
410	after 2 courses	0				0		40
411	after 2 courses	1				0		100
	after 6 courses	4				1		100
Total		16				1		972
(c) *After chemotherapy*								
395	2 weeks after 5 courses	1	2				0	100
400	6 weeks after 7 courses	0	0				0	100
402	6 weeks after 3 courses	1	0				0	100
410	1 week after 5 courses	1	3				1	100
	6 weeks after 5 courses	2	0				0	100
411	7 weeks after 8 courses	2	1				1	100
Total		7	6				2	600

found in these patients before a course of chemotherapy could be found in a group of healthy controls. During treatment (table 1b) and after treatment (table 1c) there was no significant increase in the frequency of chromosomal aberrations.

The data on the patients in group B are given in table 2. Patient 538, whose lymphocytes contained a dicentric chromosome before chemotherapy, had been given an anaesthetic a few days before the study was made. During

Table 2. Chromosome studies in lymphocytes of 12 patients in group B. (De = deletions, Da = fragmented chromosomes.)

Patient number	Treatment status	Chromosome damage: Ga	Br	Fr	Di	De	Da	No. of cells
(a) *Before chemotherapy*								
401		1	1	0	0	0	0	100
403		2	0	0	0	0	1	97
404		1	1	1	0	0	0	100
405		0	0	0	0	0	0	100
407		0	2	0	0	0	0	100
409		0	0	0	0	0	0	100
538		2	0	0	1	0	0	100
		6	4	1	1	0	1	697
(b) *During chemotherapy*								
396	after 4 courses	1	0	0	1	0	0	100
401	after 2 courses	0	0	0	0	0	0	40
403	after 3 courses	1	0	0	0	0	0	100
405	after 2 courses	2	0	3	1	1	0	100
409	after 2 courses	1	0	0	0	0	0	100
521	after 6 courses	3	0	0	0	0	0	100
	after 7 courses	2	0	0	0	0	0	100
530	after 6 courses	0	0	0	0	0	0	18
538	after 5 courses	0	0	0	0	0	0	100
		10	0	3	2	1	0	758
(c) *After chemotherapy*								
401	4 weeks after 9 courses	0	0	0	0	0	0	4
	10 weeks after 9 courses	2	0	0	0	0	0	100
403	<1 week after 12 courses	4	0	0	0	0	0	100
	8 weeks after 12 courses	1	0	2	4	0	0	100
404	4 weeks after 12 courses	2	0	0	0	0	0	100
405	1 week after 9 courses	0	0	0	0	0	0	30
	8 weeks after 9 courses	3	0	1	4	0	0	100
407	5 weeks after 8 courses	0	0	0	0	0	0	100
409	<1 week after 11 courses	2	1	1	1	0	0	100
	6 weeks after 11 courses	0	0	0	0	0	0	100
489	3 weeks after 17 courses	1	0	0	0	0	0	100
	6 weeks after 17 courses	1	0	0	0	0	0	100
521	1 week after 9 courses	5	0	0	0	0	0	100
530	8 weeks after 9 courses	4	0	0	0	0	0	100
537	1 week after 11 courses	0	1	0	0	0	0	66
	8 weeks after 11 courses	10	0	0	0	0	0	100
538	6 weeks after 9 courses	0	0	0	0	0	0	100
Total		35	2	4	9	0	0	1500

treatment the frequency of gaps and dicentric chromosomes was twice that found initially, and after treatment there were three times as many gaps and four times as many dicentric chromosomes. The increase in dicentrics was contributed by two patients, four being found in the lymphocytes of patients 403 and 405 eight weeks after completion of treatment.

The interpretation of the findings in the patients on intensive chemotherapy, table 3, is complicated by the potent effects of the irradiation they received on the chromosomes of lymphocytes. This is clearly demonstrated by patients 408 and 543. It should be noted that patient 519 had been treated intensively with drugs but not by irradiation.

Table 3. Chromosome studies in lymphocytes of patients on intensive therapy. (Ab = abnormal chromosomes, other than dicentrics or deletions; XRT = X-ray therapy; neu = neutron therapy.)

atient mber	Treatment status	Chromosome damage								No. of cells
		Ga	Br	Fr	Di	Da	Ex	En	Ab	
348	1 week after 16 courses	0	0	0	0	0	0	0	0	13
	1 week after 17 courses	2	0	0	0	0	0	0	0	100
	5 weeks after 18 courses	3	2	1	0	0	0	0	1 (Bq +)	100
	10 weeks after 18 courses	4	1	0	0	0	1	0	0	100
408	prior to therapy	0	0	0	0	0	0	0	0	100
	after 2 courses	5	1	0	0	0	0	0	0	100
	8 weeks after 25 courses + XRT	0	0	65	41	0	0	0	0 (rings)	36
	10 weeks after 25 courses + XRT	0	0	11	6	0	0	0	0	9
432	after 2 courses	2	4	0	0	0	1	0	0	84
	3 weeks after 18 courses	1	0	0	0	0	0	0	2 (minutes)	4
472	after 11 courses	1	1	1	1	0	0	0	2 (minutes)	50
	after 14 courses	0	0	0	0	1	0	0	0	50
	after 15 courses	0	0	0	0	0	0	0	0	50
	after 25 courses	16	3	0	0	0	0	0	0	100
	after 27 courses	1	1	0	0	0	0	0	0	31
515	after 13 courses	1	1	0	0	0	0	0	0	98
	after many courses + neu	6	0	3	1	0	0	0	1 (Dq +)	100
522	after 5 courses	0	0	0	1	0	0	0	0	100
	after 13 courses	2	0	2	2	0	0	0	0	100
543	after 7 courses + XRT skull	5	2	9	0	0	0	0	0	100
	after 13 courses + XRT skull	3	0	4	2	0	0	2	1 (ring)	100
519	after 6 courses	4	1	0	1	0	0	0	0	100
sger-	after 10 courses	1	0	0	0	0	0	0	1 (Dq +)	100
noma	after 11 courses	5	0	5	13	0	0	0	2 (rings)	100
	18 weeks after 11 courses	1	0	4	2	0	0	0	0	100

The observations on the effects of chemotherapy on the chromosomes of marrow (table 4) were hampered by poor yields or total failure to obtain any mitotic figures. In eight of the patients only 0–3 analysable mitoses were found in the bone marrow after treatment. In seven of these patients successful lymphocyte cultures had been obtained from samples taken on the same day.

Table 4. Chromosome studies in bone marrow of patients treated by chemotherapy.

Patient No. and group	Treatment status	Chromosome damage Ga	Br	Fr	Po	No. of cells
400 A	after 6 courses	0	0	0	0	23
402 A	after 3 courses	2	3	0	0	100
410 A	1 week after 5 courses	0	0	0	5	100
401 B	4 weeks after 9 courses	1	0	0	0	30
405 B	1 week after 9 courses	0	0	1	0	100
407 B	5 weeks after 8 courses	0	0	0	0	50
409 B	after 11 courses	1	1	0	0	100
489 B	6 weeks after 17 courses	0	0	0	0	50
530 B	2 weeks after 9 courses	0	0	0	0	29
538 B	6 weeks after 9 courses	0	0	0	0	3
408 Intensive	8 weeks after 25 courses + XRT	4	0	0	0	100
432 Intensive	3 weeks after 18 courses	0	0	0	0	13
519 Intensive	18 weeks after 11 courses	0	1	0	0	60
Total		8	5	1	5	758

Nevertheless the bone marrow chromosomes showed very few aberrations in patients in groups A and B.

Discussion

The patients in this study were all of child-bearing age so they were at risk for genetic damage to both somatic and gonadal cells. The question of the genetic risk to the progeny of patients treated in the same department has been studied by Walden and Bagshawe (1976) and it was found that although the previous reproductive function of treated patients was poorer than that of a control group because of fetal, wastage, subsequent performance was not made worse by chemotherapy. Van Thiel, Ross and Lipsett (1970) had reported normal reproductive performance in a group of patients treated for trophoblastic neoplasms, and Pastorfide and Goldstein (1973) found a higher rate of spontaneous abortions but no increase in congenital abnormalities. Thus there is no evidence that women who have had trophoblastic tumours have an increased risk of producing genetically damaged offspring following chemotherapy.

The next question is whether there is any long-term risk of a second malignancy amongst successfully treated patients. Only one such case has been detected among 400 patients followed-up at the Charing Cross Hospital, and this was an early cervical carcinoma *in situ* in a patient treated for choriocarcinoma six years earlier (Bagshawe 1977).

Because the behaviour of trophoblastic tumours can be monitored by measuring the HCG that they produce, the amount of chemotherapy is adjusted according to the response of the tumour. The prognosis for the patients in the low-risk category is extremely good, The amount of chemotherapy required to eliminate the tumour apparently does not produce a significant amount of damage in the chromosomes of lymphocytes or bone marrow. More evidence of the effects of the drugs on the chromosomes was shown in the patients in the medium-risk group, who had been given a variety of cytotoxic drugs in cycle. Still the overall picture is not of amounts of chromosome damage that give cause for concern.

There are some points of interest about the type of chromosome damage and the variation between individuals. The frequency of chromosomal gaps was increased by the treatment and this propensity of chemotherapy to increase the frequency of gaps in lymphocyte chromosomes has been observed amongst patients with a benign condition treated with chlorambucil (Reeves *et al.* 1975). The gap is undoubtedly a chromosomal lesion that, if repair is not effective, could lead to discontinuity in the genetic material. Amongst the patients in group B, two contributed the excess of dicentric chromosomes observed. This suggests considerable individual variation, in that other patients subjected to comparable amounts of cytotoxic therapy did not show much evidence of chromosome damage in lymphocytes.

Amongst the patients on intensive therapy there was much more evidence of chromosomal damage due to chemotherapy. When radiotherapy is given in addition, it is clear that this has an extremely potent effect in producing a high frequency of aberrations.

Breakage in normal bone marrow chromosomes is a rare event (Secker Walker 1971). The present study shows again the difference in the effect of clastogenic agents on lymphocytes and myeloid cells, which reflects the different kinetics of the two cell populations. As with irradiation, unstable cells are eliminated from the bone marrow because the cells are in mitotic cycle. Therefore, whilst aberrations were found in lymphocytes, they were not seen in myeloid cells in samples taken on the same day. The chemotherapy is myelodepressant, which accounts for a failure to obtain metaphases in the bone marrow aspirates of many of the patients.

This study has shown that when chemotherapy is given in the smallest possible effective dose, with planned spacing of courses, the genetic damage to somatic cells is minimised. Although cells with unstable abnormalities are eliminated and do not therefore give rise to clones of structurally abnormal cells, when the frequency of such unstable cells is increased that of cells with stable rearrangements may also rise. This was the case amongst patients with polycythaemia treated with ^{32}P who were found to have clones of cells with stable rearrangements in the bone marrow (Kay, Millard and Lawler 1966). Such clones may have a malignant potential. Therefore the lack of chromosomal damage observed amongst these patients in the low- and medium-risk categories is reassuring.

Summary

The chromosomes of the lymphocytes and bone marrow of patients treated by chemotherapy for trophoblastic tumours have been monitored for the frequency of aberrations before, during and/or after treatment. The patients were selected from three groups: low risk, medium risk and high risk. The patients in the low-risk group showed no evidence of a damaging effect of therapy on the chromosomes of the lymphocytes. Those in the medium-risk group showed increased chromosomal damage following therapy, but the increase was not judged sufficient to cause concern. Patients on intensive therapy, who had also been irradiated, showed a very high frequency of chromosomal damage in lymphocytes. Damage to bone marrow chromosomes was unremarkable in all groups. Survivors showed on the whole little evidence of genetic damage to somatic cells.

Acknowledgements

We should like to thank Professor K. D. Bagshawe for his help in the preparation of the manuscript and Mrs F. M. O'Malley for her help with the cytogenetics.

References

Bagshawe, K.D. (1976) Risk and prognostic factors in trophoblastic neoplasia. *Cancer 38*, 1373–85.

—— (1977) Lessons from choriocarcinoma, *Proc. Roy. Soc. Med. 76*, 303–6.

Court Brown, W.M. & R.Doll (1957) *Leukaemia and Aplastic Anaemia in Patients Irradiated for Ankylosing Spondylitis.* Medical Research Council Special Report Series No. 295. London: H.M.S.O.

Hossfeld, D.K., J.F.Holland, R.G.Cooper & R.R.Ellison (1975) Chromosome studies in acute leukaemias developing in patients with multiple myeloma. *Cancer Res. 35*, 2808–13.

Kay, H.E.M., R.E.Millard & S.D.Lawler (1966) The chromosomes in polycythaemia vera. *Br. J. Haematol. 12*, 340–2.

Pastorfide, G.B. & D.P.Goldstein (1973) Pregnancy after hydatidiform mole, *Obstet. Gynecol. 42*, 67–70.

Puri, H.C. & R.A.Campbell (1977) Cyclophosphamide and malignancy. *Lancet 1*, 1306.

Reeves, B.R., V.L.Pickup, S.D.Lawler, W.J.Dinning & E.S.Perkins (1975) A chromosome study of patients with uveitis treated with chlorambucil. *Br. Med. J. 4*, 22–3.

Secker Walker, L.M. (1971) The chromosomes of bone-marrow cells of haematologically normal men and women. *Br. J. Haematol. 21*, 455–61.

Van Thiel, D.M., G.T.Ross & M.B.Lipsett (1970) Pregnancy after chemotherapy of trophoblastic neoplasms. *Science 169*, 1326–7.

Tjio, J.H. & J.Whang (1962) Chromosome preparations of bone-marrow cells without prior *in vitro* culture or *in vivo* colchicine administration. *Stain Technol. 37*, 17–20.

Walden, P.A.M. & K.D.Bagshawe (1976) Reproductive performance of women successfully treated for gestational trophoblastic tumours. *Am. J. Obstet. Gynecol. 125*, 1108–14.

P. FISCHER, E. NACHEVA, J. POHL-RÜLING and P. KREPLER

Cytogenetic Effects of Chemotherapy and Cranial Irradiation on the Peripheral Blood Lymphocytes of Children with Leukaemia

The production of visible chromosome aberrations in eukaryotic cells is known to be a sensitive indicator of damage to the genetic apparatus (Kihlman 1966, Lea 1946). A direct correlation between dose and effect has been demonstrated in man for radiation-induced aberrations both *in vivo* (Fischer *et al.* 1966) and *in vitro* (Evans 1962), as well as for a large number of mutagenic and carcinogenic substances, in particular cytostatic drugs used in the treatment of malignant disease (Hampel *et al.* 1966), and in a number of non-malignant conditions (Jensen and Søborg 1966, Jensen 1967, Locher and Franz 1967, Ryan and Baker 1969). Although a significant difference in the frequencies of chemically-induced chromosome aberrations in human leucocytes in culture among donors of different age groups has been reported (Bochkov and Kuleshov 1972), the *in vivo* studies have almost exclusively been conducted on adult populations, and very little is known about the relative clastogenic effects of these agents on the chromosomes of children, in whom possible long-term mutagenic and carcinogenic hazards are of considerably greater importance.

It is self-evident that changes in treatment merely on the grounds of potentially harmful late effects cannot as yet be considered in the management of childhood leukaemia. We nevertheless felt that an investigation of the cytogenetic effects of present treatment schedules on the normal chromosomes of such patients could prove useful at some future time, when more drugs become available, and the life expectation of these children increases.

The present investigation was therefore planned to answer the following questions: Is there any increased tendency to breakage in the normal chromosomes of children with leukaemia? How great is the burden of aberrations being produced by present treatment schedules, and to what extent does cranial irradiation affect this burden?

For comparison, a group of children with a non-malignant disease, and, in cured cases, a normal life expectation, who have undergone treatment with corticosteroids and a single cytostatic drug for a restricted period of time, have been investigated.

Material and Methods

Selection of Patients

Eighteen children, aged 2 to 14, with nephrosis, nine on cortisone treatment alone, and nine cortisone-resistant cases subsequently treated with cyclophosphamide (Endoxan) made up the group of patients with non-malignant disease. 32 children with acute lymphatic leukaemia (ALL), of which 30 had received cranial irradiation in addition to intensive chemotherapy, one case of acute myeloid leukaemia (AML), and one of chronic myeloid leukaemia (CML), comprised the group of 34 children, aged 3 to 11, with leukaemia. Details of the modified Pinkels scheme, according to which the majority of patients with ALL were treated, are presented in table 1. A similar spectrum of drugs, without cranial irradiation, was used in the treatment of AML and CML.

Table 1. Modified Pinkel VII scheme of treatment in acute lymphatic leukaemia of children.[1]

Stage	Duration	Treatment	
Induction therapy	4 weeks	vincristine	1.5 mg m^{-2} weekly × 5
		prednisolone	60–90 mg m^{-2} daily × 28
		With 50000 blast cells in peripheral blood, additionally:	
		adriamycin	15 mg m^{-2} weekly
		asparaginase	10,000 units m^{-2} weekly
Cranial irradiation	21–26 days	2400 rad to cranial area and medulla oblongata	
		methotrexate	12 mg m^{-2} × 5 intrathecally
		6-mercaptopurine	25 mg m^{-2} daily p.o.
Maintenance therapy	up to 3 years	methotrexate	20 mg m^{-2} weekly p.o.
		6-mercaptopurine	50 mg m^{-2} daily p.o.
		cyclophosphamide	200 mg m^{-2} weekly p.o.
		Dosage adapted to leucocyte counts of 3–4000	
Pulses	every 3 months	vincristine	1.5 mg m^{-2} weekly × 2–3, i.v.
		prednisolone	40 mg m^{-2} daily × 14, p.o.
		methotrexate	12 mg m^{-2} intrathecally

[1] In 1974/75 another modification without cyclophosphamide was used. A few cases of AML were treated with courses of cytosine arabinoside + 6-thioguanine + prednisolone, some also with daunomycin.

Culture and Scoring Methods

Whole blood cultures using standard techniques were employed. Aliquots containing 0.4 ml of heparinised blood, 5 ml of culture medium RPMI 1640 + penicillin streptomycin and 0.1 ml PHA (Wellcome) were incubated at 37° in a closed system for 48 to 72 h. 1–2 h after addition of colcemid (0.1 $\mu g/ml^{-1}$), cultures were harvested by centrifugation, followed by 10 min hypotonic treatment (1:4 NaCl/aqua dest.) and at least two 10 min periods of fixation in methanol-acetic acid (3:1). The cell suspension was then dropped onto cold wet slides, air dried, and stained with Giemsa 10% for 7 min.

The scoring procedure was as follows: where possible 100 well-spread metaphases from each culture, considered unbroken on low-power scanning, were examined at 100 × and all deviations from the normal were noted. These included variations in chromosome number, chromatid aberrations (gaps, breaks, exchanges) and chromosome aberrations (deletions, minutes, dicentrics, rings). Abnormal monocentrics were noted but not included in the estimation of chromosome breaks. Deletions were counted as one break, and dicentrics and rings, with or without an accompanying fragment, as two breaks.

In both groups blood was cultured before treatment, as well as at various stages during treatment. However, as samples at all stages were not available in each case, and cultures were often unsatisfactory, especially with blood from the leukaemic children, results from individual patients at comparable stages have had to be pooled.

Results

Details of the cytogenetic findings are listed in table 2 for the children with nephrosis, and table 3 for the malignant cases. Dependence of aberration rates on time for both groups are presented in figure 1.

Nephrosis

Data from children with nephrosis were divided into three groups for

Table 2. Cytogenetic analysis of peripheral blood cultures in children with nephrosis treated with cortisone and cyclophosphamide.

Group[1] (no. of cases)	A (9)	B (6)	C (4)
No. of cells scored	1022	1033	400
Aneuploidy $n \neq 46$	8.2 ± 0.9	8.1 ± 0.9	5.3 ± 2.0
Polyploidy	0.9 ± 0.3	0.7 ± 0.3	0.5 ± 0.4
Chromatid aberrations (%)			
Gaps	4.8 ± 0.7	7.8 ± 0.9	5.0 ± 1.1
Breaks	1.0	2.5 ± 0.5	0.8 ± 0.4
Iso-breaks[2]	0.1 ± 0.1	—	—
Exchanges	0.1 ± 0.1	—	—
Total	5.9 ± 0.8	10.4 ± 1.0	5.8 ± 1.6
Chromosome aberrations (%)			
Deletions	—	1.4 ± 0.4	—
Minutes	—	—	—
Dicentrics + rings	—	0.1 ± 0.1	—
Total breaks[2]	0.1 ± 0.1	1.6 ± 0.4	—

[1] Group A: prednisolone 3 mg kg^{-1} d^{-1} for 4 weeks, subsequently 0.5–1 mg kg^{-1} d^{-1} for 18 to 120 months.
Group B: cyclophosphamide 75 mg m^{-2} d^{-1} + prednisolone 0.5 mg kg^{-1} d^{-1} for 1.5 to 6 months.
Group C: healed cases treated with cyclophosphamide and prednisolone, no treatment for more than 30 months.
[2] Deletions, iso-breaks = 1 break; dicentrics, rings, minutes = 2 breaks.

analysis as follows: nine patients who had received prednisolone only (group A), six cortisone-resistant cases treated additionally with cyclophosphamide for 1.5 to 6 months (group B), and four cured cases, previously treated with cortisone and cyclophosphamide, who had received no drugs for a period longer than 2.5 years before blood sampling (group C).

In group A 1022 cells were scored. The levels of gaps (4.8%) and chromatid breaks (1.0%), did not differ significantly from values found in children whose blood was cultured for possible cytogenetic abnormalities. There were no chromosome aberrations. In group B 1033 cells were scored. There was an increase in gaps (7.8%) and in chromatid breaks (2.5%). Chromosome aberrations (1.4% deletions and 0.1% dicentrics) were also present. In group C no increase in gaps and chromatid breaks could be demonstrated, and no chromosome aberrations were seen in 400 metaphases scored.

Malignant Disease

A detailed analysis of the cytogenetic findings in 34 children with leukaemia is presented in table 3. Results of successful cultures were divided into three groups, according to treatment, as follows: group I, untreated cases (10 cultures); group II, chemotherapy only (14 cultures), which included children treated under the Pinkel scheme prior to irradiation plus the patients with CML, AML, and two cases of ALL, who had received intensive chemotherapy only. Group III (41 cultures), was made up of cases of ALL at various periods after 2400 rad tele cobalt-irradiation to the cranial area. and continuous intensive chemotherapy.

In group I 1000 cells were scored. The chromatid aberration rates (4.1% gaps, 0.8% breaks) do not differ significantly from results obtained in the nephrosis groups. A small number of deletions were present, four in one thousand cells. These could have been the result of diagnostic radiology. There was thus no evidence of an increased spontaneous breakage rate in the normal chromosomes of the leukaemic children.

In group II 1400 cells were analysed at mean intervals of 1.2, 8.3 and 46 months after the beginning of chemotherapy. There was a steady increase in gaps, with mean values of 5.9, 7.3 and 12.0%, but chromatid aberration rates fluctuated from 3.4% at 1.2 months to 1% at 8.3 months and 5.3% at 46 months. Single chromosome aberrations (deletions) were present at all times (1, 0.7 and 1% respectively), but two-break aberrations, (dicentrics and minutes) were only seen after 36 months of chemotherapy.

In group III 4165 cells were scored. Findings were pooled at six increasing intervals of time after irradiation. Both types of chromatid aberration showed decreased rates immediately after irradiation; 4.4% gaps and 1.4% breaks as against 5.9% gaps and 3.4% breaks present in cultures of patients at the onset of remission and before irradiation (group II, 1.2 months).

The number of chromatid aberrations fluctuated during the first post-irradiation year, but rose steadily to 11.0% gaps and 7.6% chromatid breaks at a mean post-irradiation time of 32 months. Chromosome aberrations of

Table 3. Cytogenetic analysis of peripheral blood cultures in children with leukaemia treated with chemotherapy and cranial irradiation.

Group[1] (no. of cultures)	I (10)	II (14)			III (41)					
No. of cells scored	1000	800	300	300	500	601	764	700	900	700
Min-max (and mean) time of chemotherapy (months)	—	0.3–2 (1.2)	5–13 (8.3)	36–54 (46)	2–3 (2.8)	3–7.5 (5.3)	7–8 (7.3)	8–15 (11.7)	16–27 (22.3)	28–38 (34)
Min-max (and mean) time after irradiation (months)	—	—	—	—	0.03–0.6 (0.3)	1–4.5 (2.8)	5–6 (5.3)	6.5–12 (8.9)	13–25 (20.3)	26–36 (32)
Aneuploidy $n \neq 46$	7.3 ± 0.9	6.6 ± 0.9	6.0 ± 1.4	5.7 ± 1.4	8.8 ± 1.3	7.8 ± 1.1	8.1 ± 1.0	9.3 ± 1.2	6.6 ± 0.9	9.6 ± 1.0
Polyploidy	0.5 ± 0.2	—	1.3 ± 0.7	1.3 ± 0.7	0.8 ± 0.4	2.3 ± 0.6	0.9 ± 0.3	1.0 ± 0.4	0.9 ± 0.3	0.1 ± 0.1
Chromatid aberrations (%)										
Gaps	4.1 ± 0.6	5.9 ± 0.9	7.3 ± 1.6	12.0 ± 2.0	4.4 ± 0.9	7.0 ± 1.1	4.8 ± 0.8	5.9 ± 1.0	8.9 ± 1.0	11.0 ± 1.3
Breaks	0.8 ± 0.3	3.4 ± 0.7	1.0 ± 0.6	5.3 ± 1.3	1.4 ± 0.5	1.8 ± 0.6	3.9 ± 0.7	1.4 ± 0.5	3.0 ± 0.6	7.6 ± 1.0
Iso-breaks[2]	0.1 ± 0.1	0.1 ± 0.1	—	—	—	—	0.1 ± 0.1	—	—	—
Exchanges	0.1 ± 0.1	—	1.3 ± 0.6	—	—	—	0.3 ± 0.2	0.1 ± 0.1	0.2 ± 0.2	—
Total	5.0 ± 0.7	9.3 ± 1.1	9.6 ± 1.8	17.0 ± 2.4	5.8 ± 1.1	8.8 ± 1.2	9.0 ± 1.1	7.4 ± 1.0	12.1 ± 1.2	18.7 ± 1.7
Chromosome aberrations (%)										
Deletions	0.4 ± 0.2	1.0 ± 0.4	0.7 ± 0.5	1.0 ± 0.6	2.6 ± 0.7	0.8 ± 0.4	1.7 ± 0.5	1.1 ± 0.4	1.3 ± 0.4	2.9 ± 0.5
Minutes	—	—	—	0.7 ± 0.4	—	0.5 ± 0.3	0.5 ± 0.3	0.1 ± 0.1	0.2 ± 0.2	0.1 ± 0.1
Dicentrics + rings	—	—	—	0.3 ± 0.3	1.4 ± 0.6	2.2 ± 0.6	1.7 ± 0.5	1.1 ± 0.4	0.6 ± 0.3	0.6 ± 0.3
Abnormal monosomes	—	0.1 ± 0.1	—	—	0.5 ± 0.3	0.5 ± 0.3	0.1 ± 0.1	0.3 ± 0.2	0.3 ± 0.2	0.1 ± 0.1
Total breaks[2]	0.5 ± 0.2	1.1 ± 0.4	0.7 ± 0.5	3.0 ± 1.0	6.4 ± 1.0	6.2 ± 1.1	6.2 ± 0.9	3.7 ± 0.7	2.9 ± 0.6	3.4 ± 0.7

[1] Group I, before treatment; group II, chemotherapy alone; group III, chemotherapy + cranial irradiation.
[2] Deletions, iso-breaks = 1 break; dicentrics, rings, minutes = 2 breaks.

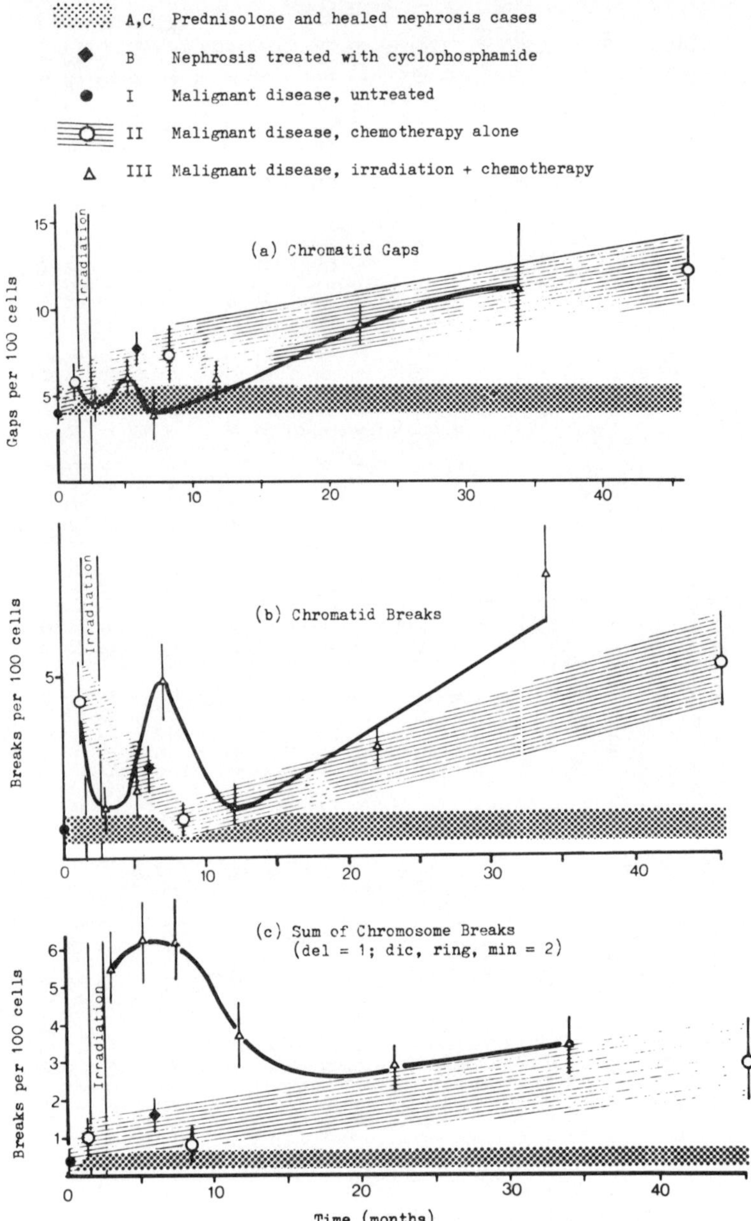

Figure 1. Aberrations in peripheral blood lymphocytes of children with leukaemia undergoing intensive chemotherapy and partial-body irradiation, and children with nephrosis treated with prednisolone and cyclophosphamide.

both types were present at all periods. Deletions were highest immediately after irradiation, 2.6% at a mean time of 0.3 months, and dicentrics at a mean time of 2.8 months post irradiation (2.2%). Aneuploidy (broken cells $n \neq 46$)

ranged from 6.6 to 9.6%. There was no dependence on disease, or time or type of treatment. However, 3 of the 32 cases of ALL were congenitally aneuploid, two with Down's Syndrome 47,XX (21 +) and one with a deletion of a G group chromosome 46,XY 21r.

The comparative effects, of chemotherapy and chemotherapy with partial body irradiation on the chromosomes of children with leukaemia, and of cortisone and short-term chemotherapy on children with nephrosis, are shown in figure 1. The dependence of the mean number of gaps with time is presented in figure 1a. In all leukaemic children the number of gaps increases after about twelve months of treatment and continues to rise with time. The relationship of chromatid break rates to type of treatment is more complex (figure 1b). Within the first twelve months of therapy there is a fluctuation in the number of chromatid breaks, but after one year the increase is steadily progressive. Chromosome aberrations are presented in figure 1c. With chemotherapy alone, a small but significant increase in the number of chromosome breaks is apparent. In children irradiated under the Pinkel VII scheme there is a six-fold increase in the number of chromosome breaks, which persists for about eight months after irradiation. About one year post-irradiation, aberration rates have dropped to levels present with chemotherapy alone.

Discussion

This study presents the first analysis of the cytogenetic effects of prolonged chemotherapy and chemotherapy combined with partial body irradiation on the lymphocyte population of *children* with leukaemia, and on the effects of cortisone and cyclophosphamide on children with nephrosis.

The cytogenetic effects of a number of drugs administered in this survey have been analysed in *adults* receiving a single preparation. Cortisone has been reported to produce no structural damage to human chromosomes (Jensen and Søborg 1966) whereas cyclophosphamide (Schmid and Bauchinger 1970), 6-mercaptopurine (Jensen 1967), arabinoside (Kihlman, Nichols, and Levan 1963) and adriamycin (Vig 1971) have been shown to cause varying degrees of damage. Reports on the effects of methotrexate (amethopterin) are conflicting. In studies on persons treated with amethopterin for long periods of time, Locher and Fränz 1967), Jensen and Søborg (1966), and Ryan and Baker (1969) reported structural damage to the chromosomes. However Sachsse and Denk (1970), Wilmanns and Martin (1968), and Vorhees *et al.* (1969) found no increase in aberrations even after prolonged administration.

In a study of the *in vitro* effects of several cytostatic drugs on human lymphocytes, Hampel *et al.* (1966) reported negative effects when both cyclophosphamide and methotrexate were added directly to the cultures. Addition to the cultures of serum from rats administered cyclophosphamide or methotrexate one hour previously resulted in the production of aberrations with cyclophosphamide, but results with methotrexate remained negative. Cyclophosphamide requires activation by the liver to the active compound and can therefore not be effective *in vitro*. Methotrexate, an inhibitor of folic acid synthesis, seems to be ineffectual as a clastogen.

Our results are in agreement with the findings reported above. Corticosteroids caused no structural defects to the chromosomes, even when given over long periods of time to children with nephrosis. The use of much larger doses, combined with the spindle-inhibitor vinblastine, also produced minimal damage to the lymphocytes in leukaemia.

In cortisone-resistant cases cyclophosphamide produced a significant but transient increase in chromatid aberrations. However, two years after the cessation of treatment, no increase over normal levels could be detected.

In the children with malignant disease undergoing intensive chemotherapy, a number of drugs were given at various times and in varying doses, and not all patients received the same treatment. It was therefore not possible to evaluate the contribution of a particular preparation, and our results represent only a mean effect of prolonged cytostatic treatment on the circulating lymphocyte.

After about one year of treatment, chromatid aberrations reach levels significantly higher than in untreated cases and continue to increase (figures 1a, b), no equilibrium between aberration production and elimination having been reached up to the end of our investigation (46 months). Preliminary investigations suggest, however, that this is not associated with an increase in sister chromatid exchanges (SCE). This would support the findings of Latt *et al.* (1975), Perry and Evans (1975), Shiraishi and Sandberg (1976), and Wolff, Rodin and Cleaver (1977), which suggest that SCEs could be the result of fundamentally different cellular events and lesions than are chromosome aberrations.

The nature of gap formation has not yet been fully explained. Bender, Griggs and Bedford (1974) proposed that they are caused by single-strand lesions in the DNA, and that all visible structural aberrations are the result of damage exclusively confined to the DNA molecule. Comings (1974) suggested that gaps could be an indication of localised defects in folding of the chromatid brought about by stereochemical changes in the protein moiety. His hypothesis is supported by Brøgger (1975), who produced gaps with β-mercaptoethanol, a substance that is known to denature protein by breaking S-S bridges, leaving the DNA unaffected. The significant increase in gaps in cultures from our leukaemic children could have been caused by either mechanism. The fluctuation in the number of chromatid breaks (figure 1b) during the first years of chemotherapy, as well as in cultures from patients receiving cranial irradiation, is possibly attributable to the intermittent nature of the treatment, in which the spectrum of drugs administered as maintenance therapy is periodically boosted by additional combinations according to the clinical condition of the individual patient (table 1). In cases treated under the Pinkel VII scheme there is, additionally, a decrease in the number of chromatid breaks during the first six post-irradiation months. This coincides with the period of maximal production of chromosome aberrations, and could signify an increased elimination of cells damaged consecutively by cytostatic drugs and irradiation.

The significant increase in chromosome aberrations in children who received chemotherapy alone is noteworthy and could be in accordance with the findings of Hsu, Pathak and Kusyk (1975), who have recently reported the presence of both chromatid and chromosome aberrations in cells pulse-treated with intercalating compounds such as adriamycin. They suggest that new damage to the chromatin may be generated by treatment with such compounds, through the subsequent action of molecules of the drug that remain in the cytoplasm of the cells and enter the nuclei at a later stage. In the lymphocyte stimulated to divide by PHA, damage caused by the incorporation of such molecules could conceivably result in the formation of chemically-induced chromosomal aberrations.

Irradiation of the cranial area in children with acute lymphatic leukaemia does not result in exposure of lymph glands or blood-forming tissues. Consequently the lymphocytes at risk are only those in the peripheral blood passing through the treatment field at the time of irradiation. Immediately after irradiation there is a six-fold increase in the number of chromosome breaks, but after about eight months this drops to values present for children who received chemotherapy alone (figure 1c).

All our data suggest, therefore, that in the treatment of ALL cranial irradiation does *not* contribute significantly to the component of chromosomal aberrations present in the peripheral blood lymphocytes after a post-irradiation period of one or two years.

Conclusions

Children with Nephrosis

(a) Prednisolone at dose levels of up to 3 mg kg^{-1} daily for 4 weeks, then reduced to 0.5–1 mg kg^{-1} daily, does not cause visible structural damage to the chromosomes, even after long periods of administration.

(b) Cyclophosphamide at doses of 75 mg m^{-2} for 1.5 to 6 months causes a small but significant increase in the number of chromatid aberrations. Thirty months after treatment, aberration rates have in most cases returned to normal levels.

Children with Leukaemia

(a) Treatment with heavy doses of corticosteroids (60–90 mg m^{-2} daily), in combination with vincristine 1–5 mg m^{-2} weekly, for about one month does not result in a significant increase in chromosomal aberrations. Prolonged, intensive chemotherapy, however, causes a significant and rising number of chromatid aberrations, both gaps and breaks.

(b) After about two years of treatment, chemotherapy alone causes a small but significant increase in the number of chromosome breaks.

(c) Radiation of 2400 rad delivered in three doses of 800 rad over a period of 21–26 days to the cranial area and medulla oblongata caused an immediate sharp rise in the number of chromosome aberrations and a concurrent drop in chromatid aberrations. In about one year, both levels revert to that present with chemotherapy alone.

(d) There is *no* evidence of an increased, spontaneous breakage rate in the normal chromosomes of children with leukaemia.

Summary

Results of a cytogenetic analysis of peripheral blood lymphocytes of children with leukaemia after massive chemotherapy and cranial irradiation, and of children with nephrosis after prednisolone therapy and cyclophosphamide, are presented. Prolonged intensive chemotherapy results in a significant rise in the number of chromatid aberrations after twelve months, and of chromosome aberrations after 24 months of therapy. After cranial irradiation, a sharp rise in chromosome aberrations is present for about three months. This drops after one year to levels present in cases with chemotherapy alone.

Acknowledgements

We are indebted to Miss Evelyn Hebrard for scoring the cases and to Mr Paul Breit for preparing the figures. Dr Margrit Kummer and sister Mary Trifter of the St Anna Children's Hospital kindly supplied us with the blood samples.

References

Bauchinger, M. & E.Schmid (1969) Cytogenetic changes in white blood cells after cyclophosphamide treatment. *Z. Krebsforsch. 72*, 77–87.

Bender, M.A., H.G.Griggs & J.S.Bedford (1974) Mechanisms of chromosomal aberration production. III. Chemicals and ionizing radiation. *Mutat. Res. 23*, 197–212.

Bochkov, N.P. & N.P.Kuleshov (1972) Age sensitivity of human chromosomes to alkylating agents. *Mutat. Res. 14*, 345–53.

Brøgger, A. (1975) Is the chromatid gap a folding defect due to protein change? Evidence from mercaptoethanol treatment of human lymphocyte chromosomes. *Hereditas 80*, 131–6.

Comings, D.E. (1974) What is a chromosome break?, in *Chromosomes and Cancer* (ed. J.German) pp.95–133. New York: Wiley.

Evans, H.J. (1962) Chromosome aberrations induced by ionizing radiations. *Int. Rev. Cytol. 13*, 221–321.

Fischer, P., E.Golob, E.Kunze-Muehl, A.Ben Haim, R.A.Dudley, T.Muellner, R.M. Parr & H.Vetter (1966) Chromosome aberrations in peripheral blood cells in man following chronic irradiation from internal deposits of Thorotrast. *Radiat. Res. 29*, 505.

Hampel, K.E., B.Kober, D.Rösch, H.Gerhartz & K.H.Meining (1966) The action of cytostatic agents on the chromosomes of human leukocytes *in vitro* (preliminary comm.). *Blood 27*, 816–23.

Hsu, T.C., S.Pathak & C.J.Kusyk (1975) Continuous induction of chromatid lesions by DNA-intercalating compounds. *Mutat. Res. 33*, 417–20.

Jensen, M.K. (1967) Chromosome studies in patients treated with azathioprine and amethopterin. *Acta Med. Scand. 182*, 445–55.

Jensen, M.K. & M.Søborg (1966) Chromosome aberrations in human cells following treatment with imuran. *Acta Med. Scand. 179*, 249–50.

Kihlman, B.A. (1966) *Actions of Chemicals on Dividing Cells.* New Jersey: Prentice Hall.

Kihlman, B.A., W.W.Nichols & A.Levan (1963) The effect of deoxyadenosine and cystosine arabinoside on the chromosomes of human leukocytes *in vitro. Brief Reports*, 139–43.

Krepler, P., A.Rosenkranz, I.Obiditsch-Mayer, L.Stockinger, R.Schatz & H.Fritzsche (1970) Cyclophosphamidtherapie des nephrotischen Syndroms im Kindesalter. *Arch. Kinderheilkunde 181*, 268–82.

Latt, S.A., G.Stetten, L.A.Juergens, G.R.Buchanan & P.S.Gerald (1975) Induction by alkylating agents of sister chromatid exchanges and chromatid breaks in Fanconi's anemia. *Proc. Natl. Acad. Sci. U.S.A. 72*, 4066–70.

Lea, D.E. (1946) *Actions of Radiations on Living Cells.* Cambridge: University Press.

Locher, H. & J.Fränz (1967) Chromosomenveränderungen bei Methotrexatbehandlungen. *Med. Welt. (N.F.) 18*, 1965–6.

Perry, P. & H.J.Evans (1975) Cytological detection of mutagen-carcinogen exposure by sister chromatid exchange. *Nature 258*, 121–5.

Ryan, T.J. & H.Baker (1969) Systemic corticosteroids and folic acid antagonists in the treatment of generalized pustular psoriasis. Evaluation and prognosis based on the study of 104 cases. *Br. J. Dermatol. 81*, 134–45.

Sachsse, W. & R.Denk (1970) Chromosomenuntersuchungen aus Lymphocyten und Knochenmark von Psoriasis-Kranken nach Langzeitbehandlung mit Methotrexat. *Arch. Klin. Exp. Dermatol. 239*, 275–81.

Schmid, E. & M.Bauchinger (1970) Inter-und intrachromosomale Verteilung von Cyclophosphamid-induzierten Chromosomendefekten beim Menschen. *Mutat. Res. 9*, 417–24.

Shiraishi, Y. & A.A.Sandberg (1976) Caffeine and sister chromatid exchange. *Proc. Japan. Acad. 52*, 379–82.

Vig, B.K. (1971) Chromosome aberrations induced in human leukocytes by the antileukemic antibiotic adriamycin. *Cancer Res. 31*, 32–8.

Voorhees, J.J., M.K.Jansen, E.R.Harrell & S.G.Chakrabarti (1969) Cytogenetic evaluation of methotrexate-treated psoriatic patients. *Arch. Dermatol. 100*, 269–74.

Wilmanns, W. & H.Martin (1968) Wirkung und Verteilung von Methotrexat im menschlichen Organismus bei der Leukämiebehandlung. *Klin. Wschr. 46*, 281–96.

Wolff, S., B.Rodin & J.E.Cleaver (1977) Sister chromatid exchanges induced by mutagenic carcinogens in normal and xeroderma pigmentosum cells. *Nature 265*, 347–9.

Chromosomal Analysis of Exposed Populations: A Review of Industrial Problems

The history of the study of chromosomal abnormalities in humans exposed to chemicals is relatively short. The first studies appear to have been carried out in 1964 by Pollini and Colombi (1964a, b) and Vigliani (1964), who published papers on the abnormalities observed in the chromosomes of peripheral lymphocytes of benzene-exposed workers. Subsequently, these observations on benzene have been extended by Forni and co-workers (1966, 1967, 1971), Tough and Court Brown (1965), and Tough *et al.* (1970). The conclusions that can be drawn from these investigations are, firstly, that benzene-exposed workers develop chromosomal aberrations. Secondly, in a few of the studies, usually when examining groups of workers exposed to low levels of benzene, there was no significant increase in the level of chromosome abnormalities. In some of the work reported, none of the other variables measured were affected, suggesting that chromosomal abnormalities may be one of the most sensitive methods of screening benzene exposed workers.

The next important observation was made by Tough *et al.* (1970), who showed that the level of chromosomal abnormalities observed in benzene workers increased with age, thereby identifying one of the confounding factors of surveys of this type.

Studies of the effects on chromosomes after human exposure to other chemicals tend to be more limited than those on benzene. Several anti-cancer drugs have been shown to produce such abnormalities after treatment. These include azothioprine (Jensen 1967), daunomycin (Whang-Peng *et al.* 1969), cytosine arabinoside (Bell *et al.* 1966), methotrexate (Jensen 1967) and 6-mercaptopurine (Bischoff and Holtzer 1967). In view of the general cytotoxic and carcinogenic effects of some of these anti-cancer drugs it is not surprising that they produce clastogenic effects. Other drugs that have been associated with an increase in chromosomal abnormalities of circulating lymphocytes are chlorpromazine, lysergide, metromidazole, perphenazine (Nielsen, Friedrich and Tsuboi 1969), phenylbutazone (Stevenson *et al.* 1971). There are conflicting reports regarding the effects of LSD (Cohen, Hirschhorn and Frosch 1967) and diazepam (Cohen, Hirschhorn and Frosch 1967, 1969; Staiger 1970) with some workers describing chromosomal abnormalities that have not been confirmed by subsequent work. Marihuana, heroin and amphetamines have also been shown to increase chromosome aberrations (Gilmour *et al.* 1971). Three drugs, namely oral contraceptives (Bishun *et al.* 1973), prednisone (Jensen 1967) and lithium carbonate (Jarvik *et al.* 1971), have been shown not to produce chromosome abnormalities in exposed populations. Relatively few surveys have been carried out on industrial populations. Probably the best, and most extensive, analyses have been carried out on benzene and these have been described above. In addition, vinyl chloride (Ducatman, Hirschhorn and Selikoff 1975; Funes-Cravioto *et al.* 1976; Purchase, Richardson and Anderson 1975, 1976; Szentesi *et al.* 1976), ethylene oxide (Kallings 1974), epichlorhydrin (Kucerova and Zhurkov 1977) and arsenic (Petves, Schmid-Ullrich and Wolf 1970) have been shown to produce clastogenic effects. The recent report that exposure to mercury results in aneuploidy and chromosomal aberrations is open to some question. Similarly the reports that exposure to lead (Schwanitz, Lehart and Beghart 1970) and spray adhesives (Cervenka and Thorn 1974) results in chromosomal aberrations have not been confirmed (O'Riordan and Evans 1974, Hook *et al.* 1974). Cadmium (O'Riordan and Evans, pers. comm.) and DDT (Rabello *et al.* 1975) exposure did not increase chromosomal aberration levels.

When reviewing the information on clastogenic effects of industrial chemicals in exposed populations, one is struck by the fact that in most cases the studies are limited to identifying a clastogenic effect. Little attempt seems to have been made to identify what factors in the industrial exposure have contributed to the increase in abnormalities, what confounding factors may be interfering with the interpretation, and what the consequences of the increase in clastogenic effects are to the population exposed.

Significance of Clastogenic Effects

Mutagenicity

If, after taking account of the relevant confounding factors, a chemical is shown to produce clastogenic effects in exposed populations, there is a clear indication that the chemical induces damage to genetic material. The consequences of genetic damage to a lymphocyte that is a non-dividing somatic cell appear to be limited. In the event that a lymphocyte is induced to divide, the majority of chromosomal abnormalities will result in the death of the daughter cells. The significance, therefore, lies in the possibility that either chromosomal aberrations or other mutations, such as point mutations, will occur in other cells in the body. It would be particularly significant if these mutations were produced in the germ cells. In the absence of any firm data on the heritable effects of chemicals in man this link must remain speculative.

One direct non-heritable consequence of germ cell chromosomal mutations would be the dominant lethal effect. Again, there is extreme paucity of data on the dominant lethal effects of chemicals in man, with only one study, that of Infante and co-workers (1976a, b, c), which claims that vinyl chloride induces fetal loss in the wives of exposed workers. This work has been criticised (Paddle 1976) and the conclusions will have to be substantiated. Thus, although chemically induced mutational effects are theoretically possible in man, we have no definitive data on which to determine their practical significance.

Carcinogenicity

From the earliest days of the study of human chromosomes, Theodor Boveri (Wolf 1974) advanced the hypothesis that chromosomal aberrations were the initiating event in the neoplastic process. Studies of tumour chromosomes since that time have demonstrated that a large number of tumours are associated with chromosomal changes. However, Levan (1969) has concluded that most chromosomal disturbances do not lead to cancer. If chromosomal abnormalities are the initiating event in the neoplastic process, one might expect to see specific karyotypes associated with specific tumour types. The only example of this is in chronic myeloid leukaemia, which is associated with the presence of the Philadelphia chromosome (Nowell and Hungerford 1960). Another suggestive example is that of venereal sarcoma of the dog, which has a common karyotypic abnormality (Nakino 1974). This case, however, is not conclusive, because it is possible that the cells in venereal sarcoma have all derived from the same stem cell. In all other tumours there is a confusing variety of chromosomal aberrations present in the tumour. Evidence is accumulating, however, that certain tumour types are associated with chromosomal abnormality patterns rather than specific karyotypic abnormalities. This does not seem to be sufficient evidence to suggest that chromosomal aberrations are the initiating step in the neoplastic process. In summary, it is possible to state that most neoplasms are associated with chromosomal abnormalities, but it is still debatable whether these abnormalities are primary or secondary phenomena, and whether they are involved in the initiation of the neoplastic process or whether they are merely associated with the progression of tumour growth.

An additional factor that needs to be taken into account is that the lesion produced in the lymphocyte by the chemical is occurring in a cell that does not become neoplastic. The exception to this is, of course, benzene, where leukaemia has been associated with benzene exposure. With other chemicals, such as vinyl chloride where the tumour originates from the hepatic endothelial cells, there can be no direct significance of the induction of chromosome abnormalities in circulating lymphocytes in the carcinogenic process. The significance of clastogenic effects to the *individual* can be deduced from the above comments and from the knowledge that there is no known effect on health that results directly from chemically induced chromosome aberrations in circulating lymphocytes. There is also evidence from people exposed to radiation, which suggests that although radiation increases chromosomal abnormalities in circulating lymphocytes, and although there is an increased risk of cancer in groups exposed to high levels of radiation, it is impossible to identify on the basis of chromosome abnormalities which individuals will eventually develop leukaemia (Nowell 1969). Drugs (Bishun *et al.* 1973), virus infections (Harnden 1974) and radiation can increase the level of chromosome abnormalities, and, therefore, an increase in levels observed on a single occasion in an individual exposed to a chemical, cannot with certainty be ascribed to the chemical exposure. One can, therefore, deduce that an increase in chromosomal aberrations observed as a consequence of a single analysis is not sufficient evidence to suggest that that individual is at risk. If, however, there is a steadily increasing level of chromosomal abnormalities on multiple consecutive samples, which can then be associated with exposure to the chemical, one may be able to conclude that the individual is exposed to too high a level of that chemical.

The significance of an increased chromosome abnormality level in the circulating lymphocytes of a *group* of exposed workers is somewhat different. If this increase is a consequence of exposure to a chemical about which nothing is known of its carcinogenicity, it is a clear indication that information on its carcinogenicity will be required urgently. In the meantime exposure to that chemical should be examined and reduced where possible. Even where the chemical concerned is an established carcinogen, the significance of clastogenic effects is tenuous. We know that there is a rough dose-response relationship for radiation-induced chromosomal abnormalities and cancer incidence (Nowell 1969). Similar dose-response relationships also probably occur for chemicals. Therefore, although individual prognosis cannot be derived from chromosomal data, an increase in chromosomal aberrations in an exposed group indicates exposure to higher levels of the chemical than in an exposed group with normal chromosomal aberration levels. This in turn suggests that the group has a higher risk of developing cancer than a group without increased levels, but the size of the increased risk will be unknown.

Confounding Factors

The results of a study on a population exposed to a chemical can be influenced by a number of factors that are unrelated to the exposure. These factors need to be identified and incorporated into the study design if the interpretation of the results is to be simple. There are several technical factors that need to be taken into account, such as length of time in culture, the variation ascribed to different individuals reading the slides (Evans and O'Riordan 1975) and possibly even the length of time between blood sampling and culturing. The increase in chromosomal abnormalities associated with age have already been mentioned (Tough *et al.* 1970). These factors are generally fairly easy to incorporate into the study design. However, it is known that virus infections (Harnden 1974) and exposure to drugs (Bishun *et al.* 1973) can also increase chromosomal aberration levels, and these may be much more difficult to include in the study design.

A recent study in our laboratory has identified three other factors that can influence the results. The first of these is the duration of exposure. In our study population exposed to vinyl chloride there was a direct correlation between the length of exposure and the level of chromosome abnormalities. The second factor is smoking. Although in the control group in this study smoking had no influence on chromosome abnormalities, in those exposed to vinyl chloride there was an increase in abnormalities associated with smoking. Probably the most difficult (third) factor to take account of is that of variations in levels of exposure. In our study there was some association between the job description (which in turn is associated with level of exposure) and the level of chromosome abnormality; thus autoclave workers appeared to have a greater increase in chromosomal abnormalities than the other groups of exposed workers. The confounding factor in this situation is high exposure over short duration. Exposure to these peaks in our study was assessed by asking the workers whether they had exposure to levels that could be smelled during the preceding twelve months. Those who recorded a history of exposure peaks had higher levels of chromosome abnormalities than those who did not, and this occurred in all job descriptions where a history of exposure peaks was identified.

The interesting feature of this observation is that it is similar to the findings reported after exposure to radiation, where exposure at a high dose rate induced more chromosome aberrations than an equivalent dose given at a low dose rate. In one study the tumour incidence in mice exposed to radiation was high when the radiation was given at a low dose rate and low when given at a high dose rate (Nowell and Cole 1965). These observations suggest that any study designed to observe effects of chemicals on chromosomes of exposed populations (particularly if they are comparing clastogenic and carcinogenic effects) must take into account smoking history and the duration, level and excursion of exposure.

One further problem encountered in studies of this type is the selection of the method for statistical analysis of the data. The data from chromosomal analysis are generated as the number of any individual abnormalities in a cell. However, the results from a worker, which probably includes data from a large number of cells, constitutes one observation for that group of workers. When comparing the results of different groups of workers it will be the variability between observations (i.e. workers) within each group that must be used for statistical analysis. It should also be remembered that the form of the data generated is such that simple statistical techniques, such as a *t*-test, are not directly applicable. An example of this problem is provided by a report on chromosomal aberrations in mercury-exposed workers in which results are expressed as percentages of abnormalities in each group, which may invalidate the later statistical analyses.

Prospects for the Future

The work described above establishes that there is a close relationship between the clastogenic and carcinogenic effects of a chemical, and that there is a dose-response relationship evident in some studies of chemical clastogens. This suggests that there may be several ways in which studies on the clastogenic effects of chemicals may be developed.

Compound identification. Established methods for identifying chemical carcinogens (chronic animal studies and epidemiology in man) are time-consuming, expensive and sometimes not very sensitive. If nothing is known about the carcinogenicity of a chemical to which people are exposed, an assessment of the priority that must be assigned to the testing programme could be made after data on clastogenic effects were obtained. The extrapolation of the results of experimental carcinogenicity studies to man is also not well understood. For example, in the case of DDT, which produces hepatic neoplasia in mice, there is no evidence for carcinogenic effects in man. Similarly, no clastogenic effects have been observed in DDT-exposed human populations (Rabello *et al.* 1975). This suggests that the study of chromosomes in exposed populations may answer the question whether chemicals of this sort are active in humans. Such studies are relatively short and cheap, and they may offer an alternative to the standard epidemiology as a means of obtaining information on biological activity directly relevant to man.

'Safe' dose. The best way of determining a safe dose of a carcinogen may well be a very careful epidemiological study. However, at low doses this is likely to take many decades and may be insensitive, particularly if the target organ has a high incidence of cancer. In the intervening period while the epidemiological data are being painstakingly accumulated, a working hygiene level needs to be established. Once it is established that a chemical is clastogenic in an exposed human population it may be possible to establish what level of exposure produces no effects on the chromosomes. The establishment of hygiene standards that reduce exposure to a level that will not produce chromosome abnormalities is in itself a reassuring step. If some assumptions are made, however, we may be able to take this further. If we assume that cancer is a consequence of somatic cell mutation, and that chromosome aberrations in lymphocytes are produced at the same dose as other types of mutations in other cells that might be involved in cancer induction, it might be reasonable to deduce that an exposure level that produces no chromosome abnormalities will also not produce cancer. There is, however, little evidence to confirm these assumptions. The consequences of this argument are, therefore, that when (a) it is established that a chemical is clastogenic and carcinogenic, and (b) the chemical is of sufficient social and economic consequence that manufacture must continue, we should explore the use of monitoring the workforce by chromosome analysis. This effort will need to be combined with epidemiology and careful assessment of exposure levels so that in due course the quantitative relationship between these different factors is established.

Population monitoring. Once it is established that a chemical is clastogenic in man, and whether or not a concept of a 'safe' dose in terms of carcinogenicity is valid, there may still be a role for continued population monitoring. The presence of clastogenic effects indicates damage to genetic material and, therefore, it is reasonable that we should strive towards hygiene practices that would avoid such an effect. Control of exposure to industrial chemicals is usually achieved by measuring atmospheric levels and ensuring that these are kept below the threshold limit value. Fixed-point monitoring of atmospheric levels throughout a plant is open to the criticism that it does not measure representative samples in terms of human exposure. In order to overcome this the relatively more cumbersome and difficult technique of personal atmospheric monitoring may be used. Although this may provide an approximate estimate of the level of exposure of an individual, there are still situations where it is not entirely satisfactory; and even if levels below the threshold limit value are achieved, this value may not be based on information on chromosomal damage. It is arguable, therefore, that when dealing with clas togenic chemicals an appropriate method of monitoring human exposure would be to examine clastogenic effects.

Individual idiosyncrasy. In any epidemiological or experimental study of the effects of carcinogens in mammals there are marked differences between the susceptibilities of individuals to the carcinogen. It is reasonable to assume that one of the major contributing factors to these differences is the individual's ability to metabolise the compound. In particular, the balance between the production of active metabolites and their detoxification may be related to individual susceptibility. Thus, in any situation in which people are exposed to clastogenic chemicals, any individual with consistently higher chromosomal aberration levels is either more susceptible to the effect of that chemical, or is prone to practices in the operation of his job that allow him to be exposed to higher levels than his colleagues. In either case steps could be taken to reduce exposure in such a way that his chromosome abnormality levels were not consistently high.

Problems for the Future

The largest technical problem facing those who wish to use the study of chromosome aberrations routinely is that it requires a high degree of technical skills and a lot of time. These factors tend to reduce its applicability for routine monitoring. One of the challenges for the future is to develop alternative methods, which will be capable of estimating chromosomal damage with the same degree of sensitivity as the classical technique but which will take a far shorter period. It has recently been suggested that analysis of sister chromatid exchange (SCE), which is much simpler and easier than conventional chromosome studies, may be useful in testing clastogenic effects. However, a major disadvantage appears to be that SCEs only remain in the lymphocytes for a relatively short period (about one to two weeks after exposure) (Stetka and Wolff 1976). In any situation where fluctuating or long-term exposures occur these flaws in the SCE technique will render it less useful than the conventional method.

The second major challenge for the future appears to me to be the identification of confounding effects. Until we know with great accuracy what factors other than exposure to the chemical influence chromosome abnormality levels we will be unable to provide optimally designed experimental studies on human populations. If techniques are improved and confounding factors are identified, we may have a technique that is of immense importance in controlling exposure to carcinogenic and mutagenic chemicals.

Summary

The most obvious aspect of detecting clastogenic effects in the lymphocytes of exposed populations is that genetic damage has been induced in somatic cells, but as such the consequences to the individual or the population exposed remain uncertain. The consequence of germ cell mutations to succeeding generations is understood, but has not yet been demonstrated unequivocally as causally related to chemical exposure. Potentially the most useful aspect of clastogenic effects may derive from the growing evidence that cancer may be induced by somatic mutations, and that many carcinogens are clastogenic. In the near future, therefore, emphasis should be on the utility of studying abnormal aberration levels as a means of solving problems associated with potential and actual occupational carcinogens.

Any monitoring programme has to take account of a number of confounding factors, such as virus infection, drug therapy, and X-rays. Synergism between the effect of chemicals and of smoking are also important. Once these have been adequately identified, studies should be designed to take account of them. Studies on exposed populations can be used for compound identification, population monitoring, determination of 'safe' doses and identification of individual idiosyncrasy. In considering the results obtained in these studies there is a clear distinction between the significance to the individual and to the exposed group.

A major constraint in applying chromosomal analysis techniques to exposed populations is the expense and the resources required. Simpler effective alternatives (such as the sister chromatid exchange technique) will have to be developed and evaluated if this type of monitoring is to become widely used.

References

Bell, W.R., J.J.Whang, P.P.Carbone, G.Brecher & J.B.Block (1966) Cytogenetic and morphologic abnormalities in human marrow cells during cytosine arabinoside therapy. *Blood 27*, 771–81.

Bischoff, R. & H.Holtzer (1967) The effect of mitotic inhibitors on myogenesis *in vitro. J. Cell Biol. 36*, 111–27.

Bishun, N.P., D.C.Williams, J.Mills, N.Lloyd, R.W.Raven & D.V.Parke (1973) Chromosome damage induced by chemicals. *Chem. Biol. Interactions 6*, 375–92.

Cervenka, J. & H.Thorn (1974) Chromosomes and spray adhesives. *New Engl. J. Med. 290*, 543.

Cohen, M.M., K.Hirschhorn & W.A.Frosch (1967) *In vivo* and *in vitro* chromosomal damage induced by LSD–25. *New. Engl. J. Med. 277*, 1043–9.

—— (1969) Cytogenetic effects of tranquilising drugs *in vivo* and *in vitro. J. Am. Med. Ass. 207*, 2425–6.

Ducatman, A., K.Hirschhorn & I.J.Selikoff (1975) Vinyl chloride exposure and human chromosome aberrations. *Mutat. Res. 31*, 163–8.

Evans, H.J. & M.L.O'Riordan (1975) Human peripheral blood lymphocytes for the analysis of chromosome aberrations in mutagen tests. *Mutat. Res. 31*, 135–48.

Forni, A. (1966) Chromosome damages due to chronic exposure to benzene, in *Proc. XV Int. Congr. Occupational Health, II/1*, pp.437–9. Vienna: Wiener Medizinische Akademie.

Forni, A. & L.Moreo (1967) Cytogenetic studies in a case of benzene leukaemia. *Europ. J. Cancer 3*, 251–5.

Forni, A., E.Pacifico & A.Limonta (1971) Chromosome studies in workers exposed to benzene or toluene or both. *Arch. Environ. Health 22*, 373–8.

Funes-Cravioto, E., B.Lambert, J.Lindsten, L.Ehrenberg, A.T.Natarajan & S.Osterman-Golkar (1976) Chromosome aberrations in workers exposed to vinyl chloride. *Lancet 1*, 459.

Gilmour, D.G., A.Bloom, K.Lele, E.Robbins & C.Maximillian (1971) Chromosomal aberrations in users of psychoactive drugs. *Arch. Gen. Psychiat. 24*, 268–72.

Harnden, D.G. (1974) Viruses, chromosomes and tumours: the interaction between viruses and chromosomes, in *Chromosomes and Cancer* (ed. J.German), pp.151–90. New York: Wiley.

Hook, E.B., N.H.Hatcher, P.S.Brinson, O.J.Stanecky, L.Fisher, G.Feck & P.Greenwald (1974) Negative outcome of a blind assessment of the association between spray adhesive and human chromosome breakage. *Nature 249*, 165–6.

Infante, P.F., J.K.Wagoner & A.J.McMichael (1976a) Genetic risks of vinyl chloride. *Lancet 1*, 734–5.

Infante, P.F., J.K.Wagoner, A.J.McMichael, R.J.Waxweiler & H.Falk (1976b) Genetic risks of vinyl chloride. *Lancet 1*, 1289–90.

Infante, P.F., J.K.Wagoner & R.J.Waxweiler (1976c) Carcinogenic, mutagenic and teratogenic risks associated with vinyl chloride. *Mutat. Res. 41*, 131–42.

Jarvik, J.F., N.P.Bishun, H.Bleweiss, T.Kato & E.Moralishvilli (1971) Chromosome examinations in patients on lithium carbonates. *Arch. Gen. Psychol. 24*, 166–8.

Jensen, M.K. (1967) Chromosome studies in patients treated with azathioprine and amethopterin. *Acta Med. Scand. 182*, 445–55.

Kallings, L.Q., quoted in Ehrenberg, L., K.D.Hiesche, S.Osterman-Golkar & I.Wennberg (1974) Evaluation of genetic risks of alkylating agents: tissue doses in the mouse from air contaminated with ethylene oxide. *Mutat. Res. 24*, 83–103.

Kucerova, M. & V.S.Zhurkov (1977) Population cytogenetic study of mutagenic effect of epichlorhydrin. *Mutat. Res. 46*, 227–8.

Levan, A. (1969) Chromosomal abnormalities and carcinogenesis. *Frontiers Biol. 15*, 716–31.

Makino, S. (1974) Cytogenetics of canine venereal tumours: world wide distribution and a common karyotype, in *Chromosomes and Cancer* (ed. J.German), pp.335–72. New York: Wiley.

Nielsen, J., U.Friedrich & T.Tsuboi (1969) Chromosome abnormalities in patients treated with chlorpromazine, penphenazine and lysergide. *Br. Med. J. 3*, 634–6.

Nowell, P.C. (1969) Biological significance of induced human chromosome aberrations. *Fed. Proc. 28*, 1797–1803.

Nowell, P.C. & L.J.Cole (1965) Hepatomas in mice: incidence increased after gamma irradiation at low dose rates. *Science 148*, 96.

Nowell, P.C. & D.A.Hungerford (1960) A minute chromosome in human CML. *Science 132*, 1497.

O'Riordan, M.L. & H.J.Evans (1974) Absence of significant chromosome damage in males occupationally exposed to lead. *Nature 247*, 50–3.

Paddle, G.M. (1976) Genetic risks of vinyl chloride. *Lancet 1*, 1079.

Petves, J., K.Schmid-Ullrich & U.Wolf (1970) Chromosome aberrations in human lymphocytes in the case of chronic arsenic injuries. *Dt. med. Wschr. 95*, 79–80.

Pollini, G. & R.Colombi (1964a) Il danno chromosomico midollare nel' anemia aplastica benzolica. *Med. Lavoro 55*, 241–55.

—— (1964b) Il danno chromosomico dei linfocifi nell' emopatia benzinica. *Med. Lavoro 55*, 641–54.

Purchase, I.F.H., C.R.Richardson & D.Anderson (1975) Chromosomal and dominant lethal effects of vinyl chloride. *Lancet 2*, 410.

—— (1976) Chromosomal effects in peripheral lymphocytes. *Proc. Roy. Soc. Med. 69*, 290–1.

Rabello, M.N., W.Becak, W.E. de Alweida, P.Pigati, M.T.Ungaro, T.Murata & C.A.B.Pereira (1975) Cytogenetics study on individuals occupationally exposed to DDT. *Mutat. Res. 28*, 449–54.

Sato, H., E.Pergamen & V.Nair (1973) LSD in pregnancy: chromosomal effects. *Life Sci. 10*, 773–9.

Schwanitz, G., G.Lehnart & E.Beghart (1970) Chromosomenschaden bei beruflicher bleiblastung. *Dt. Med. Wschr. 95*, 1636–41.

Staiger, G.R. (1970) Studies on the chromosomes of human lymphocytes treated with diazepan *in vitro*. *Mutat. Res. 10*, 635–44.

Stevenson, A.C., J.Bedford, A.G.S.Hill & F.H.F.Hill (1971) Chromosomal studies in patients taking phenylbutazone. *Ann. Rheumatoid Dis. 30*, 487–500.

Stetka, D.G. & S.Wolff (1976) Sister-chromatid exchange as an assay for genetic damage induced by mutagen-carcinogens. I. *In vivo* test for compounds requiring metabolic activation. *Mutat. Res. 41*, 333–42.

Szentesi, I., E.Hornyak, G.Ungvary, A.Gzeizel, Z.Bognar & M.Timar (1976) High rate of chromosomal aberration in PVC workers. *Mutat. Res. 37*, 313–16.

Tough, I.M. & W.M.Court Brown (1965) Chromosome aberrations and exposure to ambient benzene. *Lancet 1*, 684.

Tough, I.M., P.G.Smith, W.M.Court Brown & D.G.Harnden (1970) Chromosome studies on workers exposed to atmospheric benzene. *Eur. J. Cancer 6*, 49–55.

Verschaeve, L., M.Kirsch-Volders, G.Susanne, C.Greotenbriel, R.Haustermans, A.Lecomte & D.Roossels (1976) Genetic damage induced by occupationally low mercury exposure. *Environ. Res. 12*, 306–16.

Vigliani, E.C. & G.Saita (1964) Benzene and leukaemia. *New York J. Med. 271*, 872–6.

Whang-Peng, J., B.G.Leventhal, J.W.Adamson & S.Perry (1969) The effect of daunomycin on human cells *in vivo* and *in vitro*. *Cancer 23*, 113–21.

Wolf, U. (1974) Theodor Boveri and his book 'On the problem of the origin of malignant tumours', in *Chromosomes and Cancer* (ed. J.German), pp.3–20. New York: Wiley.

A.T.NATARAJAN, P.P.W.VAN BUUL and T.RAPOSA

An Evaluation of the Use of Peripheral Blood Lymphocyte Systems for assessing Cytological Effects induced in vivo by Chemical Mutagens

Studies on the frequencies of chromosome aberrations in PHA-stimulated human peripheral blood lymphocytes have proved to be one of the reliable methods of quantitatively assessing human exposure to ionising radiations. It is obvious, therefore, to consider this system for monitoring human exposure to chemical mutagens. It was indeed found, for example, that workers occupationally exposed to vinyl chloride had a higher frequency of chromosome aberrations in their peripheral blood lymphocytes than the controls (Funes-Cravioto *et al.* 1975). However, a detailed analysis of the frequencies of aberrations in blood lymphocytes from patients treated with massive doses of potent cytostatic agents (which efficiently induced chromosome aberrations under *in vitro* conditions or *in vivo* bone marrow cells in experimental animals) indicated that the lymphocyte system is very inadequate for monitoring human populations exposed to chemical mutagens (Schinzel and Schmid 1976). The present paper summarises our experience with this system, and includes (a) an analysis of the frequencies of chromosome aberrations in workers exposed to low levels of vinyl chloride; (b) an analysis of frequencies of sister chromatid exchanges (SCEs) in blood lymphocytes from patients treated with cytostatic agents, during treatment as well as a few weeks after recovery; and (c) an analysis of chromosome aberrations and SCEs in lymphocytes of rats following *in vivo* treatment with directly and indirectly acting alkylating agents.

Studies on Chromosome Aberrations in Vinyl Chloride Workers

In our earlier investigation on the workers exposed to vinyl chloride (VC) in Sweden (Funes-Cravioto *et al.* 1975), it was found that out of the six cases analysed, only four workers had elevated frequencies of aberrations, and that there was no correlation between the duration of exposure (9 to 29 years) and the frequencies of chromosome aberrations. It was estimated that the air concentration of VC monomer was 20 to 30 p.p.m. Further analysis of these six cases showed that the two with the highest frequencies of aberrations were regular cigarette smokers. Based on this experience, the survey in

Holland was planned on a larger scale, taking into consideration possible factors that may influence the results. The first group of workers analysed were those exposed to VCM, for 2 to 5 years, at an air concentration of around 1 p.p.m. Any over-exposure is detected individually by pocket monitors. Complete medical records of all individuals under study are available. Controls were selected taking into consideration age, sex and smoking habits. 90 individuals have so far been analysed, comprising 33 VC workers and 57 controls. All the samples were coded and the code was broken only after the analysis of these 90 cases. 100 cells were scored from Giemsa stained preparations from 48-h cultures in each case. The results are presented in table 1. There appears to be no strongly significant difference between the control and VC groups with regard to the occurrence and frequencies of chromosome aberrations, $0.01 > P < 0.05$. The frequencies of aberrations varied from 0-3/100 cells in the VC group and 0-7 in the control group. It should be pointed out that it is difficult to choose appropriate controls for such studies, and our controls were taken partly from administrative offices and partly from other chemical plants.

Table 1. Frequencies of chromosome aberrations in the lymphocytes of workers exposed to vinyl chloride (VC). all scoring was done with 48-h cultures, and 100 cells were scored in each case.

	No. analysed	No. with aberrations	Aberrations per 100 cells			
			gaps	breaks	exchanges	total
VC group	33	10 (30.3%)	0.36	0.15	0.03	0.54
Control group	57	33 (57.8%)	0.73	0.46	0.04	1.23

Patients Treated with Cytostatic Drugs

Since it was reported by Schinzel and Schmid (1976) that the stimulated lymphocytes from the patients treated with cytostatics did not always contain elevated frequencies of chromosome aberrations, we decided to confine our analysis to the frequencies of SCEs. As we were interested in detecting the effects of the treatments after a period of recovery, all blood samples were taken about two months after the termination of the therapy. The results from a group of 5 patients treated with cyclophosphamide (Cyc), and one with thiotepa (T.tepa), will be considered here. The control group consisted of patients of similar age suffering from some kinds of anaemia, but who had not received any therapy with cytostatic drugs. Since the patients were relatively old (see table 2) and not healthy, it was difficult to culture the lymphocytes for two cycles in BUdR (10 μg ml^{-1}) (which is necessary for the FPG technique, Perry and Wolff 1974), but this was overcome to some extent by increasing the fetal calf serum to 20% in the medium. The results are presented in table 2. The control group had an average of 10.3 SCE per cell, a frequency that is within the normal range obtained in different laboratories. The two patients suffering from acute idiopathic thrombocytopaenic purpura (ITP) had consistently lower frequencies of SCEs (20 and 50 cells were analysed in cases 2 and 5 respectively, see table 2). In the treated group, two patients (7 and 8) had significantly higher frequencies of SCEs. The increase was contributed to by some cells containing higher numbers of SCEs. For example, in patient 7, 8 cells out of 40 had more than 30 SCEs per cell, the highest recorded being 85. It can be concluded that some lymphocytes containing DNA lesions can circulate in the blood for at least two months following treatment, though this fraction can be very low.

Table 2. Frequencies of sister chromatid exchanges in patients treated with alkylating agents. Blood samples from the treated patients were taken at least two months after the last treatment. All scoring was done on 72-h cultures, the slide being processed by FPG technique. 25–50 cells were scored in each case.

Patient	Age	Disease[1]	Treatment	SCEs per cell
1	30	none	none	10.3 ± 0.6
2	33	ITP	,,	8.7 ± 0.7
3	36	CHE	,,	11.0 ± 1.7
4	60	PA	,,	11.6 ± 1.0
5	71	ITP	,,	6.6 ± 1.2
6	74	MP	,,	11.2 ± 1.2
7	18	AML	Cyc + Vin + CCNU	21.0 ± 2.7
8	62	LS	,,	15.6 ± 3.9
9	68	LRM	,,	7.2 ± 0.5
10	72	SAA	T.tepa	9.9 ± 0.7
11	76	RCS	Cyc + Vin	10.1 ± 0.9

[1] ITP, acute idiopathic thrombocytopaenic purpura; CHE, congenital hereditary elliptocytosis; PA, paraproteinic anaemia; MP, myeloproliferative disorder; AML, acute myelogenous leukaemia; LS, lymphoblastic sarcoma; LRM, lymphoreticular malignancy; SAA, sideroachrestic anaemia; RCS, reticular cell sarcoma; CCNU, 1-(2-chloroethyl)-3-cyclohexyl-1-nitrosourea.

In order to evaluate the effects of cyclophosphamide in the human peripheral lymphocyte system during therapy, a 76-year-old patient suffering from reticular cell sarcoma (RCS), was monitored during therapy. Successful cultures were made on days 13, 16, 18 and 20 of treatment with a daily dose of 100 mg of cyclophosphamide. A subsequent culture was made on the 45th day, i.e. 25 days after the termination of therapy. The results are presented in figure 1. The frequencies of SCEs were 23.4, 15.6, 16.2, 12.3 per cell on days 13, 16, 18 and 20 of treatment respectively. The frequency came down to 6.3 after 25 days recovery. Due to the difficulties in culturing these lymphocytes, only 10 to 15 cells could be analysed in each case.

From these preliminary studies it can be concluded that exposure to cyclophosphamide (or any other alkylating agents) can be monitored by

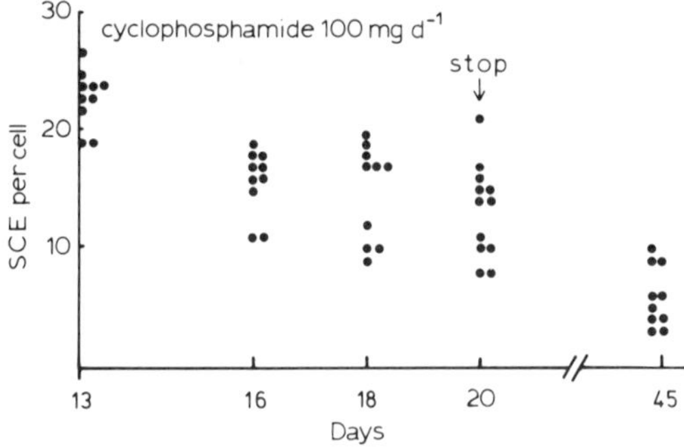

Figure 1. Sister chromatid exchanges per cell during and following treatment with cyclophosphamide.

determining the frequencies of SCEs in the peripheral blood lymphocytes during the treatment. But this comes back to control level, after the completion of treatment, in a few weeks time. This human situation is very similar to the results reported in rabbits (Stetka and Wolff 1976). It should be pointed out that in one case, in which we detected few cells with high frequencies of SCEs, we had the possibility of analysing large numbers of cells, most probably due to the fact that the patient in question was very young (18 years). Thus, it seems that in studies planned to evaluate the *in vivo* effects of chemical mutagens in humans, it may be more profitable to screen younger patients than older ones.

Experiments with Laboratory Animals

Although many tissues, such as bone marrow and spermatogonia (Schmid *et al.* 1971, Tates and Natarajan 1976, Tates *et al.* 1977), can be studied to assess the cytological effects of chemical carcinogens and mutagens administered *in vivo* in experimental animals, peripheral lymphocytes have two advantages: the animal need not be killed for screening, and several blood samples can be taken from the same animal over a period of time. Lilly, Bahner and Magee (1975) reported the usefulness of the rat peripheral lymphocyte system to demonstrate the chromosome-breaking effects of dimethyl nitrosoamine (DMN), a compound that needs metabolic activation to become effective. Though one can screen such compounds for cytological effects *in vitro* by using rat liver microsomes for activation (Natarajan *et al.* 1976), *in vivo* studies are still important in view of the fact that the *in vitro* systems do not have the natural detoxification mechanisms in operation. We have demonstrated earlier the effects of DMN and diethyl nitrosoamine (DEN) in *in vitro* experiments (Natarajan *et al.* 1976). In order to assess the validation of the lymphocyte system in experimental animals, we are investigating the effects of chemical compounds *in vivo*. The compounds included in this study were DMN, DEN (carcinogens that need metabolic activation) and methyl nitrosourea (MNU, a directly acting carcinogen). In all experiments rats were used, and frequencies of both chromosome aberrations and SCEs were determined. In spite of very high doses (300 mg kg^{-1}), no effect of DEN could be detected in the peripheral blood lymphocytes for SCEs or chromosome aberrations. Positive results were obtained for DMN and MNU (table 3): in both cases a dose-dependent increase of chromosome aberrations and SCEs was found. Much higher doses of DMN than MNU were required to produce similar effects, though both are methylating agents. This may be due to the fact that DMN needs to be activated in the liver and the reactive moiety does not reach the target cells in sufficient quantities to produce an effect. The lack of activity of DEN, *in vivo*, can only be attributed to the very short life of the active form or a very effective deactivation system present in the animal.

Table 3. Frequencies of chromosome aberrations and sister chromatid exchanges induced by DMN, DEN and MNU in rat peripheral blood lymphocytes treated *in vivo*. Chromosome aberrations were scored with 48-h cultures and SCEs with 56-h cultures. 100 cells (50 from each animal) were scored for chromosome aberrations and 25 cells were scored for SCEs in all cases. In calculating aberration frequencies only breaks and exchanges were included.

	Dose (mg kg^{-1})	Aberrations per 100 cells	SCE per cell
Control		0.0	5.8 ± 0.3
DMN	60	0.0	8.9 ± 1.1
	120	6.0 ± 2.4	12.6 ± 1.2
	240	9.0 ± 2.7	21.6 ± 2.2
MNU	20	6.0 ± 2.4	21.4 ± 2.9
	40	26.0 ± 4.3	30.0 ± 2.8
	80	22.0 ± 4.1	33.6 ± 2.4
DEN	300	0.0	6.8 ± 0.5

General Conclusions

From the data presented above, the following conclusions can be drawn with regard to the utility of peripheral lymphocyte systems for monitoring populations exposed to chemical mutagens, as well as for screening chemicals for their *in vivo* mutagenic effects:

(1) For low chronic exposures this system does not seem to be sensitive in detecting mutagenic effects. This can be due to (a) resistance of G_0 cells to effects produced by chemical mutagens, (b) prolonged repair time available for these cells before actual monitoring takes place, and (c) constant replenishment with new lymphocytes. The third possibility seems unlikely, as it is known that following whole-body irradiation, aberrant cells persist at least up to three years in man (Preston, Brewen and Gengozian 1974).

(2) Peripheral lymphocytes may still prove to be a good technique for assessing *in vivo* effects of chemical mutagens, especially those that need

metabolic activation. However, this may produce false negatives, e.g. DEN, which is effective after *in vitro* activation and carcinogenic *in vivo*.

Summary

The utility of peripheral blood lymphocyte systems for assessing cytological effects of chemical mutagens under *in vivo* conditions has been evaluated. Three different approaches have been used, namely the determination of: (a) frequency of chromosome aberrations in a group of workers exposed to low and chronic exposure of vinyl chloride with appropriate controls; (b) frequencies of SCEs in patients exposed to high doses of cytostatic drugs during and after therapy; and (c) frequencies of both SCEs and chromosome aberrations in rats treated with directly and indirectly acting alkylating agents. For low chronic exposures, the lymphocyte system does not seem to be a sensitive indicator of chromosome damaging effects. However, for *in vivo* screening of chemical mutagens, it may still prove to be a useful technique for directly acting agents. For compounds that used metabolic activation, this method may generate false negatives in some cases.

Acknowledgements

We would like to acknowledge the help of Drs E. A. Loeliger, C. A. Hartgrink-Groeneveld, H. L. Haak and Ms C. T. Krösschell in providing blood samples; the excellent technical assistance of Aya den Hertog, Johan Goudzwaard and Matty Meijers; and the warm encouragement of Prof. Dr F. H. Sobels.

References

Funes-Cravioto, F., B.Lambert, J.Lindsten, L.Ehrenberg, A.T.Natarajan & S.Osterman-Golkar (1975) Chromosome aberrations in workers exposed to vinyl chloride. *Lancet 1*, 459.

Lilly, L.J., B.Bahner & P.N.Magee (1975) Chromosome aberrations induced in rat lymphocytes by N-nitroso compounds as a possible basis for carcinogen screening. *Nature 258*, 611–12.

Natarajan, A.T., A.D.Tates, P.P.W. van Buul, M.Meijers & N. de Vogel (1976) Cytogenetic effects of mutagens/carcinogens after activation in a microsomal system *in vitro*. I. Induction of chromosome aberrations and sister chromatid exchanges by diethylnitrosamine (DEN) and dimethylnitrosamine (DMN) in CHO cells in the presence of rat-liver microsomes. *Mutat. Res. 37*, 83–90.

Perry, P. & S.Wolff (1974) New Giemsa method for the differential staining of sister chromatids. *Nature 251*, 156–8.

Preston, R.J., J.G.Brewen & N.Gengozian (1974) Persistence of radiation-induced chromosome aberrations in marmoset and man. *Radiat. Res. 60*, 516–24.

Schinzel, A. & W.Schmid (1976) Lymphocyte chromosome studies in humans exposed to chemical mutagens. The validity of the method in 67 patients under cyto-static therapy. *Mutat. Res. 40*, 139–66.

Schmid, W., D.T.Arakaki, N.A.Breslau & J.C.Culbertson (1971) Chemical mutagenesis. The Chinese hamster bone marrow as an *in vivo* test system. I. Cytogenetic results on basic aspects of the methodology, obtained with alkylating agents. *Humangenetik 11*, 103–18.

Stetka, D.G. & S.Wolff (1976) Sister chromatid exchanges as an assay for genetic damage induced by mutagens-carcinogens. I. *In vivo* tests for compounds requiring metabolic activation. *Mutat. Res. 41*, 333–42.

Tates, A.D. & A.T.Natarajan (1976) A correlative study on the genetic damage induced by chemical mutagens in bone-marrow and spermatogonia of mice. I. CNU-ethanol. *Mutat. Res. 37*, 267–77.

Tates, A.D., A.T.Natarajan, N. de Vogel & M.Meijers (1977) A correlative study on the genetic damage induced by chemical mutagens in bone-marrow and spermatogonia of mice. *Mutat. Res.* (in press).

F. FUNES-CRAVIOTO, C. ZAPATA-GAYON, B. KOLMODIN-HEDMAN, B. LAMBERT, J. LINDSTEN, E. NORBERG, M. NORDENSKJÖLD, R. OLIN and Å. SWENSSON

Chromosome Aberrations and Sister Chromatid Exchange in Laboratory and Factory Workers and their Children

Two-and-a-half years ago we accidentally discovered that there was an increased frequency of chromosome aberrations in cultured lymphocytes obtained from some technicians working in a laboratory performing hormone analysis. The results were confirmed by an analysis of repeated blood samples from the same technicians. These observations prompted us to study a larger group of subjects from the same laboratory. The mean frequency of chromosome aberrations in this group was found to be significantly increased. The study has then gradually been extended to include personnel from other chemical laboratories, and very similar results have been obtained. Furthermore, an analysis of the frequency of sister chromatid exchange (SCE) has been added for some of the subjects studied.

Many organic solvents were commonly used in the different laboratories. We therefore tried to find subjects who had been exposed to such compounds but who had a different type of work, and a group of workers from a roto-printing factory, who had been exposed mainly to benzene and toluene, was included. This group was also found to have an increased frequency of chromosome aberrations.

Since several of the female technicians with an increased frequency of chromosome breaks had been working in the laboratories while they were pregnant, it was also considered of interest to study a group of their children. These children were found to have a mean frequency of chromosome aberrations as high as that of their mothers, and a significantly increased frequency of SCE in comparison with control children.

Up until now a total of 75 adults and 14 children, together with 53 control adults and 14 control children, have been examined. The present paper constitutes a report of the results of this study, and a discussion of the problems involved in the interpretation of the significance of these results.

Materials

Altogether 156 subjects were studied, 89 'exposed' and 67 controls (table 1). The exposed subjects (75 adults and 14 children) belonged to eight

Table 1. Number of subjects studied with regard to the frequency of chromosome aberrations and sister chromatid exchange.

Group (see text)	Number of subjects			
	Chromosome aberrations only	SCE only	Chromosome aberrations + SCE	Total
Exposed				
A + B + C	25	2	10	37
D	7	—	—	7
E	9	—	—	9
F	14	—	—	14
G	4	—	4	8
children	10	—	4	14
total	69	2	18	89
Control				
adults	32	11	10	53
children	5	7	2	14
total	37	18	12	67
Total	106	20	30	156

different groups. All the exposed adults were interviewed concerning smoking and drinking habits, use of drugs, details of their work and symptoms of disease, e.g. past illnesses such as hepatitis and infections, symptoms of blood disorders and fatigue. Haemoglobin content, counts of red and white blood cells as well as thrombocytes, differential white blood cell count and liver tests (SGOT, SGPT, alkaline phosphatase and bilirubin) were carried out in the majority of cases.

Subjects Analysed for Chromosome Aberrations

The number of subjects in each group and their mean age is given in table 2.

Group A. Fourteen clinically healthy technicians and research workers (10 women and 4 men) from the laboratory performing hormone analysis in which the high frequency of cells with chromosome breaks was originally discovered. Radioimmunoassays were commonly used and there was a heterogenous exposure to a number of organic solvents, mainly chloroform and toluene. A mixture of 3% benzene and 97% cyclohexane was openly used for column chromatography in small rooms, without hoods. There had been a heavy exposure to benzene during the 1960s.

Measurements of the air concentration of benzene and cyclohexane were carried out in 1975 and 1976. Air was sampled during work hours in charcoal tubes, eluted and analysed by a gas chromatographic procedure. The mean concentrations were 0.5 p.p.m. for benzene and 176 p.p.m. for cyclohexane, i.e. well below the threshold limit values (TLV). There was no undue radioactivity.

Six apparently healthy women had a low number of granulocytes (less

Table 2. Mean age and frequency of abnormal cells and chromosome aberrations in groups of control and exposed subjects.

Group	No. of subjects	Age range (years)	Mean age	Total cells analysed	Abnormal cells			Total chromosomes analysed	Breaks		
					Total	Frequency range (%)	Mean frequency		Total	Range, per 100 cells	Mean, per 100 cells
Exposed											
A	14	24–60	32.2	1 400	138	3–16	9.86	64 400	172	3–20	12.29
B	15	24–36	30.0	1 500	122	2–17	8.13	69 000	141	2–22	9.40
C	6	29–56	41.5	600	38	4–9	6.33	27 600	59	5–17	9.83
D	7	25–49	33.0	700	54	2–22	7.71	32 200	66	2–29	9.43
E	9	23–30	26.5	900	68	3–17	7.55	41 400	85	3–21	9.44
F	14	23–54	37.2	1 400	108	2–15	7.71	64 400	124	2–17	8.86
G	8	54–65	61.3	800	76	4–17	9.50	36 800	95	6–17	11.87
adults	73	23–65	36.0	7 300	604	2–22	8.27	335 800	742	2–29	10.16
children	14	0.02–10.5	4.3	1 500	137	3–17	9.79	69 000	161	4–21	11.50
Control											
adults	42	18–63	27.7	4 200	200	0–20	4.76	193 200	233	0–21	5.55
children	7	0.16–11.0	4.4	800	17	1–5	2.43	36 800	21	1–7	3.00
total	49	0.16–63	24.4	5 000	217	0–20	4.34	230 000	254	0–22	5.08

than 2000 per mm^3). The values were normal on a second occasion in four of them.

Group B. Fifteen clinically healthy technicians (13 women and 2 men) working in another laboratory analysing hormones. Radioimmunoassays and a variety of organic solvents were used. Benzene had been used in large amounts openly and without the use of hoods during the 1960s.

Measurements of the air concentrations of benzene, chloroform, cyclohexane, diethyl ether, iso-octane, methanol and toluene were made in 1975 using the same procedure as for group A. The benzene concentrations ranged between 0.5-0.9 p.p.m. during a morning shift, toluene between 5-10 p.p.m., and diethyl ether between 200-385 mg m^{-3}. The concentrations of all the compounds analysed were below TLV with the exception of a single value of 175 p.p.m. of chloroform, i.e. eight times the TLV, which was recorded in connection with washing of glassware. There was no undue radioactivity.

Two of the women had initially moderately low white blood cell counts. These values were normal on a second occasion when the subjects were still actively working in the laboratory. One of the men, who never used alcohol, had slightly elevated values of bilirubin and transaminases in the serum. His working conditions were changed and the values then normalised. Another man who regularly drank wine, also had slightly elevated levels of transaminases.

Group C. Six female technicians from yet another laboratory performing hormone analysis. They were mainly exposed to chloroform, toluene and diethyl ether. Radioimmunoassays were commonly used. The air concentration of solvents was not measured. All the subjects were clinically healthy. No abnormal routine tests were noted.

Group D. Seven males working as organic chemists at the Royal Institute of Technology. The exposure ranged from 2-28 years. Four of the men were exposed to benzene. Tetrahydrofurane was also a commonly used solvent. Two of the subjects had sleep disturbances not obviously connected with work. Two of the men, who had no excessive intake of ethyl alcohol, had slightly elevated serum levels of γ-glutamyl transpeptidase. The white blood cell counts were normal.

Group E. Nine female technicians from four different chemical laboratories. Three had previously been exposed to benzene. All were exposed to toluene, methanol, chloroform and xylene. No air concentrations of solvents were measured. All the subjects were clinically healthy. No abnormal routine tests were noted.

Group F. Fourteen men working in a roto-printing factory. They had been exposed to toluene since around 1950. Before 1958 the toluene was probably contaminated by a low percentage of benzene. The exposure ranged between 1.5-26 years. The air concentration of toluene was measured for 2 months in 1975 and showed time weighted average values of 100-200 p.p.m. with occasional rises up to 500-700 p.p.m. when the workers had to carry out repair work inside the machines.

Most of the subjects often experienced headache, fatigue and sometimes vertigo, nausea and a feeling of drunkness. However, they were all clinically healthy and had normal laboratory test values.

Group G. Eight men working in a roto-printing factory. They were exposed to benzene during the 1940s for 2-10 years. In 1950 benzene was substituted by toluene. The exposure to benzene as well as to toluene ranged from 2-26 years. The air concentration of benzene was not measured. The concentrations of toluene were the same as under F.

Five of the subjects had suffered from leucopenia during the 1940s and therefore had not been allowed to work for 0.5-1 years. In 1975 and 1976 they were clinically healthy and had a normalised white blood cell picture.

Group H. Fourteen children (7 girls and 7 boys) between 4 days and 11 years of age. Their mothers belonged to groups A and B and had been working in the laboratories during pregnancy. All the children were apparently healthy except a 7-year-old girl with idiopathic thrombocytopaenia, which had apparently started in connection with a virus infection at the age of 5. She received no drug treatment at the time of investigation. No routine blood tests were carried out on the children.

Controls. The control group comprised 42 adults (19 women and 23 men) and 7 children (3 girls and 4 boys) all with normal karyotypes. Ten of the adults (all women) had just begun medical studies, 10 male students were analysed in connection with medical examination before military service, 11 (5 women and 6 men) belonged to the administrative staff of the different laboratories investigated, 7 (2 women and 5 men) had previously had a malformed child or repeated spontaneous abortions, 2 men were electricians, 1 was a female nurse and 1 a female student technician. Three of the children were hospitalised for cystoscopy because of different urological disorders. Two children were referred for chromosome analysis as 'funny looking kids', one was the sister of a child with an unbalanced translocation, and one was investigated for suspected lactose intolerance, which could not be verified. Apart from the exceptions mentioned all the control subjects were apparently healthy.

A more detailed interview was made with 24 of the adult control subjects after they had been selected as controls. Not surprisingly it was then noted that the majority of these subjects had been submitted to diagnostic X-ray examinations (generally of the lungs and teeth) on one or more occasions. Many had been taking different kinds of drugs for various disorders, some of the women used contraceptive pills, many used deodorant sprays, paints, etc., and many were smokers. It proved to be very difficult to evaluate all these kinds of exposure, and the controls were therefore not matched with the exposed subjects with regard to all these variables.

Subjects Analysed for Sister Chromatid Exchange (SCE)

Twelve female technicians from groups A, B and C, 4 children of two of these technicians working during pregnancy, and 4 males from group G were studied with regard to the frequency of SCE. With the exception of two of the technicians all subjects were also analysed with regard to the frequency of chromosome aberrations.

Altogether 21 apparently healthy adult control subjects were studied. Eight (all females) had just begun medical studies, 8 (7 males, 1 female) were blood donors and 5 were male students drafted for military service. Ten of these subjects were also studied with regard to frequency of chromosome aberrations.

Nine children were included as controls. They were submitted to a paediatric department for various reasons such as suspected malabsorption (5 boys), androgenital syndrome (1 boy), unilateral renal agenesis and urethral valve (1 boy) and cow milk protein sensitivity (1 girl). Diagnostic X-rays had been performed in five of the children prior to blood sampling for the SCE study. Chromosome aberrations were studied in only two of these children.

Methods

The frequency of chromosome aberrations was analysed in cells from conventional lymphocyte cultures incubated for 72 hours (medium: 80% Parker 199, 20% of the subject's own plasma). A micromethod was used for the study of the children and the plasma of the medium was then substituted by fetal calf serum. Generally 100 cells were analysed in each case and only cells with the normal number of 46 chromosomes were included. The analysis was carried out blindly on coded slides. Agreement between two independent observers was required for scoring the cells as normal or abnormal. The criteria used for the analysis and classification of the chromosome aberrations were the same as described previously (Funes-Cravioto *et al.* 1974). Gaps were not included as aberrations.

The analysis of the number of SCE was carried out on cultured lymphocytes as described previously (Latt 1973, Wolff and Perry 1974, Lambert *et al.* 1976). The analysis was made directly from the microscope. Twenty cells with the normal number of 46 chromosomes were examined in each case. All slides were coded and analysed blindly.

Results

Chromosome Aberrations

Two variables were used for the analysis, i.e. the frequency of cells with chromosome aberrations and the frequency of chromosome breaks per 100 cells. Similar results were obtained for both these variables.

The distribution of subjects within the exposed and control groups of adults showed considerable overlap (figure 1). The mean frequency of chromosome breaks per 100 cells varied between 8.9-12.3, and the frequency of abnormal cells varied between 6.3-9.9% in the exposed groups of adults (tables 2 and 3). The corresponding values for the control adults were 5.6 and 4.8, respectively. There was no significant heterogeneity between the exposed groups (one-way analysis of variance, $F=0.89$, $P=0.51$), and the data were

Table 3. Frequency of different types of chromosome aberrations in groups of control and exposed subjects.

Group	No. of cells analysed	Chromatid aberrations				Chromosome aberrations								Total aberrations	
		Breaks		Exchanges		Breaks		Dicentrics		Rings		Other			
		no.	per cell	no.	per cell	no.	per cell	no.	per cell	no.	per cell	no.	per cell	no.	per cell
Exposed															
A	1400	55	0.040	3	0.002	89	0.064	3	0.002	6	0.004	2	0.001	158	0.113
B	1500	54	0.037	1	0.001	72	0.049	4	0.003	1	0.001	2	0.001	134	0.089
C	600	16	0.027	2	0.003	25	0.042	2	0.003	2	0.003	2	0.003	49	0.082
D	700	32	0.046	0	0.000	26	0.037	3	0.004	0	0.000	1	0.001	62	0.089
E	900	33	0.037	0	0.000	44	0.050	1	0.001	2	0.002	1	0.001	81	0.090
F	1400	40	0.028	4	0.003	68	0.049	3	0.002	0	0.000	1	0.001	116	0.083
G	800	23	0.029	1	0.001	64	0.080	0	0.000	1	0.001	1	0.001	90	0.112
children	1500	83	0.055	0	0.000	72	0.048	1	0.001	1	0.001	1	0.001	158	0.105
Controls															
adults	4200	75	0.018	7	0.002	111	0.026	7	0.002	8	0.002	1	0.000	209	0.050
children	800	12	0.015	0	0.00	7	0.009	1	0.001	0	0.000	0	0.000	20	0.025

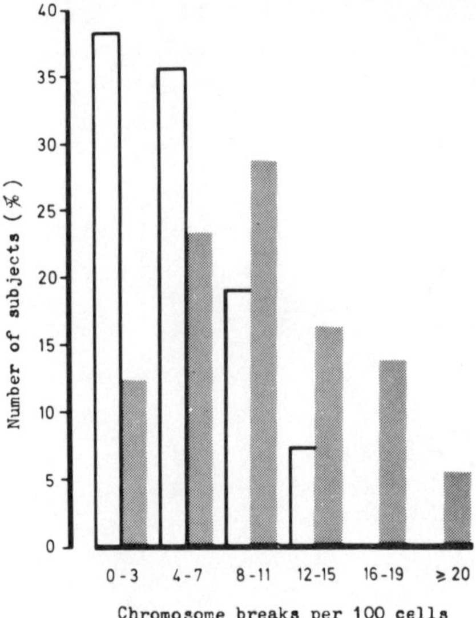

Figure 1. Distribution of exposed (screened columns) and control (open columns) adult subjects according to the frequency of chromosome breaks per 100 cells.

therefore pooled. The difference between the exposed and control adults was statistically significant (median test, $P < 0.001$).

The exposed children had a mean frequency of chromosome breaks per 100 cells of 11.5 and a mean frequency of abnormal cells of 9.8%. The corresponding values for control children were 3.0 and 2.4, respectively. The difference between the exposed and control children was statistically significant (median test, $P < 0.01$).

As seen from figure 2 there was a correlation between the frequency

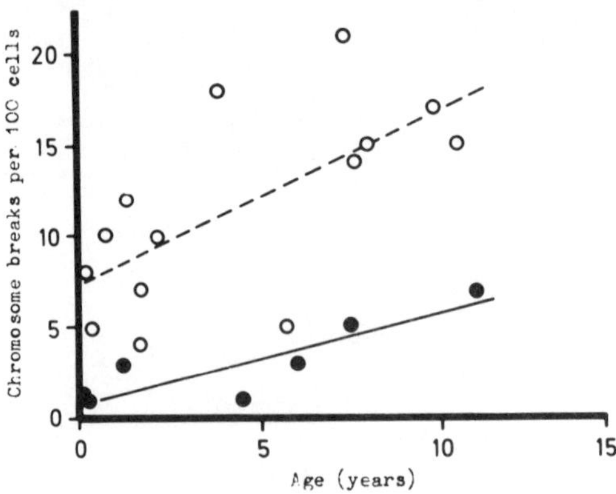

Figure 2. The correlation between chronological age and number of chromosome breaks per 100 cells in the exposed (open circles) and control (solid circles) children.

of chromosome breaks and age for both the exposed ($r=0.66$, d.f.$=12$, $P<0.01$) and control ($r=0.86$, d.f.$=5$, $0.05>P>0.01$) children. No age effect was found in the adult exposed and control groups.

The correlation between the frequency of chromosome breaks and years of exposure in the exposed adults was $r=0.36$ (d.f.$=32$, $0.05>P>0.01$; information available for 34 subjects).

There was no correlation between the frequency of chromosome breaks in the exposed mothers and their children. Furthermore, there was no correlation with smoking habits (information available for 26 subjects). There were no significant sex differences.

The higher number of chromosome aberrations in the exposed group was mainly due to an equal increase in the frequency of chromatid and isochromatid breaks.

Seven subjects belonging to groups A and B were studied on repeated occasions. The results obtained are difficult to interpret, but are included for the sake of completeness (table 4). Three of the subjects were studied after they had been on leave and all then showed somewhat lower values.

Table 4. Results of repeated analysis of the frequency of chromosome breaks per 100 cells in seven subjects.

Subject	Breaks per 100 cells	Status
64474	12	working
	19	working, 1 month later
	9	after 6 months leave
73874	12	working
	16	working, 1 month later
	10	after 20 days leave
67474	13	working
	6	after 80 days leave
74174	20	working
	12	working, 10 months later
71174	>24[1]	working
	>12[1]	working, 16 months later
67374	5	at beginning of employment
	6	working, 12 months later
70674	14	working
	15	working, 17 months later

[1] Both cultures contained cells with pulverised chromosomes.

Sister Chromatid Exchange

The results are presented in table 5. The mean frequency of SCE was 13.1 in the whole adult control material and 8.4 for the control children. The difference is statistically significant ($P<0.01$, Mann-Whitney's U-test). No sex or age differences were found within the adult control group.

There was a significant increase of SCE in the female technicians ($P < 0.001$), but not in the group of four printers, in comparison to the adult controls. The exposed children revealed a significantly higher SCE frequency than the control children ($P < 0.001$, Mann-Whitney's U-test).

Subjects with a high frequency of SCE did not have a high frequency of chromosome aberrations. In fact there was a negative correlation ($r = -0.62$, $0.1 > P > 0.05$) in the group of 10 female technicians in which both variables were analysed.

Table 5. Frequency of sister chromatid exchange in groups of exposed and control subjects.

Group (see text)	No. of subjects	Age range	Mean age	Mean SCE per cell per individual: range	Mean SCE per cell per individual: group mean
Exposed					
adults					
A+B+C	12	23–41	30.3	12.1–24.5	19.7
G	4	54–63	60	6.7–12.9	10.3
children	4	0.4–3.75	1.9	14.9–16.6	15.8
Control					
adults 1	15	19–49	23.8	7.8–19.7	13.5
adults 2	6	50–66	56	8.4–17.0	12.2
children	9	0.2–4.5	2.4	7.1–11.3	8.4

Discussion

The present study has shown that subjects working in some chemical laboratories and in the roto-printing industry, as well as children of women working in these laboratories during pregnancy, have an increased frequency of chromosome breaks and SCE. Age-dependent differences in the frequency of these cytogenetic changes were also observed. The frequency of SCE was significantly lower among non-exposed children than among non-exposed adults. Furthermore, a positive correlation was noted between the frequency of chromosome breaks and chronological age both in the control and exposed groups of children but not in the adults.

As mentioned in the introduction, the present study gradually evolved from the original, accidental observation of a high frequency of chromosome aberrations in some laboratory technicians. The cause of the observed cytogenetic changes was not known at the beginning of the study and is still not known. The selection of exposed and control subjects can therefore be criticised. Nevertheless it seems reasonable to assume that some factor(s) in the working environments studied do induce chromosome breaks and SCE. Unfortunately, we were not able to study any of the subjects before and after employment, in order to prove the causal relationship between the working environment and cytogenetic changes. This would have been of great value and might also have given some information regarding the time required for the induction of the cytogenetic changes.

Since the cytogenetic changes were observed not only in lymphocytes from adults but also in their children one might postulate that the causative agents, be they environmental factors or their active metabolites, either are able to pass the placenta or occur in increased amounts in the homes of the exposed mothers. The positive correlation between the frequency of chromosome aberrations and age in the two groups of children may therefore be due to continuous exposure during infancy and childhood, or to latent DNA damage in stem cells introduced during pregnancy. The finding of a higher level of chromosome breaks in the children of exposed mothers than in control children suggests that stem cell damage during pregnancy may have occurred. If so, an additional explanation for the age related differences within the group of exposed children may be that the exposure levels in the laboratories were higher earlier, and have gradually declined during the last ten years.

Even if many of the subjects working in the laboratories had been using radioactive isotopes, there are no reasons to believe that ionising irradiation would be the common causative factor. Chemical agents seems much more likely in view of the fact that the roto-printing workers, as well as several subjects among the controls, also demonstrated high frequencies of chromosome breaks and SCE that did not appear to be related to previous X-ray exposure. Furthermore, no undue radioactivity could be measured in the laboratories.

Our observation of a higher frequency of chromosome breaks and SCE in adult controls compared to control children may be due to a continuous exposure to mutagenic agents in the common environment. In fact, the mean frequency of chromosome breaks in the control groups was rather high in comparison with the results obtained in some, but not in all, previously published studies (Court Brown *et al.* 1966; Court Brown, Jacobs and Tough 1967; Bauchinger *et al.* 1976; Kučerová 1976; Luchnik and Sevankaev 1976; Szeutesi *et al.* 1976). However, it is almost impossible to sort out one or a few causative agents from all potentially mutagenic compounds that occur in both the common working environments. Furthermore, different compounds may constitute the aetiological factors in the different environments. The cytogenetic changes may also depend on interaction between several agents, or even between agents and host factors, e.g. DNA repair functions, enzyme induction and metabolism.

Our primary interest was focused on organic solvents, because these agents were common to all working environments studied, and chromosome aberrations have previously been reported in subjects exposed to e.g. benzene, toluene and vinyl chloride (Ducatman, Hirschhorn and Selikoff 1975; Forni *et al.* 1971; Forni, Pacifido and Limonta 1971; Funes-Cravioto *et al.* 1975; Hartwich and Schwanitz 1972; Purchase, Richardson and Anderson 1976; Snyder and Kocsis 1975; Tough *et al.* 1970). However, those who had been exposed most heavily to organic solvents did not necessarily demonstrate the highest frequency of cytogenetic changes. Furthermore, there was a low correlation between the frequency of chromosome breaks and the number of years that the subjects had been employed. It should also be pointed out that the subjects included in the study had been exposed to a number of other compounds, but were selected because they had been working with organic solvents.

The considerable variation between subjects with regard to the cytogenetic changes could indicate varying degrees of exposure to the mutagenic compounds. On the other hand, the finding that the mean frequency of chromosome breaks in the different exposed groups did not vary too much (9.4-12.3 per 100 cells) suggests a more complicated explanation. Perhaps an equilibrium between the degree of exposure, the turnover of cells and the degree of damage is reached with time in the exposed groups. This hypothesis is to some extent supported by the finding of a positive correlation between the frequency of chromosome breaks and chronological age in the children but not in the adults. Different subjects might also respond differently when exposed to the same environment. Genetically determined qualitative and quantitative differences between individuals with regard to the metabolism of mutagenic compounds and DNA repair could, for instance, play a role in this context.

The chromosome aberrations observed were mainly isochromatid and chromatid breaks, i.e. aberrations that may be caused by latent DNA damage but that also could have been induced during the cultivation of the cells *in vitro*. Sister chromatid exchanges detected by the technique used in this work arise during the *in vitro* cultivation, but may also have ocurred because of remaining DNA damage from previous exposure. Since the cells were incubated with the subject's own plasma the mutagenic compounds could have been present in the blood specimens obtained, and caused the DNA damage during *in vitro* cultivation. The increased frequency of aberrations and SCE noted in the children of exposed mothers suggests, however, that the DNA damage was induced earlier but had remained latent in the lymphocytes for several years. If the aberrations were induced during the cultivation of the cells *in vitro*, one would expect to find lower frequencies of aberrations when the exposed subjects had been on leave, while this would not be the case if the aberrations were due to latent damage. The frequency of aberrations was somewhat lower in those three subjects who had been on leave (20 days–6 months). This material is much too small to allow any definite conclusions, but they do not contradict the hypothesis that latent damage of DNA leading to isochromatid and chromatid breaks can be induced a long time before the cell divides.

The present technique for the study of SCE is relatively new and so far there are no reports of *in vivo* induced SCE in human subjects. However, several mutagenic agents, predominantly alkylating agents and cytotoxic drugs, have been shown to induce SCE when added to human lymphocyte cultures *in vitro* (Latt 1974, Beek and Obe 1975, Kato 1977), and increased frequencies of SCE have been observed after clinical treatment with cytotoxic drugs (Perry and Evans 1975, and our own unpublished observations). The present results suggest that SCE may be used as a sensitive indicator of cytogenetic changes due to environmental exposure. There was, however, a considerable heterogeneity in the frequency of SCE within the groups of exposed subjects, as well as among controls. The reason for this heterogeneity is not known at present, but it should be noted that the frequency of SCE is measured against a background of bromodeoxyuridine-induced SCE (Lambert *et al.* 1976). Possible explanations for the observed heterogeneity may be individual differences in the uptake of bromodeoxyuridine or host-dependent interactions between other SCE-inducing agents and bromodeoxyuridine.

Some results indicate that SCE may arise by mechanisms different from those that cause chromosome aberrations (Wolff, Rodin and Cleaver 1977). If this is true, one should not necessarily expect a positive correlation between alterations in the frequencies of chromosome aberrations and SCE. In fact, our preliminary results suggest that there may be a negative correlation, since several subjects with high aberration frequencies displayed a low number of SCE, and vice versa. However, further and extended studies of SCE of human subjects exposed *in vivo* are needed in order to obtain more information about causative agents, individual differences and the relation between SCE and chromosome aberrations.

The significance of the observed cytogenetic changes in relation to the subject's health is so far not known. All the people studied were apparently healthy. The total number of individuals studied was relatively small, especially with regard to SCE, and most were relatively young. Therefore, no conclusion can be drawn from the present study regarding the correlation between chromosome breaks and SCE in lymphocytes from the peripheral blood and the health of the subjects. It might, however, be mentioned that an increased incidence of malignant lymphomas and leukaemia has been reported among Swedish chemistry graduates (Olin 1976). A larger group of at-risk individuals than chemistry graduates has to be considered if further studies can demonstrate that there is a relationship between the cytogenetic changes observed in the present study and certain malignant disorders.

Summary

Cultured lymphocytes from 73 subjects working in chemical laboratories and the printing industry were found to have a significantly increased frequency of chromosome aberrations, mainly chromatid and isochromatid breaks, in comparison with 49 control subjects (42 adults and 7 children). An increase of the same magnitude was also found in 14 children, aged 4 days–11 years, of 11 women who had been working in the laboratory during pregnancy. A significant correlation between age and frequency of chromosome aberrations was noted both for the exposed and control children, but not for the adults. The frequency of sister chromatid exchange was significantly increased in a group of 12 technicians working in laboratories performing hormone analysis. Four children of two female technicians working during pregnancy also had a significantly increased frequency of sister chromatid exchange. There was a considerable overlapping between exposed and control groups with regard to frequency of chromosome aberrations as well as sister chromatid exchange. The cause and biological significance of these findings are not yet known.

Acknowledgements

The authors would like to thank all the exposed and control subjects for their participation in the present study. The financial support for the study was provided by the Swedish National Board of Safety and Health (026 5531/75) and the Swedish Medical Research Council (3681). One of us (C.Z-G.) was a research fellow of the National Council of Science and Technology, Mexico, during the course of the study.

References

Bauchinger, M., E.Schmid, H.J.Einbrodt & J.Dresp (1976) Chromosome aberrations in lymphocytes after occupational exposure to lead and cadmium. *Mutat. Res. 40*, 57–62.

Beek, B. & G.Obe (1975) The human leukocyte test system. VI. The use of sister chromatid exchanges as possible indicators for mutagenic activities. *Humangenetik 29*, 127–34.

Court Brown, W.M., K.E.Buckton, P.A.Jacobs, I.M.Tough, E.V.Kuenssberg & J.D.E. Knox (1966) Chromosome studies on adults. *Eugenics Laboratory Memoirs XLII*. Cambridge: University Press.

Court Brown, W.M., P.A.Jacobs & I.M.Tough (1967) Some types of information obtainable from chromosome studies on defined population groups, in *Human Radiation Cytogenetics* (eds H.J.Evans, W.M.Court Brown & A.S.McLean) pp.115–21. Amsterdam: North-Holland.

Ducatman, A., K.Hirschhorn & I.J.Selikoff (1975) Vinyl chloride exposure and human chromosome aberrations. *Mutat. Res. 31*, 163–8.

Forni, A.M., A.Cappellini, E.Pacifido & E.Vigliana (1971) Chromosome changes and their evolution in subjects with past exposure to benzene. *Arch. Environ. Health 23*, 385–91.

Forni, A.M., E.Pacifido & A.Limonta (1971) Chromosome studies in workers exposed to benzene or toluene or both. *Arch. Environ. Health 22*, 373–8.

Funes-Cravioto, F., N.N.Yakovienko, N.P.Kuleshov & V.S.Zhurkov (1974) Localization of chemically induced breaks in chromosomes of human leucocytes. *Mutat. Res. 23*, 87–105.

Funes-Cravioto, F., B.Lambert, J.Lindsten, L.Ehrenberg, A.T.Natarajan & S. Osterman-Golkar (1975) Chromosome aberrations in workers exposed to vinyl chloride. *Lancet 1*, 459.

Hartwich, G. & G. Schwanitz (1972) *Dtsch Med. Wschr. 97*, 45.

Kato, H. (1977) Mechanisms for sister chromatid exchanges and their relation to the production of chromosomal aberrations. *Chromosoma 59*, 179–91.

Kučerová, M. (1976) Cytogenetic analysis of human chromosomes and its value for the estimation of genetic risk. *Mutat. Res. 41*, 123–30.

Lambert, B., K.Hansson, J.Lindsten, M.Sten & B.Werelius (1976) Bromodeoxyuridine-induced sister chromatid exchanges in human lymphocytes. *Hereditas 83*, 163–73.

Latt, S.A. (1973) Microfluorometric detection of deoxyribonucleic acid replication in human metaphase chromosomes. *Proc. Nat. Acad. Sci. U.S.A. 70*, 3395–9.

—— (1974) Sister chromatid exchanges, indices of human chromosome damage and repair: detection by fluorescence and induction by mitomycin C. *Proc. Nat. Acad. Sci. U.S.A. 71*, 3162–6.

Luchnik, N.V. & A.V.Sevankaev (1976) Radiation-induced chromosomal aberrations in human lymphocytes. I. Dependence on the dose of gamma-rays and an anomaly at low doses. *Mutat. Res. 36*, 363–78.

Olin, R. (1976) Leukaemia and Hodgkin's disease among Swedish chemistry graduates. *Lancet 2*, 916.

Perry, P. & H.J.Evans (1975) Cytological detection of mutagen-carcinogen exposure by sister chromatid exchange. *Nature 258*, 121–5.

Purchase, I.F.H., C.Richardson & D.Anderson (1976) Chromosomal effects in peripheral lymphocytes. *Proc. Roy. Soc. Med. 69*, 290–2.

Snyder, R. & J.J.Kocsis (1975) *CRC Critical Reviews in Toxicology.*

Szeutesi, I., E.Hornyak, G.Ungvary, A.Czeizel, Z.Bognaz & M.Timaz (1976) High rate of chromosomal aberrations in PVC workers. *Mutat. Res. 37*, 313–16.

Tough, I.M., P.G.Smith, W.M.Court Brown & D.G.Harnden (1970) Chromosome studies on workers exposed to atmospheric benzene: the possible influence of age. *Europ. J. Cancer 6*, 49–55.

Wolff, S. & P.Perry (1974) Differential Giemsa staining of sister chromatids and the study of sister chromatid exchanges without autoradiography. *Chromosoma 48*, 341–53.

Wolff, S., B.Rodin & J.E.Cleaver (1977) Sister chromatid exchanges induced by mutagenic carcinogens in normal and xeroderma pigmentosum cells. *Nature 265*, 347–9.

N.P.BOCHKOV and K.N.YAKOVENKO

Sensitivity of Human Lymphocyte Chromosomes to Ethylenimine Derivatives at different Periods during Cell Culture

Experimental data on the cytogenetic effects of chemical mutagens at different cell cycle stages are sparse and fragmentary. For a long time chemical mutagens were considered to be able to damage chromosomes only at the synthetic stage, that is during the period of replication (Rieger and Michaelis 1962, Kihlman 1966). However, it was convincingly proved in plants that a number of chemicals might damage chromosomes at a presynthetic period (Dubinina and Dubinin 1967, 1968), and that some mutagens have their effect only at the postsynthetic stage (Kihlman and Odmark 1965).

There has been little work on the sensitivity of human chromosomes to the action of chemical mutagens during the cell cycle, and limited evidence that the dynamics of chromosomal aberration frequencies during the cell cycle depend on the specificity of the chemical. Induction of chromosomal aberrations is possible at different stages of the cell cycle, and not only during the synthetic period (Brewen & Christie 1967, Funes-Cravioto and Yakovenko 1972). Systematic data on this problem are certainly necessary both for understanding mechanisms of chemical mutagenesis and for the characterisation of cell cultures as a system for testing environmental factors for their mutagenic activity.

In the present work the dependence of chromosomal aberration frequency in cultured human lymphocytes upon time of addition of several ethylenimine derivatives has been studied.

Materials and Methods

Experiments were carried out on cultured human peripheral blood according to the generally accepted method. Ethylenimine derivatives were utilised as mutagens in the following doses: phosphamide 60 $\mu g\ ml^{-1}$, thiophosphamide 20 and 30 $\mu g\ ml^{-1}$, dipin 40 $\mu g\ ml^{-1}$ and fotrin 40 $\mu g\ ml^{-1}$. The tested mutagens contained different numbers of alkylating groups and active mutagenic centres (Yakovenko, Azhayev and Bochkov 1974a, b). Chemicals were added for 1 h and cultures were then washed three times in tenfold Hank's solution and fresh medium was added. Mutagens were added to separate cultures every 4 h from culture initiation to 104 h of culture, i.e. at hours

0, 4, 8, etc., of culture. The cells were harvested from 56 h onwards, in 8-h intervals up to 108 h of culture. On the whole, the investigation included 126 variants of experiments with the analysis of 17 150 metaphases. From 100 to 150 cells were investigated for each variant. All slides were coded for scoring the frequency of chromosome aberrations.

Results and Discussion

In the first series of experiments the effects of different ethylenimine derivatives added from 0 to 52 h with harvesting at 56 h of cultivation were compared. All the variants of the experiment with each mutagen were carried out with the blood of the same individual.

In table 1 the data on the frequency of aberrant cells are given. Table 2 gives the number of damaged chromosomes per 100 cells. As may be seen from these tables, the maximum sensitivity to the action of all the mutagens was observed at 28–32 h, in other words 24–28 h before harvesting. The cytogenetic activity of dipin did not seem to differ from the beginning of culture up to 32 h. Beyond this time the frequency of chromosome aberrations decreases: the minimum sensitivity for all the tested mutagens was observed at 48–52h.

The results of the experiment given in table 1 are also presented in figure 1 as step functions, each step reflecting a relative frequency of aberrant cells differing statistically from the neighbouring ones. The variants of the experiments that did not differ from each other statistically were united in one step.

It may be seen that dipin and fotrin, which are multicentric mutagens in contrast to the monocentric phosphamide and thiophosphamide, gave smaller fluctuations of aberration frequencies during the most sensitive first cell cycle

Table 1. Frequency of aberrant cells depending on time of addition of the mutagen.

Time of addition	Phosphamide ($60\ \mu g\ ml^{-1}$)	Thiophosphamide ($20\ \mu g\ ml^{-1}$)	Dipin ($40\ \mu g\ ml^{-1}$)	Fotrin ($40\ \mu g\ ml^{-1}$)
0	27.3	32.5	57.3	59.3
4	35.3	31.2	54.7	45.3
8	35.3	33.0	61.3	48.0
12	30.1	42.6	43.3	56.0
16	41.3	48.1	52.0	64.7
20	34.7	50.3	53.3	58.0
24	29.3	55.4	45.3	60.0
28	47.2	41.8	52.7	74.0
32	46.0	31.9	48.7	71.3
36	36.0	24.5	39.3	58.7
40	19.3	22.3	24.0	59.3
44	10.8	14.0	16.0	28.0
48	5.0	9.2	2.7	5.3
52	9.3	13.5	7.3	14.0

Table 2. Number of damaged chromosomes per 100 cells depending on the time of addition of the mutagen.

Time of addition	Phosphamide (60 μg ml^{-1})	Thiophosphamide (20 μg ml^{-1})	Dipin (40 μg ml^{-1})	Fotrin (40 μg ml^{-1})
0	39.3	47.5	108.0	113.3
4	64.0	44.0	81.3	60.7
8	66.0	42.0	100.7	73.3
12	61.0	72.9	63.3	86.7
16	90.0	87.2	78.7	112.7
20	56.7	100.1	92.7	99.3
24	55.3	122.3	70.0	101.3
28	93.1	74.0	78.7	130.0
32	90.0	56.6	75.3	117.3
36	59.3	38.4	55.3	89.3
40	23.3	29.0	26.7	84.0
44	11.7	18.0	17.3	32.0
48	6.4	10.1	3.3	6.7
52	11.3	15.0	7.3	14.0

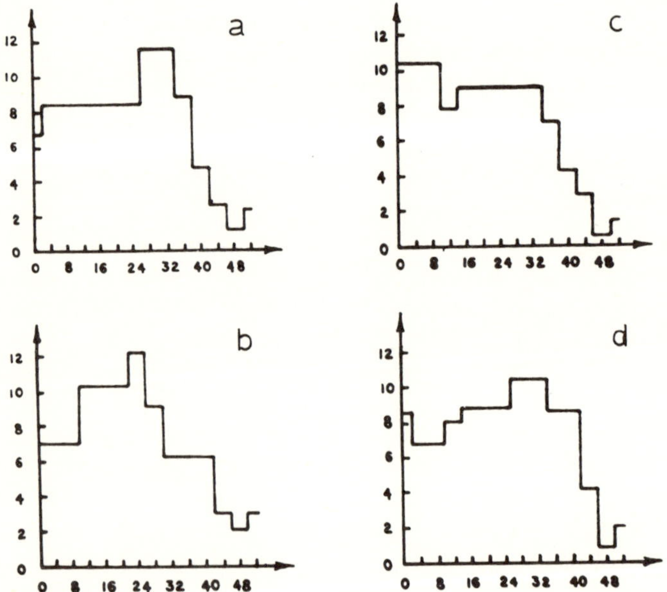

Figure 1. Dependence of the frequency of aberrant cells induced by ethylenimine derivatives on the time of exposure to mutagens: (a) phosphamide, (b) thiophosphamide, (c) dipin, (d) fotrin. Abscissae: time of addition of mutagen. Ordinates: effect expressed as percentage of the sum of effects over all times of exposure.

period (up to 32 h). This would imply that even the derivatives of the same chemical produce different effects during the cell cycle. To study the mutagenic activity of a chemical, it is therefore reasonable to expose cultures to the compound for 24–28 h before harvesting.

In the second series of experiments, the investigation of the action of one mutagen was more detailed in order to trace time regularities of the sensitivity of chromosomes to mutagens over several cell cycles.

The cytogenetic effect produced by thiophosphamide at a concentration of 30 μg ml^{-1} given for 1 h every 4 h from 8 to 104 h of culture, with harvesting every 8 h from 60 to 108 h, was studied. Thus, 70 different combinations of additions of mutagen and harvesting of cultures were analysed.

The aberration data from this experiment are given in table 3. The columns of this table are equivalent to the curves 'time of exposure *vs* effect', and the lines correspond to the curves 'time of harvesting *vs* effect' at different moments of exposure to the mutagen.

Table 3. Frequency of aberrant metaphases (%) following exposure to 30 μg ml^{-1} of thiophosphamide depending on time of addition and time of harvesting.

Time of addition	Duration of cultivation (h)						
	60	68	76	84	92	100	108
8	56.0	49.3	34.0	24.0	19.3	15.3	15.3
16	54.0	46.7	26.0	13.3	13.3	10.0	14.0
24	81.0	68.7	40.0	26.0	16.0	18.7	16.7
32	72.7	51.3	33.3	29.0	15.3	12.7	16.0
40	50.0	73.3	55.0	30.0	24.0	20.7	16.7
48	16.0	39.3	48.0	37.3	22.0	20.0	16.7
56	4.7	16.7	38.0	66.7	56.7	30.7	15.0
64		7.3	28.0	49.0	52.0	33.3	24.0
72			5.3	28.7	56.0	60.0	39.0
80				6.7	20.7	55.3	53.0
88					9.3	27.3	48.0
96						9.3	24.0
104							19.0

The curves 'time of exposure *vs* effect' are described for all the harvestings by correlation equations of the fourth degree. They are presented in figure 2. In table 4 the coefficients of the approximating polynomials and the results of statistical processing of the conformity between the experimental and calculated data are given.

As distinct from the curves 'time of exposure *vs* effect', the shape of the curves 'time of harvesting *vs* effect' also depends upon the time of exposure to the mutagen. The histograms of these curves are given in figure 3, where we may see that with exposure to the mutagen before 32 h of cultivation, and with the prolongation of cultivation time before the harvesting, the frequency

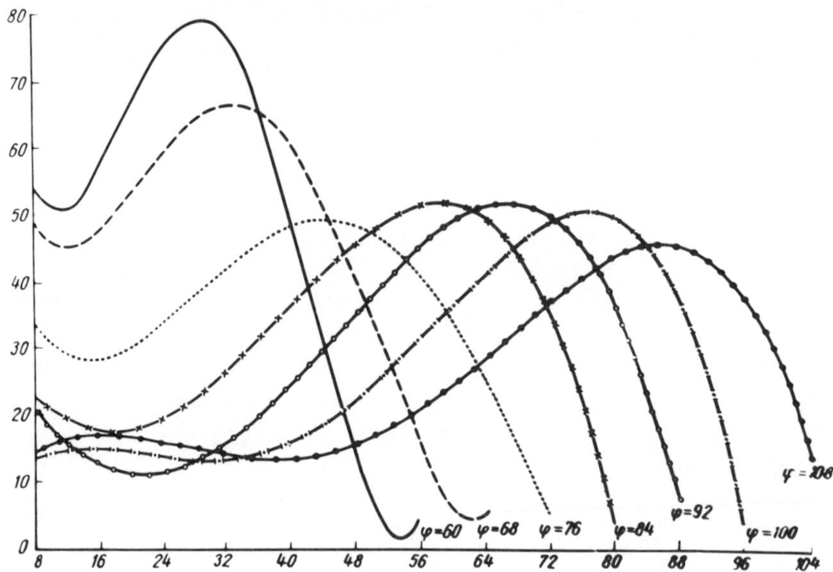

Figure 2. The curves 'time of adding thiophosphamide *vs* effect' at different times of harvesting lymphocyte cultures. Abscissa: time of addition of mutagen. Ordinate: percentage of aberrant metaphases. φ is time of harvesting in hours from the beginning of culture.

of aberrant metaphases decreases. Beginning from 40 h cultivation, these curves change their form: at first, the effect increases and then, on reaching the peak, starts decreasing. Taking into account the changing shapes of the curves 'time of harvesting *vs* effect' the conclusion might be drawn that the curves 'time of exposure *vs* effect' evidently carry more valuable information about the changing of chromosome sensitivity in different periods of lymphocyte culture.

We may see in table 3 and figure 2 that every curve 'time of exposure *vs* effect' is at first characterised by a zone of relatively constant sensitivity followed by a maximum after which a pronounced decrease in effect begins.

Table 4. Coefficients of polynomial approximations and statistical results (X = hour of addition/8).

Time of harvest	Order of the member of equation							
	X^0	X^1	X^2	X^3	X^4	χ^2	n	P
60	+123.62	−128.69	+74.27	−15.190	+0.9800	1.05	6	0.98
68	+88.43	−69.47	+36.08	−6.453	+0.3584	8.17	7	0.30
76	+59.12	−39.19	−15.72	−2.070	+0.0816	4.58	8	0.80
84	+37.25	−18.68	+4.79	−0.111	−0.0215	10.98	8	0.20
92	+34.75	−17.44	+3.12	+0.122	−0.0259	13.32	9	0.14
100	+5.64	+13.38	−6.31	+1.084	−0.0543	16.18	11	0.14
108	+3.36	+16.45	−6.47	+0.915	−0.0392	12.50	12	0.40

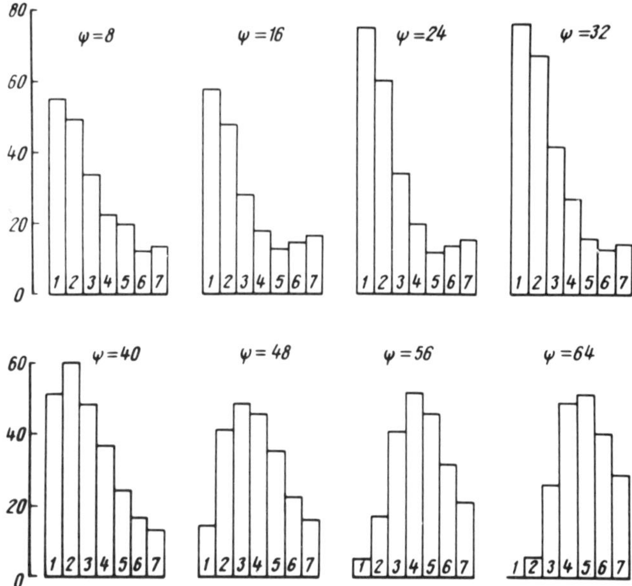

Figure 3. Dependence of 'time of harvesting *vs* effect' on different times of addition of thiophosphamide. Abscissae: time of harvesting. Ordinates: Percentage of aberrant metaphases. φ is time of addition of mutagen. Figures inside the histograms are times of harvesting in hours from the beginning of culture: 1 = 60, 2 = 68, 3 = 76, 4 = 84, 5 = 92, 6 = 100, 7 = 108.

The minimal sensitivity is observed 4 h before harvesting, which means a relative stability of the G_2 stage on exposure to chemical mutagens, as distinct from the variability encountered in radiation effects.

In table 5 are given the results of mathematical analysis of the curves 'time of exposure *vs* effect' at different harvestings. Attention is called to the fact that the distance from the point where the convex curve changes into a concave one (point of inflexion) to the point of harvesting, within the limits of precision of the measurement, is a constant value, namely 44.1 ± 2.7 hours. The point of inflexion is the beginning of the presynthetic stage of the last mitosis before harvesting. This definition is in accord with the estimation of the point of inflexion that was found in the first series of experiments, although with different concentration of thiophosphamide. As for the position of the maximum, as may be seen in table 5, when harvesting at 60–76 h it is located further from harvesting (32–35 h before harvesting) than at later times (about 24 h before harvesting). It is too difficult to judge if this is an experimental error or a characteristic feature of the first post-stimulation mitotic cycle of lymphocytes.

Mathematical analysis of the curves 'time of exposure *vs* effect', the results of which are given in table 5, permits every curve to be partitioned into three sections: (1) from the moment of stimulation to the point of inflexion; (2) from the point of inflexion to the maximum; and (3) from the maximum

Table 5. Mathematical analysis of the curves 'time of exposure *vs* effect' at different harvestings.

Time of harvest	Distance from point of inflexion to harvesting (h)	Distance from maximum to harvesting (h)
60	41.3	31.7
68	47.0	35.3
76	48.1	31.9
84	44.8	25.6
92	45.4	25.3
100	41.1	23.4
108	41.3	23.1
Evaluation		
Mean	44.14 ± 2.71	28.04 ± 4.50
Dispersion	7.32	20.05
Asymmetry	+0.109	+0.371
Excess	−1.857	−1.473
96% confidence limits		
Mean	38.83-49.45	19.26-36.82
Asymmetry	−2.013- +2.013	
Excess	−3.310- +3.310	

Precision of the evaluations: point of inflexion, ± 6.14%; position of the maximum, ± 15.98%.

to the harvesting. The advantage in this partitioning lies in the fact that the boundaries of these sections, and their quantitative characteristics, might be determined by formalised mathematical methods. In particular, the differentiation of the polynomial approximating the experimental curve 'time of exposure *vs* effect' permits us to determine the point of inflexion and the maximum, whereas its integration enables us to determine the area under the curve in each area.

Using the notion of the mean value of the function on a certain section, an average sensitivity of every section might be calculated on all the curves 'time of exposure *vs* effect', taken at different harvestings. The results of these calculations are given in table 6. The data in this table show that the mean sensitivity of the section from culture stimulation to the point of inflexion decreases with increasing time of harvesting, implying the elimination of the damaged cells during culture. The mean sensitivity of the section from the point of inflexion to the maximum is practically the same in all the harvestings, from 76 h onwards. The mean sensitivity of the section from beyond the initial maximum is a constant value for all the curves 'time of exposure *vs* effect'.

Cytogenetic analysis of the efficiency of a fixed concentration of thiophosphamide and a variable time of exposure to the mutagen and time of harvesting, shows that lymphocytes stimulated by PHA are asynchronised in their time of exit from the G_0 phase. Thus even in those cases when the inter-

Table 6. Mean sensitivity of sections of the curves 'time of exposure *vs* effect', as percentages of aberrant metaphases. (Section 1, from stimulation to the point of inflexion; section 2, from the point of inflexion to the maximum; section 3, from the maximum to the time of harvesting.)

Time of harvest	Section of curve 1	2	3
60	54.78	73.46	39.97
68	48.27	62.85	38.14
76	31.10	45.30	34.17
84	22.70	45.62	37.22
92	18.37	43.66	37.28
100	18.18	40.08	36.57

val between the exposure to the mutagen and time of harvesting is 100 h, a proportion of cells are seen to enter into their first mitosis in culture.

After the G_0 stage lymphocytes have an average generation time of about 44 h. Since departure from the G_0 stage is evidently a random process, with big dispersion, then during the first 24 h of cultivation the population of lymphocytes becomes very heterogeneous in its composition in relation to different stages of interphase. In connection with this, the notions 'first', 'second', etc., mitosis lose their distinctness, except for those cases when the effects of mutagenic factors are limited by the resting stage. Even in this case, however, we may speak only of the majority of cells beginning one or another division.

Thus, in the case of the action of the mutagen on a proliferating culture, not only the ordinal number of mitoses after stimulation should be taken into consideration, but also its ordinal number after exposure to the mutagen. In other words, we suggest a two-measure characteristic of a cell population, taking for the initial point of calculation the moment of stimulation, and, for the final point, the moment of harvesting the culture of lymphocytes. Mitotic cycle (M) is defined by two indices, the first of which indicates what mitotic cycle after stimulation the cell was in at the moment of exposure to the mutagen, and the second what mitotic cycle after stimulation the cell was in at the moment of harvesting the culture. For example, if the cell was exposed to the mutagen in the first interphase after stimulation, and this cell is harvested in the first mitosis, it will be attributed to subpopulation $M_{1,1}$; if a cell is harvested in the second mitosis, it should be attributed to subpopulation $M_{1,2}$, and so on. There might also be situations when cells are exposed to the mutagen when they are in the second interphase after stimulation. In this case the index $M_{2,2}$ means that cells were exposed to the mutagen in the second interphase and harvested in the second mitosis after stimulation, but in the first mitosis after the exposure to the mutagen. Index $M_{2,3}$ is a symbol of a subpopulation that was exposed to the action of the mutagen in the second interphase and harvested in the third mitosis after stimulation, but in the second mitosis after the exposure to the mutagen. From this point of view, in the second series of the experiments, the investigated range of time of adding the mutagen and the harvesting of the cultures overlaps three mitoses: $M_{1,1}$, $M_{1,2}$, and $M_{2,2}$.

It is difficult as yet to evaluate exactly the sensitivity of separate stages of interphase to ethylenimine derivatives. It may only be asserted that the G_2 stage is most resistant and the G_1 stage most sensitive. However, because of mitotic cycle asynchronism within the cell population, precise data on the sensitivity of certain periods of the interphase is of purely theoretical interest. In practice, the most practical parameter to consider is the mean sensitivity of the section of the curve 'time of exposure *vs* effect', from the point of inflexion to the time of harvesting.

Summary

The dependence of the frequency of chromosome aberrations in a culture of human lymphocytes from time of addition of four ethylenimine derivatives: phosphamide (60 μg ml^{-1}), thiophosphamide (20 and 30 μg ml^{-1}), dipin (40 μg ml^{-1}), and fotrin (40 μg ml^{-1}), has been investigated. The mutagens tested differ both in number of alkylating groups and in number of active mutagen centres. The chemicals were added in the culture for one hour from the zero time of cultivation of lymphocytes, which had not been stimulated, to 104 h. The cells were harvested from 56 to 108 h of culture.

The mutagens were cytogenetically effective at all the stages of the cell cycle. Maximal sensitivity was observed at 28–32 h of culture, i.e. 24–28 h before harvesting, with the exception of dipin where sensitivity was practically constant from zero to 32 h. After the maximum, the sensitivity of chromosomes to all the studied mutagens sharply decreased and reached the minimum at 48–52 h.

In the case of thiophosphamide, the curves 'time of exposure *vs* effect' are shown to have the same shape, independent of the time of harvesting the culture, described by a fourth order polynomial, while the curves 'time of harvesting *vs* effect' have essentially different shapes depending upon the moment of mutagen action.

For assaying mutagenicity in terms of chromosome aberrations a two-measure characteristic of a cell population is suggested. The first of these indicates what cycle after lymphocyte stimulation the cell is in at the moment of exposure to the mutagen, and the second what cycle after stimulation the cell is in at the moment of harvesting the culture.

References

Brewen, J.G. & N.T.Christie (1967) Studies on the induction of chromosomal aberrations in human leukocytes by cytosine arabinoside. *Exp. Cell Res. 46*, 276–91.

Dubinina, L.G. & N.P.Dubinin (1967) Evidence for chromosome damage by alkylating compounds at presynthesis phase of the cell cycle. *A.S.R. U.S.S.R. 175* (*1*), 213.

——(1968) News on the effects of alkylating compounds on chromosome mutations. *Genetika 4* (*2*), 5–24.

Funes-Cravioto, F. & K.N.Yakovenko (1972) Location of TIO-TEF-induced chromosomes at different periods of culturing peripheral blood lymphocytes. *Genetika 8* (*6*), 131–7.

Kihlman, B.A. (1966) *Actions of Chemicals on Dividing Cells.* New York, London: Prentice-Hall.

Kihlman, B.A. & G.Odmark (1965) Deoxyribonucleic acid synthesis and the production of chromosomal aberration by streptonigrin, s-ethoxycaffeine and 1,3,7,8-tetramethyluric acid. *Mutat. Res. 2*, 494–505.

Rieger, R. & A.Michaelis (1962) Die Auslösung von Chromosomen aberrationen bei Vicia faba durch chemische Agenzien: eine Überzicht. *Kulturpflanze 10*, 212.

Sinkus, A.G. (1972) Cytogenetic effect of rubomidin C in the culture of human lymphocytes. Chromosome damage at stage G_2 of the mitotic cycle. *Genetika 8* (*6*), 138–41.

Yakovenko, K.N., S.A.Azhayev & N.P.Bochkov (1974a) Cytogenetic effect of ethylenimine derivatives in the culture of human lymphocytes. Communication I. Experimental data and mathematical test of quantitative regularities. *Genetika 10* (*10*), 135–43.

——(1974b) Cytogenetic effect of ethylenimine derivatives in the culture of human lymphocytes. Communication II. Mathematical model of the effects of different concentrations of dipin and fotrin. *Genetika 10* (*11*), 138–46.

The Action of Anthracyclines on the Somatic Chromosomes of Man

The anthracyclines claim their significance as antibiotics with anticancer properties against a variety of tumours: the best studied member of the group, adriamycin (ADM), has potentials against several carcinomas, soft tissue sarcomas, pediatric solid tumours, malignant lymphomas and acute leukaemia, among others (Carter 1975). These chemicals are made up of an aglycon chromophore linked to an amino sugar (figure 1), and can inhibit nucleic acid synthesis and bind to nucleic acids through intercalation (see reviews by DiMarco, Arcamone and Zunino 1975; Vig 1977). Whether these drugs bind to a specific kind of DNA is not clear but increasing ΔTm is observed with increasing A:T content of DNA (DiMarco, Arcamone and Zunino 1975). Interestingly though, the alternate dG-dC sequence binds ADM ten times better than alternate dA-dT (Tsuo and Yip 1976).

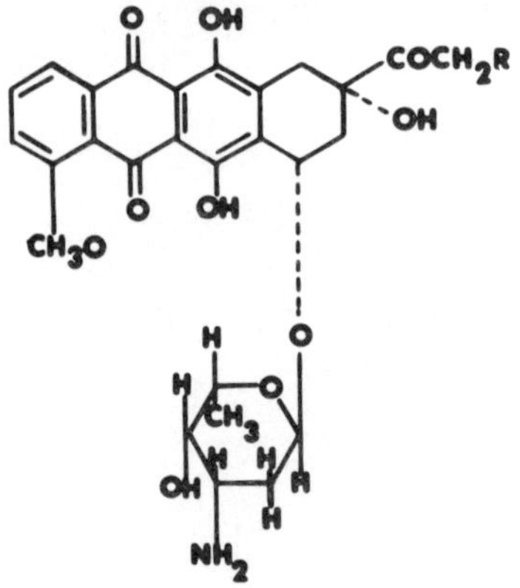

Figure 1. The structure of adriamycin (R=OH) and daunomycin (R=H).

Anthracyclines are capable of inhibiting cell division at several points along interphase. CHO cells show a critical time in S phase, as do human lymphocytes (Krishan and Frei 1976). G_1 and G_2 stages are also inhibited (Byfield 1976; Clarkson and Humphrey 1977; DiMarco, Arcamone and Zunino 1975; Hittelman and Rao 1975; Vig 1977). However, it may not be possible to draw general conclusions from the data so far available. Additionally, the binding of anthracyclines to non-histone proteins (100 times greater than other proteins) may suggest an effect on gene expression, thus pointing to a dual effect of the molecule at DNA template level as well as at regulatory level (Kikuki and Sato 1976).

Even though several modifications of the basic molecule are available, any detailed studies on the effects on chromosomes are limited to adriamycin (ADM) and daunorubicin (or daunomycin, DNR). Nonetheless, the genetic effects of these chemicals are in line with the radiochemical data, which show that the largest amount of DNR is taken up by the nuclei, with some labelling in the mitochondrial fraction (see DiMarco, Arcamone and Zunino 1975). The fact that ADM can cause both single-strand and double-strand breaks in DNA (Byfield 1976) suggests the involvement of DNA in chromosome breakage induced by anthracyclines.

Materials, Cells and Methodologies

The sources of anthracyclines used, especially the equivalents of DNR, were rubidomycin by the French group (in experiments by de Grouchy and de Nova 1967), the rubomycin C of Russians (in studies by Sinkus 1972) and daunomycin by Pharmitalia Pharmaceuticals (used mostly by American and Italian workers). ADM, however, was obtained from Pharmitalia. In most experiments with chromosome aberrations, peripheral leucocytes in whole blood microcultures were utilised. The PHA-stimulated cultures were exposed to the chemical at different post-culture times. The continuously treated cultures were exposed for 1, 2 or 3 days (e.g. de Grouchy and de Nova 1967; Massimo, Dagna-Bricarelli and Fossati-Guglielimoni 1970; Vig *et al.* 1968), whereas in pulse treatments the exposure period varied from 1 to 4 h (e.g. Sinkus 1972a; Vig 1970, 1971a, b) with recovery periods ranging from 2 h (Sinkus 1972a) to 64 h (Vig 1971b). For studies with human embryonic fibroblasts (Sinkus and Kuliev 1972) the suspension of cells obtained from muscle tissues of 6 to 9-week-old embryos were treated with rubomycin C for 2 h and chromosome preparations were made by treating cells with colchicine after 24 and 72 h.

In the case of *in vivo* studies carried out with DNR (Massimo *et al.* 1970, Sinkus 1972, Whang-Peng *et al.* 1969) and ADM (Massimo, Dagna-Bricarelli and Cherchi 1972), lymphocytes taken from patients treated with therapeutic doses were cultured at various periods after the drug was administered; however, bone marrow preparations were processed for chromosome analysis immediately after taking the samples. This unavoidable difference in technique for obtaining chromosomes from lymphocytes and bone marrows may be responsible for the differences observed in the frequency of aberrations.

The Nature and Localization of Aberrations

In vitro Studies

The clastogenic potentials of anthracyclines were first indicated in 1963 by DiMarco and associates (DiMarco 1968) as blocking of mitotic processes and anomalous scattering of chromosomes. However, Ostertag and Kersten found no effect on chromosome structure of HeLa cells treated with the drug (see Vig 1977). The details of the nature of aberrations induced by anthracyclines were provided independently by de Grouchy and de Nova (1967) and Vig *et al.* (1968) using human lymphocytes and medullary cells in the former studies and human leucocytes in the latter. But, whereas in one laboratory chromosome aberrations were observed after treating the cells with doses of DNR as low as 0.02 μg ml^{-1} for 24 h (Vig *et al.* 1968), in other studies as much as 0.6 μg ml^{-1} of this antibiotic were used for 2 or 3 days (de Grouchy and de Nova 1967). The responses to such differences in dose are still unclear in view of the fact that, in our laboratory, treatments of leucocytes with doses like 0.4 or 0.6 μg ml^{-1} for only a few hours are usually sufficient to kill the cells.

Aberrations induced with DNR are of both chromatid and chromosome types and include fragments as well as exchanges. However, in cells treated at low doses with brief recovery periods, most aberrations can be of deletion type without much rejoining (e.g. as for ADM; Newsome and Singh 1977). Small marker chromosomes, similar to Ph^1 chromosomes, have also been observed (de Grouchy and de Nova 1967) and may simply result from loss of a chromosome fragment originating close to the centromere. In some cases, incomplete or 'sub-chromatid' reunions were observed, along with what appears to be simple adhesion, both of non-sister chromatid ends of two chromosomes (Vig *et al.* 1968) similar to what is reported after cells are treated with streptonigrin (see Vig 1977) or for embryonic cells close to senescense (Benn 1976). The nature and origin of such 'exchanges' is a matter of conjecture.

In the case of ADM, the early reports were also with human lymphocytes in culture. Massimo, Dagna-Bricarelli and Fossati-Guglielimoni (1970) showed the inhibition of blastogenesis and cellular alterations apparently resulting from chromosome damage in cells treated with 0.05 or 0.1 μg ml^{-1} of the chemical added throughout a 72 h culture period. Concentrations of 1 μg ml^{-1} were lethal or completely inhibited cell division. At about the same time, Vig (1971) reported chromosome aberrations in cells treated for only 6 h with doses as low as 0.02 to 0.15 μg ml^{-1}. In these studies some cells were totally demolished or pulverised, whereas in others chromosome morphology was deformed.

Molecular alterations in DNR do not always retain the chromosome-breaking properties of the parent molecule. Thus, the N-acetyl derivative of DNR, for example, is ineffective in inducing aberrations. DiMarco (1968) suggested that the lack of chromosome-breaking property of this derivative was due to its large size and consequent reduced capacity to bind to DNA—one of the mechanisms postulated to be responsible for chromosome breakage induced by anthracyclines.

The effect of anthracyclines appears to be cell cycle specific, the least sensitive stages being G_0 and mitosis. The G_0 resistance may be due to resistance of chromatin to the anthracycline molecule, impermeability of the cell membrane or lack of some molecular environment conducive to the action of the drug in these physiologically relatively inert cells. The resistance of mitotic chromosomes could be due to their compact nature. The evidence, using *in vitro* cultures of human leucocytes, is derived from studies in which treatment of cells for the first few hours of culturing produced none or only a few aberrations. In similar studies, when cells were treated during early G_1 (about 10–16 h post-culture), the anthracycline treatments did produce aberrations (Vig 1971b). However, cells treated in early G_1 start expressing aberrations only after long recovery periods (30–60 h), whereas those treated in early S or late G_1 seem to constitute a uniformly more sensitive population and to express a high frequency of aberrations even in the first sample of analysable metaphases. These differences may, in part, be attributable to differential sensitivity of G_1 and S populations to progression through cell cycle.

In the production of chromosome aberrations, the S phase appears to be much more sensitive to the action of anthracyclines—both DNR (Vig 1971b) and ADM (Vig 1971a)—than other phases even though DNR prolongation of G_2 cells, as for example in CHO cultures (Hittelman and Rao 1975), is about three times greater than that of S. However, a high frequency of aberrations in S should not be correlated with the synthesis of DNA *only*, since chromosomal proteins synthesise at about 2.5 times the rate of their synthesis in G_1 (see Vig 1977 for details). Also, the sensitivity of G_2 cells, however slight, has been demonstrated by studies with human lymphocytes (Sinkus 1972) and CHO cells (Hittelman and Rao 1975).

The effect of anthracyclines in induction of aberrations is interesting in that in appropriately treated cell populations, not only is there a non-delayed type of effect, but also the aberrations keep appearing for several cell cycles after the drug is removed. Thus, in one study, Sinkus and Kuliev (1972) set up experiments to find the time for which this residual effect persists after pulse treatment with DNR for 2 h at concentrations of 0.02 and 0.05 μg ml^{-1}. In this case, the chromosome aberrations in cells cultured from 6 to 9-week-old human embryos persisted for at least 4–6 passages. In pulse-treated human leucocytes also, such aberrations appear to last for several cell cycles (Vig 1970, 1971b).

In most of these studies, using brief exposures or long recovery, primary types of aberrations were of chromatid-type exchanges along with chromosome-type (=isochromatid?) fragments. One also finds cells carrying both chromatid aberrations and dicentrics. It is unlikely that the observed high frequency of 'twin' fragments is due to breakage in the two sister chromatids in neighbouring regions (Kusyk and Hsu 1976). The possibility exists that a chromosome in early S is composed of parts that are single stranded (G_1-like) and some that are already functionally bistranded. The data of Reddy and Miller (1977) support this conclusion. An alternate possibility is that sensitivity of G_1 chromosomes results in the production of chromosome breakage followed by the rejoining of some of the broken ends after replication of chromatin. Chromatid aberrations, however, may also originate if only a 'half chromatid' (a single DNA helix) is affected. Nonetheless, chromosome breakage at the centromere can originate at either G_1, S or G_2.

A rather rare type of chromosome aberration is a triradial, interpreted to be the result of isochromatid-chromatid type exchange. In anthracycline treated cells, e.g. those with ADM or DNR, such aberrations can be seen more frequently than has been the experience with other chemicals or radiations. Some of the triradials have all three ends free, whereas in others a pair of sister chromatids is seen in reunion. These triradials usually have only one centromere in the whole complex and are less frequent than dicentric triradials (see table 5 in Vig 1971b). Some of these monocentric triradials appear to originate from rearrangements within one chromosome. This interpretation has been supported by the occurrence of rare cells with monocentric triradials

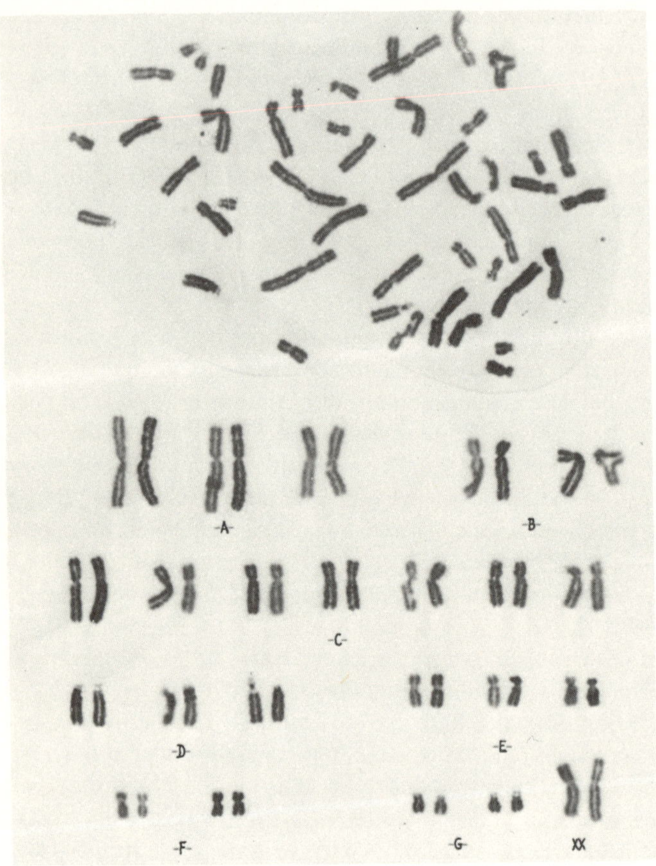

Figure 2. An ADM-treated cell showing a monocentric triradial as the only exchange. Karyotyping assigns this chromosome to group B, probably no.5, suggesting an intrachromosomal origin for this aberration.

as the only type of exchange and a lack of any chromosome-type of aberration. For example, when a karyotype is prepared on the cell shown in figure 2, the triradial is interpreted to have originated by intrachromosomal rearrangement of one of the chromosomes 5.

Anthracyclines induce chromosome breakage in human leucocytes nonrandomly. The positions of breaks along the length of the chromosome (or genome) are similar for ADM and DNR. Similarly, non-randomness is maintained between chromosomes. Thus, in one study (Vig *et al.* 1970) DNR-induced breaks were found at a frequency higher than expected in chromosome pairs 1, 2, 4/5, 6–12 and 13/15, whereas others showed deficiency. The Y chromosome was the least affected, followed by 19/20, 21/22, 3 and 16/18. A study with continuously treated or pulse-treated leucocytes with ADM also gave closely parallel results (Vig 1971a). It is rather interesting that anthracycline-induced chromosome aberrations in man are not confined primarily to known heterochromatic blocks, whereas those induced in mouse leukaemic cells (Cornforth, unpublished) do show a preponderance in the A:T-rich, constitutively heterochromatic regions. Whether the localisation of aberrations has anything to do with the selective production of a 'banding' effect by anthracyclines is not known.

Some chemicals frequently induce chromatid exchanges that involve homologues. In the case of mitomycin C (MMC), such breakage and rejoining takes place at corresponding regions, producing quadriradials as observed in human leucocytes (Shaw and Cohen 1965) and *Vicia faba* (Rao and Natarajan 1967). Whereas anthracyclines induce frequent exchanges between homologues, these rarely involve corresponding positions on these chromosomes. In one study 37.5% of all exchanges induced by ADR involved homologues or apparent homologues, but only 8 out of a total of 120 such exchanges could meet the criterion of region-specific reciprocal rejoining (Vig *et al.* 1970). As with other chemicals, chromatid exchanges induced by anthracyclines are more frequently of asymmetrical type. In one study, when cells were treated with 0.02 μg ml^{-1} of ADM for 4 h, 63% of all exchanges were of U-type. These results are similar to those obtained for DNR (see Vig 1977).

Interactive response with other chemicals. In preliminary studies with DNR (Vig *et al.* 1968) the addition of arginine resulted in an increase in mitotic index as well as in chromosome aberrations. This increase in the frequency of aberrations appeared to be due to an increase in the frequency of cells showing multiple aberrations. The data indicated that arginine somehow increased the chances of survival and/or progression to metaphase of badly damaged cells otherwise destined to die during that cell cycle. The addition of this amino acid had no effect on the pattern of distribution of aberrations (Vig *et al.* 1970). On the other hand, when cells treated with anthracyclines are treated with inhibitors of protein synthesis, e.g. puromycin, a reduction in the frequency of aberrations is noticeable (Vig 1970). The reduction affects exchanges very drastically, sometimes reducing the rejoining to only 12% of the value obtained without addition of puromycin. Contrary to this, similar exposure to FUdR of the cells treated with DNR did not affect the overall frequency of aberrations, nor was there any noticeable effect on the ratio between exchanges and free fragments. Contrary to these results, an addition of deoxyribose cytidine (2×10^{-4}M) to ADM-treated cells did increase the frequency of all types of aberrations found in human leucocytes (Vig 1973). Even though this synergism is not limited to deoxyribose cytidine, it was found to be more consistent (about 30% higher than in the material treated with ADM alone) than with other nucleosides. This increase in total aberrations is not at the expense of exchanges creating more free fragments. Rather, the data indicate that the synergistic effect is real and is due to some mechanism that facilitates the action of ADM on chromatin. A hypothesis has been proposed that such an effect originates at the points of chromatin-nuclear membrane attachment sites, which are postulated to be the weak spots along the length of chromatin (Vig 1975).

It has been known for some time that the rate of cell killing by ADM increases with hyperthermia to 43°C compared to that observed at 37°C (Hahn, Braun and Har-Kedar 1975) apparently due to increased uptake of ADM at higher temperatures. Such a mechanism should also be responsible for an increase in the frequency of chromosome aberrations. The cells were therefore treated with 0.1 and 0.2 μg ml^{-1} of ADM for 1 h at 37°C and higher temperatures and recovered after various intervals. The data indicate a significant increase in the frequency of chromosome aberrations of both chromatid and isochromatid (or chromosome) type (Vig, unpublished). The cells treated at 37°C, but given a post-ADM treatment at 43°C, failed to show an increase in the frequency of aberrations. Also, the treatment of cells with ADM at 4°C followed by 37°C incubation showed a drastic reduction in the frequency of aberrations compared to those observed in the material treated at 37°C, even though, as previously, post-treatment at 4°C did not affect the frequency of aberrations when ADM was in contact with the cells at 37°C. These data permit the conclusions that ADM affects the cells immediately, and that post-ADM treatments do not affect the enzymatic machinery associated with production of aberrations once aberrations or lesions are induced by anthracyclines. The data from studies on prematurely condensed chromosomes (PCC) carried out with CHO cells by Hittelman and Rao (1975) strongly support the interpretation.

In vivo Studies

In vivo studies have been carried out with human bone marrow cells taken from patients on chemotherapy using DNR only (Ballerini 1970, Whang-Peng *et al.* 1969). In the latter case, 7 patients with acute lymphocytic or granulocytic leukaemia given DNR at the rate of 80 to 420 mg m^{-2} showed aberrations in up to 90% of their cells 'right after treatment *in vivo*'. These aberrations disappeared in 1 to 2 weeks after treatment was stopped. This observation seems to be logically related to the known half-life ($\simeq$55 h) of the drug in human plasma and correlates well with similar observations made for *in vitro* cultures (Sinkus 1972). Structural aberrations in these cells were similar to those found in leucocytes treated *in vitro*. Besides, in patients treated with DNR (Ballerini) or ADM (Massimo, Dagna-Bricarelli and Cherchi 1972), aneuploidy appears to be a common feature after long exposures.

The treatment of haematological patients with DNR also produces chromosome aberrations in lymphocytes (Sinkus and Orlova 1970). However, a rather low frequency of cells (between 5 and 9.2% after treatment *vs* 1.2% before treatment) show aberrations. These include dicentric rings, indicating possible survival of an interarm intrachromosomal aberration to be followed by sister reunions in the same or the following cell cycle. However, the differences in the frequency of aberrations observed between lymphocytes in Sinkus' study and bone marrow in another (Whang-Peng *et al.* 1969) may reflect a lack of quantitative relationship between cells from various organs. In other species, e.g. rats (Jensen and Philip 1971), a 12 to 16 times higher frequency of metaphase cells with aberrations has been realised in the bone marrow cells than in cultured leucocytes. It is not known if these differences originate from causes like differential permeability to the anthracycline molecule (as postulated for DNR resistant DC-3F and CLM-7 lines of Chinese hamster cells; see Vig 1977), or if they result from procedural differences necessary to obtaining good metaphases. However, there have been reported differences in the effectiveness of DNR on leucocytes from various individuals. De Grouchy and de Nova (1967) had demonstrated a differential effect of the antibiotic on medullary cells in comparison to that on leucocytes. Similarly, quantitative effects of anthracyclines regarding induction of aberrations differ for normal *vs* leukaemic cells in mice (Stromskaya and Pogosianz 1971), with the latter being about 3 to 4 times more sensitive to DNR but not to urethane.

Karyotypic Profile Alterations

Effects of anthracyclines on the induction of numerical inequalities among the daughter cells are known. Massimo, Dagna-Bricarelli and Cherchi (1972), for example, reported induction of polyploidy and aneuploidy induced in human lymphocytes *in vitro* by ADM. Similar results have been previously reported by Whang-Peng *et al.* (1969) for DNR. Hasholt, Visfeldt and Dano (1971), on the other hand, reported a rather interesting phenomenon of karyotypic changes induced by DNR as associated with resistance to the drug. When a sub-line of *Ehrlich ascites* tumour, made resistant to the drug by long-term treatment, was studied karyotypically, a change from near tetraploidy in the original line to hyperdiploidy, along with the appearance of marker chromosomes, in the resistant line was found. On reversion from DNR resistance to sensitivity, the karyotype also changed to that of the original tumour. However, the rate of growth in the resistant hyperdiploid lines was slower than those in the original tetraploid sensitive ones.

Induction of Sister Chromatid Exchanges

Anthracyclines, being potent inducers of chromosome aberrations, including exchanges, should be capable of inducing sister chromatid exchanges (SCE). In a study documenting the relationship between chromosome aberration induction and induction of SCE, Perry and Evans (1975), and more recently Neustad (pers. comm.), have shown that only fractions of concentrations needed for the production of traditional exchanges can induce SCE in high frequencies. Even though the details of data are not available they reported a 'significant increase in SCE . . . observed in blood cells sampled 24 h after injection' with adriamycin, 'at dose levels which produce only a negligible frequency of chromosomal aberrations in these cells'. Similar comparisons are now available for other organisms, e.g. CHO (1975) and mouse (Cornforth, unpublished). However, a relationship between induction of SCE and traditional exchanges does not necessarily hold for every chemical; nor is there a definite correlation between regions of exchange induction and those for induction of SCE.

Summary

Anthracyclines constitute a group of promising anticarcinogens. The basic structure of the molecule is an aglycon chromophore linked to an amino sugar. These drugs may inhibit nucleic acid synthesis and induce single- as well as double-strand breaks in the DNA. The anthracyclines, particularly daunomycin and adriamycin, are known to have potential for breaking chromosomes. Both chromatid- and chromosome-type aberrations are produced, sometimes in the same cell. The most sensitive stage of the cell cycle to the induction of chromosome aberrations is S, with early G_1 and G_2 cells being only slightly sensitive. The G_0 cells and chromosomes in prophase or metaphase appear to lack anthracycline sensitivity.

Anthracyclines appear to have a residual effect lasting three to four cell cycles. A large number of chromatid exchanges along with 'chromosome-type' fragments are the predominant type of aberrations. Rarely, intra-chromatidal-chromosomal aberrations, giving rise to monocentric triradials, are seen, suggesting that the entire chromosome may not replicate at the G_1–S border.

Anthracycline-induced aberrations are non-randomly distributed, but are not preferentially localised in C-chromatin. Addition of arginine to DNR-treated cells increases mitotic index as well as aberration yield, whereas the frequency of exchanges is reduced with the addition of puromycin. Hyperthermia and deoxyribose cytidine show strong synergism with adriamycin. ADM and DNR can induce chromosome aberrations *in vivo*, as well as *in vitro*. Numerical inequalities induced in cells treated with these agents indicate effects on spindle fibres or centromeres. Karyotypic alterations associated with resistance-susceptibility to anthracyclines have been observed in Ehrlich tumour cells.

Adriamycin has been shown to be very effective in inducing sister chromatid exchanges both *in vitro* and *in vivo.*

References

Ballerini, G. (1970) Alterations of karyotypes by antimitotic treatment. *Haematologica 55*, 225–31.

Benn, P.A. (1976) Specific chromosome aberrations in senescent fibroblast cell lines derived from human embryos. *Am. J. Hum. Genet. 28*, 465–73.

Byfield, J.E. (1976) Molecular origin of cell death induced by cyclophosphamide, adriamycin and X-rays. *Proc. Am. Ass. Cancer Res. 17*, 27.

Carter, S.K. (1975) Adriamycin—a review. *J. Natl Cancer Inst. 55*, 1265–74.

Clarkson, J.M. & R.M.Humphrey (1977) The effect of adriamycin on cell cycle progression and DNA replication in Chinese hamster ovary cells. *Cancer Res. 37*, 200–5.

DiMarco, A. (1968) Mechanism of action of daunomycin. *A. Ge. Me. Ce. 17*, 102–21.

DiMarco, A., F.Arcamone & F.Zunino (1975) Daunomycin (daunorubicin) and adriamycin and structured analogues: biological activity and mechanism of action, in *A Mechanism of Action of Antimicrobial and Antitumor Agents* (eds J.W.Cockran & F.H.Hahn) pp. 101–28.

Grouchy, J. de & C. de Nova (1967) Cytogenetic effect of rubidomycin (or daunomycin). *Ann. Genet. 11*, 39–44.

Hahn, G.M., J.Braun & I.Har-Kedar (1975) Thermochemotherapy: synergism between hyperthermia (42–43°) and adriamycin (or bleomycin) in mammalian cell inactivation. *Proc. Nat. Acad. Sci. U.S.A. 72*, 937–40.

Hasholt, L., J.Visfeldt & K.Dano (1971) Karyotypic profile alterations in *Ehrlich ascites* tumor cells during development of resistance to daunorubicin. *Acta Pathol. Microb. Scand. 79*, 665–75.

Hittelman, W.N. & P.N.Rao (1975) The nature of adriamycin-induced cytotoxicity in Chinese hamster cells as revealed by premature chromosome condensation. *Cancer Res. 35*, 417–20.

Jensen, M.K. & P.Philip (1971) The cytogenetic effects of rubidomycin. *Mutat. Res. 12*, 91–6.

Kikuki, H. & S.Sato (1976) Binding of daunomycin to nonhistone proteins from the rat. *Biochem. Biophys. Acta 434*, 509–12.

Krishan, A. & I.Frei (1976) Effect of adriamycin on the cell cycle traverse and kinetics of cultured human lymphoblasts. *Cancer Res. 36*,143–50.

Kusyk, C.J. & T.C.Hsu (1976) Adriamycin-induced chromosome damage: elevated frequencies of isochromatid aberrations in G_2 and S Phase. *Experientia 32*, 1513–14.

Massimo, L., F.Dagna-Bricarelli & A.Fossati-Guglielimoni (1970) Effects of adriamycin on human lymphocytes stimulated with PHA *in vitro. Europ. J. Clin. Biol. Res. 7*, 793–9.

Massimo, L., F.Dagna-Bricarelli & M.G.Cherchi (1972) Effects of adriamycin on blastogenesis and chromosomes of blood lymphocytes: *in vitro* and *in vivo* studies. *Int. Symp. Adriamycin*, 35–46.

Newsome, Y.L. & D.N.Singh (1977) Cytological effects of adriamycin on human peripheral lymphocytes. *Acta Cytol. 21*, 137–40.

Perry, P. & H.J.Evans (1975) Cytological detection of mutagen-carcinogen exposure by sister chromatid exchanges. *Nature 258*, 121–5.

Rao, R.N. & A.T.Natarajan (1967) Somatic association in relation to chemically induced chromosome aberrations in *Vicia faba. Genetics 57*, 821–35.

Reddy, M.D. & M.W.Miller (1977) X-ray induction of chromosome and chromatid aberrations in some cells after treatment with hydroxyurea. *Experientia 33*, 321–2.

Shaw, M.W. & M.M.Cohen (1965) Chromosome exchanges in human leukocytes induced by mitomycin C. *Genetics 51*, 181–90.

Sinkus, A.G. (1972a) Cytogenetic action of rubomycin C in human lymphocyte cultures, chromosome aberrations at the G_2 stage of the mitotic cycle. *Genetika 8* (*6*), 138–41.

—— (1972b) Chromosomal damage on cultured human lymphocytes taken from patients after rubomycin C treatment. *Tsitologiya 14*, 1184–7.

Sinkus, A.G. & A.M.Kuliev (1972) Chromosome injuries in some cell generations after treatment of human cell cultures with rubomycin C. *Bull. Exp. Biol. Med. 74*, 1460–2.

Sinkus, A.G. & R.S.Orlova (1970) Chromosome changes in lymphocyte cultures of patients treated with mitomycin C. *Tsitologiya 12*, 352–6.

Stromskaya, T.P. & H.E.Pogosianz (1971) Effect of rubidomycin C and urethane on chromosomes of normal and leukemic mouse cells. *Genetika 7* (*11*), 57–63.

Tsuo, K.C. & K.F.Yip (1976) Effect of deoxyribonucleases on adriamycin-polynucleotide complex. *Proc. Am. Ass. Cancer Res. 17*, 13.

Vig, B.K. (1970) Alterations in the pattern of daunomycin induced chromosomal aberrations by inhibitors of protein and DNA synthesis, *Mutat Res. 9*, 607–14.

—— (1971a) Chromosome aberrations induced in human leukocytes by the antileukemic antibiotic adriamycin. *Cancer Res. 31*, 32–8.

—— (1971b) Nature of chromosome aberrations induced in pre-DNA synthesis period in human leukocytes by daunomycin. *Mutat. Res. 12*, 441–52.

—— (1973) Synergism between deoxyribose cytidine and adriamycin in causing chromosome aberrations in human leukocytes. *Mutat. Res. 21*, 163–70.

—— (1975) Chromatin-nuclear membrane attachment in relation to DNA replication and chromosome aberrations: A new hypothesis. *J. Theoret. Biol. 54*, 191–9.

—— (1977) Genetic toxicology of mitomycin C, actinomycins, daunomycin and adriamycin. *Mutat. Res. 49*, 189–238.

Vig, B.K., S.B.Kontras, E.F.Paddock & L.D.Samuels (1968) Daunomycin-induced chromosomal aberrations and the influence of arginine in modifying the effect of the drug. *Mutat. Res. 5*, 279–87.

Vig, B.K., L.D.Samuels & S.B.Kontras (1970) Specificity of daunomycin in causing chromosome aberrations in human leukocytes. *Chromosoma 29*, 62–73.

Whang-Peng, J., B.G.Leventhal, J.W.Adamson & S.Perry (1969) The effect of daunomycin on human cells *in vivo* and *in vitro*. *Cancer Res. 23*, 113–21.

M.T. DOLOY, R.LE GO, M.HARDY and M.REILLAUDOU

Chromosome Abnormalities produced in Human Lymphocytes by Organic Extracts from River Water

To determine the eventual toxicity of drinking water for man, it appeared to be of interest to investigate the effects of agents found in it. This study was made possible by a grant from the Ministry of the Quality of Life. The feeble effects observed with the usual method of lymphocyte cultures treated with the organic extracts led us to search for optimal experimental conditions to increase the effect observed, while maintaining the mitotic activity. The influence of the pollutant concentration in the culture and the pollutant-lymphocyte contact time were given special emphasis.

This report compares the effects observed in several systems of human lymphocyte cultures.

Materials and Methods

Water Organic Extracts

Chloroform extracts of organic substances from river water were provided by IRCHA (Institut National de Recherche Chimique Appliquée). After complete evaporation, the organic substances in each sample were weighed and dissolved in a very small volume of 0.1N NaOH before being introduced into the cultures. After preliminary studies, the concentration of the extract in the culture was fixed at 200 $\mu g\ ml^{-1}$ of nutritive medium. The effect observed at lower concentrations was less, and at higher concentrations the cytotoxic effect was often too high to permit chromosomal analysis.

Benzopyrene

3,4-benzopyrene in dimethylsulphoxide (DMSO) was introduced into the cultures at concentrations of 1 or 10 $\mu g\ ml^{-1}$.

Culture Techniques

Standard culture technique. The lymphocytes were cultured for 2 or 3 d using the Moorhead technique. For a test phase the lymphocytes were cultured for 4 or 5 d. The micro-pollutant was added either at the beginning of the culture time (2-d cultures), or after 24 h incubation (longer cultures).

Long-lasting culture technique. The lymphocytes were isolated from the blood by gravity sedimentation followed by filtration on a nylon wool column. We noted that the lymphocyte cultures were considerably altered if the nylon wool column filtration was omitted. After 15 min contact with the nylon wool, the effluent cells were centrifuged (100 to 150 g) for 7 min. The lymphocytes represented at least 80% of the nucleated cells. The cells (10^6 per ml) in a final volume of 5 ml were cultured in a medium containing: 4 ml McCoy medium, 1 ml human AB (Rh +) serum, 0.05 ml heparin (Liquemine Roche), 0.075 ml phytohaemagglutinin (PHA Wellcome) diluted at 1/100. 200 units ml^{-1} of penicillin and 200 $\mu g\ ml^{-1}$ of streptomycin were also added to the medium. The PHA concentration in this medium is 100 times lower than in the standard medium.

The cells were incubated for 4 or 7 d at 37°C. Every 3 or 4 d, 3 ml of supernatant medium was exchanged for fresh medium. The mitotic activity was very low because of the low concentration of PHA. This stage was called the latent period. In the absence of PHA the cultures responded poorly. A normal mitotic index could be recovered by introducing PHA at the usual concentration of 0.015 ml per ml of medium. 3 d after mitotic stimulation the lymphocytes were treated according to standard methods for chromosomal observations (colcemid addition, hypotonic treatment, Giemsa staining). Several cultures were obtained from one human blood sample. The cultures inoculated with the pollutants were compared with control cultures and 100 metaphases were observed microscopically for each culture. The karyotype was established for each abnormal cell. For each cell the chromosomes were scored, analysed according to the Denver classification, and the different kinds of chromosomal anomalies were studied, i.e. the number of abnormal chromosomes; types of chromatid lesions (gap, breaks, interchanges, pulverisation); and chromosomal abnormalities (fragments, translocations, dicentrics, rings).

For each extract, only a small quantity of organic material was available, which made the comparison of the results difficult; for this reason benzopyrene was chosen as a reference pollutant.

A statistical analysis for the evaluation of possible significant chromosomal abnormalities was performed by the statistical unit of INSERM (URS-INSERM, Villejuif, by P.Lazar and D.Hemon). A weight (W) was defined for each type of anomaly. The value chosen for the weight is the logarithm to the base 10 of the inverse of the frequency of occurrence (f_t) in the set of control cultures, i.e. $W = \log 1/f_t$. The presence of a type of abnormality in a culture gives information about the cytogenetic efficiency of the pollutant. This information (i) is assumed to be equal to the product of its observed frequency (f_{obs}) by its weight, i.e. $i = f_{obs} \times \log 1/f_t$. The information ($I$) included in the whole set of observations from one culture is equal to the sum of the individual information (i) from each type of abnormality. For this estimation some abnormalities that may be produced by an artifact are excluded from the score (e.g. chromatid gaps and hypodiploidy).

Application of the Student and Fisher test allowed comparison of the data from the treated culture and data from the control culture on the same blood sample. The variance used for this test is equal to the mean variance on the I-values derived from all the control cultures.

Results

In the 'classical' cultures incubated more than 3 d, the quality of the cells is somewhat unsatisfactory and the mitotic index is poor. On the other hand, in the 'long-lasting cultures' starting after 4-d latent period, the cell quality is adequate for the cytogenetic study. However, in the case of the 7-d or more latent period, the quality of the cells again is unsatisfactory.

Table 1. Comparison between the control cultures.

Rest phase (d)	0	0	4
Mitotic stimulation (d)	2	3	3
No. of observed cells	2856	3142	3595
Percentage of:			
gaps	2.49	4.74	3.62
iso-chromatid gaps	0.39	0.54	0.7
chromatid breaks	0.42	0.25	1.14
exchanges	—	—	—
pulverisation	—	—	—
fragments	0.42	1.08	0.86
translocations	—	0.12	—
dicentrics	—	0.03	0.028
rings	—	—	—
hyperdiploidy	0.04	0.25	0.03

The frequency of abnormalities in the control cultures, either classical 2- or 3-d or long-lasting 4-d latent period, are similar (table 1). The comparison between the benzopyrene-treated cultures shows more abnormalities in the cultures with a 4-d latent period than in the 2 or 3-d classical cultures (table 2).

Table 2. Comparison between the benzopyrene-treated cultures.

	Control			Low Dose			High Dose			
est phase (d)	0	0	4	0	0	4	0	0	0	4
itotic stimulation (d)	3	4	3	3	4	3	2	3	4	3
o. of observed cells	357	100	300	361	100	400	200	291	100	403
ercentage of:										
gaps	4.76	9	1.67	1.94	12	5	1	2.4	11	7.75
isochromatid gaps	0.56	—	0.33	0.55	1	1	—	0.7	—	2.5
chromatid breaks	1.68	—	1.33	2.5	7	3.75	—	2.75	1	6
exchanges	—	—	—	—	—	—	—	—	—	1
pulverisation	—	—	—	—	—	—	—	—	—	—
fragments	0.84	—	0.33	1.66	—	4.75	3	5.5	1	13
translocations	—	—	—	—	—	—	1	—	—	0.25
dicentrics	—	—	—	—	—	—	—	—	—	0.5
rings	—	—	—	—	—	—	—	—	—	—
hypodiploidy	3.1	5	4.67	7.76	6	7.5	—	4.8	8	10.25
hyperdiploidy	0.56	—	—	0.28	1	—	—	0.7	2	1

In the case where the same pollutant has been introduced in the 2-d classical and the 3-d classical cultures there were more abnormalities in the 3-d culture (table 3). In this experiment, the cell pollutant contact time was the same: the pollutant was introduced at the start of the 2-d culture and introduced after 1 d in the 3-d cultures.

Table 3. Comparison between cultures treated with the same pollutant (2 and 3 days).

Rest phase (d)	0	0
Mitotic stimulation (d)	2	3
No. of observed cells	900	900
Percentage of:		
gaps	3.9	5.67
isochromatid gaps	0.22	0.56
chromatid breaks	0.33	1
exchanges + pulverisation	0.11	—
fragments	1.67	2.4
translocations	0.22	0.11
dicentrics + rings	—	0.22
hyperdiploidy	0.33	0.78

The results of the experiment in which the same pollutant was introduced both in the 3-d classical and the long-lasting 4-d latent period cultures, showed more abnormalities in the long-lasting cultures (table 4).

Initially, the pollutants were introduced in the 2-d classical cultures. Only rarely the Student-Fisher test allowed the detection of an apparent cytogenetic

Table 4. Comparison between cultures treated with the same pollutant (with or without rest phase).

Rest phase (d)	0	4
Mitotic stimulation (d)	3	3
No. of observed cells	900	800
Percentage of:		
gaps	6.89	7.75
isochromatid gaps	1.11	1
chromatid breaks	4.78	5.13
exchanges	0.11	0.13
pulverisation	—	—
fragments	0.89	2.38
translocations	—	0.13
dicentrics	—	—
rings	—	0.25
hypodiploidy	7	12.63
hyperdiploidy	0.11	0.5

effect by comparing one culture only with its own control. The comparison between the whole set of treated cultures and the whole set of controls shows a significant difference. This indicates that the cytogenetic effect was too weak to be demonstrated for 100 cells. In the 3-d classical cultures, a greater percentage of pollutants showed a significant effect. This effect was again more evident in the long-lasting cultures (table 5).

Table 5. Probability P for the cultures treated with water extract to belong to the control culture population.

Rest phase (d)	0		4	
Mitotic stimulation (d)	3		3	
Treated cultures	No.	%	No.	%
$P > 0.05$	33	75	44	*58*
$0.05 > P > 0.01$	5	11	7	9
$0.01 > P > 0.001$	—	—	11	*14*
$P < 0.001$	6	14	14	18

These results with different pollutants support the preceding results with benzopyrene and the intercomparisons on the same pollutants by the different culture methods.

Discussion

Because only a very small quantity of the different pollutants was available, the comparison between the different culture methods was difficult. The differences are mainly apparent in the benzopyrene results.

The advantage of the long-lasting culture technique has been demonstrated. However, these results require confirmation by testing with other types of pollutants. The long-lasting culture technique allows the lymphocyte contact to be increased, so that the pollutant can be active either directly on the molecular structures of the nucleus, or through the mechanisms of the intermediary cellular metabolism. The fact that this long-lasting contact does not impair the ability of the cells to undergo mitosis is also of interest.

From a statistical point of view, the choice of the weighting factor for the different types of abnormalities does not appear to be satisfactory. In fact, the choice of the logarithm of the inverse of the frequency in the control culture results in a minimisation of the importance of the occurrence of the rare types of abnormalities with respect to the more frequent ones. However, at the time, this statistic was chosen because no other form of weight estimation seemed more appropriate.

Conclusion

The long-lasting culture technique described in this paper appears to be a method of choice for studies on the cytogenetic effect of pollutants.

The Induction of Non-Disjunction in Mammalian Oogenesis

Chromosome studies on more than 5000 consecutive newborn babies (for review see Nielsen and Sillesen 1975) revealed a considerable incidence of numerically unbalanced karyotypes at birth. Approximately 4–5 babies among 1000 carry such an aneuploidy, which is thought to originate in most cases from meiotic non-disjunction. Animal data from this stage of development are lacking so far. At an earlier stage of ontogenesis, i.e. during the early post-implantation development, the incidence of aneuploidy is even considerably higher (Boué and Boué 1973) and one can also observe trisomies other than from chromosomes 13, 18, 21 and from aneuploidies of the sex chromosomes (Creasy, Crolla and Alberman 1976). Data on spontaneous abortions suggest that the incidence of aneuploid fetuses may be estimated at about 5% at this early stage of development. This estimated incidence agrees with the numbers observed among induced abortions at a similar stage of gestation (Yamamoto *et al.* 1975).

How many aneuploidies are present at the very early beginning of ontogenesis (at the time of fertilisation) in man? There must be at least around 5%, whereas Boué and Boué suspect that a more realistic incidence would be 50% of all conceptions.

Studies using marker chromosomes in informative families having a mongoloid child, which were confounded by a statistical bias (Langenbeck *et al.* 1976), revealed that the majority of trisomies 21 originate from maternal non-disjunction during the first meiotic division. Animal models are the best suited experimental systems to study the influence of certain mutagens on meiotic non-disjunction, which is obviously not feasible in man. Some new data will be presented below on the spontaneous incidence of non-disjunction in different mouse strains (Hansmann 1977), and data from the literature will be reviewed on mutagen-induced non-disjunction.

Materials and Methods

8 to 12-week-old female mice, inbred (C3H), F_2 hybrids (101 × C3H), and from a randomly bred NMRI colony, were either mated to vasectomised males for an indicator of ovulation, or treated with hormones, PMS and HCG, 48 h later for stimulated ovulation. A low dose (1.5 i.u. PMS and 1.0 i.u. HCG)

and a high dose of hormones (10 i.u. PMS and 10 i.u. HCG) were chosen to study the influence on the first meiotic division of hormonally stimulated ovulation. The females were sacrificed on the morning of the vaginal plug or 15 h after the application of HCG.

Chromosome preparations were made from ovulated oocytes according to Röhrborn and Hansmann (1971), which is a modification of Tarkowsky's method (1966). Na-citrate at a concentration of 1.3% was chosen instead of 1.5%. The slides were coded and the chromosomes analysed microscopically at the stage of metaphase II. Chromosome counts of less than 20 might at least partly be due to preparational loss and were therefore discarded.

Results

First results from three mouse strains of this long-term study on non-disjunction in mammals are presented in table 1. No significant difference between these three strains was observed. For simplification, the data of each strain are added together in table 1. No significant increase of hyperploid oocytes was observed after hormonally stimulated ovulation.

Table 1. Spontaneous non-disjunction during the first meiotic division in three mouse strains.

Strain	Ovulation[1]	Hormone dose	No. of oocytes	No. hyperploid
C3H	spontaneous	—	413	1
	induced	1.5/1.0	386	1
		10/10	451	2
(101 × C3H)F_1	spontaneous	—	159	1
	induced	1.5/1.0	138	1
		10/10	103	1
NMRI	spontaneous	—	438	1
	induced	1.5/1.0	509	1
		10/10	423	1
Total			3020	10

[1] Either mated with vasectomised males or stimulated with pregnant mare serum (PMS) and human chorionic gonadotropin (HCG) 48 h later.

Discussion

The present study on a large number of ovulated oocytes from three different strains shows that non-disjunction is a rare event during the first meiotic division in mouse oogenesis. Our data are compared to previous results in table 2.

The low incidence in mouse oocytes and also in those of the Chinese hamster (Hansmann, Neher and Röhrborn 1974) is in apparent contrast to the situation in man. In man, one can estimate that incidence figures are higher

Table 2. Chromosome studies on unfertilised oocytes.

Author	Species	Strain	Method[1]	Treatment	No. of oocytes	No. hyperploid
Röhrborn & Hansmann (1971)	mouse	$(101 \times C3H)F_1$	PMS/HCG	control	81	5
				trenimone	107	20
Uchida & Lee (1974)	mouse	(C3H × ICR/Swiss	*in vitro*	control	1054	0
				X-rays	1149	6
Hansmann (1974)	mouse	C3H	PMS/HCG	control	335	8
				amethopterin	450	44
				cyclophosphamide	136	8
Röhrborn & Hansmann (1974)	mouse	C3H	spontaneous	control	128	3
				norethisterone-acetate	229	13
Hansmann, Neher & Röhrborn (1974)	Chinese hamster		PMS/HCG	control	220	0
				trenimone	137	3
Reichert, Hansmann & Röhrborn (1975)	mouse	NMRI	PMS/HCG	control	143	0
				X-rays	204	6
Shimada *et al.* (1976)	mouse	ddY	PMS/HCG	control	198	0
				cadmium	271	3
Martin, Dill & Miller (1976)	mouse	CBA	*in vitro*	60–150 days	166	0
				151–240 days	101	6
				241–330 days	146	0
Becker & Schöneich (1976)	mouse	NMRI	spontaneous	control	410	2

[1]Methods: *in vitro*, culture of preovulatory oocytes up to metaphase II; spontaneous : mating vasectomised males with females without hormonal pretreatment; PMS/HCG, application of pregnant mare serum (PMS) and human chorionic gonadotropin (HCG) to stimulate ovulation.

by several factors as already indicated. The reasons for this exceptional frequency of non-disjunction in man may be many. Different chromosome number and morphology, or functional aspects of repetitive DNA, might also be involved. A possible complication for mutagenicity studies arises from this species-specific incidence of spontaneous non-disjunction. The effect found in mouse and other rodents need not reflect the situation in man and may reflect differences due to a different metabolism and repair system. From teratogenesis studies it is known that strains with a higher spontaneous incidence of malformations are more susceptible to the actions of teratogens (e.g. Fraser *et al.* 1954). It is quite clear that mechanisms of teratogenicity and mutagenicity are different, but we should envisage that mouse data on non-disjunction may also be an under-representation of possible events in human germ cells.

A significant increase of non-disjunction was observed in ovulated mouse oocytes after exposure to X-rays in two groups (Uchida and Lee 1974; Reichert, Hansmann and Röhrborn 1975). It appears that some chemical mutagens are more potent in increasing the events of non-disjunction and this was observed with two alkylating agents and an antimetabolite (Röhrborn and Hansmann 1971, Hansmann 1974). In addition a gestagen applied in a high dose, and for a longer period, increased non-disjunction after treatment was stopped (Röhrborn and Hansmann 1974). No significant increase of non-disjunction, but arrest of meiosis, was observed after treatment with cadmium (Shimada, Watanabe and Endo 1976). Further studies are needed, using the same mutagens and set-up but with lower concentrations, and using a number of species.

Previous studies (see table 2) have shown that a short period before ovulation is highly sensitive for induced non-disjunction. More information is required on the sensitivity of earlier stages in oogenesis, especially the long-lasting dictyotene, and also stages during the prenatal phase of oogenesis. Studies on transplacentally induced mutagenesis are to be encouraged and these studies may also bring new information on those mechanisms involved in the process of non-disjunction.

Chromosome studies on ovulated oocytes are expensive and time consuming, but they have the advantage that non-disjunction is observed immediately after the first meiotic division, long before well-known processes like selection against unbalanced karyotypes start to act. Another way to analyse spontaneous and induced non-disjunction is the study of cleavage stages beginning with the pronucleus stage (e.g. Hansmann 1974). Early embryos after implantation are also very well suited, and in combination with banding techniques they also give us information on structural anomalies. We should keep in mind, however, that the further removed the analysis from the time of induction, the less efficient are we in detecting induced changes. Analysis nearer to the time of induction may involve technically more complicated procedures, but will yield far more information.

Summary

The incidence of non-disjunction during the first meiotic division was studied by analysing more than 3000 oocytes at metaphase II from three different mouse strains, after spontaneous or hormonally induced ovulation. No increase of hyperploid oocytes was observed after the application of a low and a high dose of pregnant mare serum and human chorionic gonadotropin for stimulated ovulation. The overall incidence of hyperploidy was about 0.3%, indicating that non-disjunction during the first meiotic division in mouse oocytes is a very rare event.

This is in apparent contrast to the human situation, where studies on newborns and abortions reveal that the incidence figures of aneuploidy are higher by several factors. At least in Down's syndrome it could be shown that most non-disjunctional events occurred during the maternal first meiotic division. The reasons for this exceptional situation in man compared to mouse are known.

In earlier studies with mice and Chinese hamsters it could be shown that, under certain circumstances, and depending on the stage of meiosis treated, X-rays and chemical mutagens can increase the incidence of spontaneous non-disjunction.

Acknowledgements

The study is supported by the Deutsche Forschungsgemeinschaft. Thanks are due for the technical assistance of Mrs Gebauer, Miss Hagemann and Mrs Schindler, and to Mrs Loddenkemper for her help in preparing the manuscript.

References

Becker, K. & J.Schöneich (1976) Frequencies of spontaneous meiotic nondisjunction in mouse oocytes. EEMS, 6th Annual Meeting, Gernrode 1976.

Boué, J. & A.Boué (1973) Anomalies chromosomiques dans les avortement spontanés, in *Chromosomal Errors in Relation to Reproductive Failure* (eds A.Boué & C.Thibault) pp.29–56. Paris: INSERM.

Creasy, M.R., J.A.Crolla & E.D.Alberman (1976) A cytogenetic study of human spontaneous abortions using banding techniques. *Hum. Genet. 31*, 177–96.

Fraser, F.C., H.Kalter, B.E.Walker & T.D.Fainstad (1954) The experimental production of cleft palate with cortisone and other hormones. *J. Cell Comp. Physiol. 43*, Suppl.1, 237–59.

Hansmann, I. (1974) Chromosome aberrations in metaphase II-oocytes. Stage sensitivity in the mouse oogenesis to amethopterin and cyclophosphamide. *Mutat. Res. 22*, 175–91.

——(1977) Influence of hormonally stimulated ovulation on spontaneous nondisjunction during first meiotic division in 3 mouse strains (in prep.).

Hansmann, I., J.Neher & G.Röhrborn (1974) Chromosome aberrations in metaphase II-oocytes of Chinese hamster (*Cricetulus griseus*). I. The sensitivity of the preovulatory phase to triaziquone. *Mutat. Res. 25*, 347–59.

Langenbeck, U., I.Hansmann, B.Hinney & V.Hönig (1976) On the origin of the supernumerary chromosome in autosomal trisomies, with special reference to Down's syndrome. A bias in tracing nondisjunction by chromosomal and biochemical polymorphisms. *Hum. Genet. 33*, 89–102.

Martin, R.H., F.J.Dill & J.R.Miller (1976) Nondisjunction in aging female mice. *Cytogenet. Cell Genet. 17*, 150–60.

Nielsen, J. & I.Sillesen (1975) Incidence of chromosome aberrations among 11,148 newborn children. *Hum. Genet. 30*, 1–12.

Reichert, W., I.Hansmann & G.Röhrborn (1975) Chromosome anomalies in mouse oocytes after irradiation. *Humangenetik. 28*, 25–38.

Röhrborn, G. & I.Hansmann (1971) Induced chromosome aberrations in unfertilized oocytes of mice. *Humangenetik. 13*, 184–98.

—— (1974) Oral contraception and chromosome segregation in oocytes of mice. *Mutat. Res. 26*, 535–44.

Shimada, T., T.Watanabe & A.Endo (1976) Potential mutagenicity of cadmium in mammalian oocytes. *Mutat. Res. 40*, 389–96.

Tarkowsky, A.K. (1966) An air drying method for chromosome preparations from mouse eggs. *Cytogenetics 5*, 394–400.

Uchida, I. & C.P.V.Lee (1974) Radiation induced nondisjunction in aged mice. *Nature 250*, 601–2.

Yamamoto, M., R.Fujimori, T.Ito, K.Kamimura & G.Watanabe (1975) Chromosome studies in 500 induced abortions. *Humangenetik. 29*, 9–14.

R. LE GO

Image-Processing Automation for Chromosome Analysis

Our research is especially devoted to biological dosimetry and its relation to chromosome analysis. For this purpose we employ the Giemsa stain exclusively.

In order to accelerate the work of the technicians in cytogenetics, we have built two systems. The first is a metaphase finder, and the second is an electronic and informatics assembly whose purpose is to provide a computer-aided karyotyping system.

Metaphase Finder

The metaphase finder (Eidomat) is incorporated in a Zeiss photomicroscope for which a motorised stage has been built and calibrated for the x, y coordinates. The sensor is a set of photodiodes that scan the microscope field along the vertical diameter. The linear matrix is composed of 40 elements in this first version of the instrument, but will have 512 photodiodes in the second generation version already under construction.

The analogue signals emitted by the sensors are sent to a logic assembly through a double threshold and a chopper. The 'white' threshold eliminates the background level of the signal. The 'black' threshold eliminates the too high amplitudes of the signal that corresponds to such objects as dark nuclei; this double thresholding results in the transformation of the chopped signal into a linear matrix of 0s and 1s. The 1 values correspond to a mean grey level, which is designed to correspond to the mean grey level of a correct metaphase. The 0 values represent the areas where the colour is too white or too black. The binary values from the current position of the diodes, and from the three preceding steps of chopping, are memorised on line.

The logic assembly must compare the adjacent 1s, in the horizontal and vertical direction, to find sub-matrices of 1s only, of 2×2 or 3×3 or 4×4. Each such sub-matrix of local correct grey level is called a locus. When such a locus is found, the locus counter is incremented. The system operates on line and is designed to recognise a predicted number of adjacent loci during a limited number of steps. When this occurs the image is presumed to offer the 'good' grey level on a 'good' surface, and a decision signal is emitted. All the counters are then set to zero. The system is made 'blind' during a certain number of steps to eliminate the immediately adjacent area. The decision signal will pick up on line the x, y coordinates of the stage at the point of recognition. These values are sent to a mini-computer associated with the system and stored in memory.

An interactive program has assigned the slide area to be scanned to the computer. When this area has been explored, the system stops and is set to phase 2. During this phase the computer pilots the stage to locate the first memorised image under the objective and allows the operator to keep or cancel this image. This is done immediately by pressing one of the two appropriate buttons. The second image is then immediately presented, and so on until the last image is reached. At this time the computer has marked in memory only a list of the good metaphases on the slide.

The system is operated through a computer program activated by a monitor and a number of commands. For instance, ETA will allow the setting of a number of parameters for the scan: area, overlap or spacing between the x-lines, number of loci assigned to a good metaphase, etc., and identity parameters of the slide. LIS produces the listing of the good metaphase addresses. PRU punches this list together with the identity parameters of the slide so that the paper tape will be linked to the slide for a new examination. The computer, fed with the paper tape, will relocate the good metaphases on the same slide. CHE, n locates the *n*th metaphase of the list and CHE, 0 sets the stage at home position.

The accuracy of this metaphase finder is about 15–25%. Thus for about 100 locations, from 15 to 25 are actually good metaphases depending on the quality of the smear, staining, etc. The time required to scan the useful surface of a smear 2.5 × 6.0 cm is about 5 to 6 min, i.e. 0.3 min cm^{-2}. At this speed it is acceptable to spend 5 min more to cancel the non-relevant images in the smear.

A special feature has been included in the system with a view to the possibility of exploring very poor smears with rare metaphases. The stage is moved very slowly so that the moving field can be observed on a microscope screen. When a good metaphase is passing across the field the operator can choose the address himself by pressing a button. Thus he can be sure that all of the examinable metaphases will be located in memory and on the associated paper tape.

Karyotyping by Pattern Recognition

The second system, called ASTI (Automatic System for the Treatment of Images) comprises an image numeriser, an automaton and a computer program. The numeriser is composed of a Zeiss photomicroscope with a plumbicon camera (Thomson) mounted on it, an A to D converter with a grey-level threshold and a 32K MOS memory. Symmetrically the digital content of the memory is sent through a D to A converter to a video display. The result of the conversion can be continuously monitored on the screen and the different corrections of light, focussing, thresholding, etc., made on line. When the best possible image has been obtained on the screen, the conversion result is stored in memory at a speed of 150 ms. The communication between the microscope and the memory is then cut and set between the memory and the automaton. A control panel and a joy-stick put the automaton in conversational mode with the operator.

The automaton ASTI has several functions. It displays the memorised digital image on the monitor screen and it isolates connected images. That is, it separates the individual image of each chromosome and calculates a number of parameters on each of the isolated images: area, weighted area or mass, frame coordinates in the whole set of data, location of the first line and last line touching the image frame. Each sub-image scissored from the data set is framed by the minimum rectangle and this vignette, composed of at most 1000 points (grey values), can be sent to the mini-computer.

The automaton also permits a number of corrections to be made on the original metaphase image through the use of a 'vector function' activated by the joy-stick. The vector function appears as a small line segment or vector, and the joy-stick is used to position it at will. When the vector touches a blob on the screen the image is indicated to the automaton, which can then perform different functions on it.

The image can be sent to the computer with its hard-calculated parameters. It can also be marked, that is a given logical level is linked to it, and it can be linked to other images marked by the vector. Thus together they can constitute a subset of the metaphase, for instance, the A-class, and so on. The image designated by the vector can be masked; in this case it will virtually vanish from the memory and will be ignored by the automaton when it sends successively and automatically the framed images to the computer. The image can be cut by the vector so that a frontier is created at the location of the vector creating two individually separated items. This permits the separation of touching chromosomes.

These different functions allow the cleaning up of the metaphase image so that it will contain only pertinent and well separated chromosome images. This cleaning up takes about one or two minutes for the operator.

The automaton can now be set to the automatic isolation of images. The computer will then receive consecutively the vignettes containing the isolated chromosomes and apply to each of them the various algorithms of pattern recognition. At the present time the available software provides three functions. It can edit a facsimile of the microscopic metaphase image. Each chromosome is drawn in the same situation in the document as it had in the original metaphase. An order number is assigned to each chromosome. It can edit a propositional karyotype in which the vertically oriented images are drawn by groups A to G. In this document, there will be a number of misclassifications. These may be due to many causes, such as scattering of chromosomes, staining artifacts, touching arms, and little G. Therefore the third step in karyotyping is an interactive program by which, through the use of a special language, the operator may command, e.g., an exchange between images a and b, or the addition of image c to the B-group, and so on. Thus the definitive classification will correspond to his own conception of the particular karyotype.

The time required for the different steps described is approximately as follows: setting the appropriate light intensity, focussing and entry of the image into memory, 30 s; clearing the metaphase image, 1–2 min; drawing the metaphase facsimile, 30 s; drawing the propositional karyotype, 3 min; correcting the misclassifications, 3–5 min depending on the number of classification errors; drawing the definitive karyotype, 3 min. In fact, 6.5 min is the fixed time for this operation, and the variable part can run from 4 to 7 minutes. That is, 10 to 13 minutes for the set of documents. This running time is obviously too long for a really operational application of semi-automatic karyotyping. For this reason the next stage of our work is to shift the system to a higher level computer, a MITRA-125 (SEMS) having 64K words. This should allow us to produce about one hundred karyotypes daily.

A.V.CARRANO, J.W.GRAY and M.A.VAN DILLA

Flow Cytogenetics: Progress towards Chromosomal Aberration Detection

Over the past several years it has become increasingly evident that mankind is being exposed to a wide variety of clastogenic agents, i.e. physical or chemical agents that are capable of breaking chromosomes. This realisation has placed a considerable strain on the scientists responsible for determining the genetic consequences of such exposure, largely because conventional methods of chromosomal aberration analysis are ill-suited to large scale population studies. They are subjective, slow and tedious. Semi-automated and automated systems have been proposed as means of assisting or supporting the cytogeneticist for aberration scoring (Castleman *et al.* 1976; for a comprehensive review see Mendelsohn 1976) but, in general, they lack the high throughput or accuracy necessary for population monitoring.

For the past three years we have been developing the instrumentation and biological methodology for the analysis of isolated metaphase chromosomes by flow cytometry (Gray *et al.* 1975a, b; Carrano *et al.* 1976; Carrano, Van Dilla and Gray 1977). This technique permits the examination of approximately 1000 individual metaphase chromosomes per second. The parameter quantified is the amount of fluorescence emitted by each stained chromosome as it crosses a beam of intense laser light. The data output is a distribution of chromosome frequency versus fluorescence intensity which, for ethidium bromide stained chromosomes, constitutes a DNA flow karyotype. This approach offers the advantages of objectivity and speed; to establish a single karyotype the time required from initiation of data collection through data analysis is approximately 20 minutes. We describe below the results of our continuing efforts to apply this technology to the analysis of chromosomal aberrations in a model system, the Chinese hamster cell line M3-1. Consideration is given to both homogeneous aberrations, i.e. events that are present and identical in every cell (e.g. heritable translocations), and heterogeneous aberrations, i.e. events that are usually not heritable and occur at random in a cell population (e.g. dicentrics and deletions).

Materials and Methods

Cell Culture and Chromosome Isolation

The Chinese hamster M3-1 cells are cultured in 650 cm^2 surface area glass roller bottles in 150 ml of Minimal Essential Medium supplemented with NCTC-135 (final concentration 10%), fetal calf serum (final concentration 15%), and gentamycin and chlorotetracycline as antibiotics. Under these conditions, the generation time is 13 h. For chromosome isolation, cells are seeded in the same roller bottles at a density of 5×10^4 per cm^2 and grown for approximately 44 h. At this time the medium is removed and fresh medium containing colcemid (0.037 $\mu g\ ml^{-1}$ final concentration) is added for a period of 5 h. Following mitotic cell accumulation the roller bottles are rotated at 300 r.p.m. for 6 min on a culture rotator (Talandic Corporation, Pasadena, California U.S.A.). The medium is then decanted, the detached cells are counted and aliquots of 5×10^6 cells are pelleted in a centrifuge. Approximately 2×10^7 cells (mitotic index > 90%) are harvested from each roller bottle.

The chromosome isolation procedure has been adopted from Wray and Stubblefield (1970). The mitotic cell aliquots are resuspended in potassium chloride (0.075M) at 4°C for 30 min, pelleted, and resuspended in 0.5 ml of pH 6.7 chromosome isolation buffer consisting of 0.1 mM piperazine-N, N′-bis 2-ethanesulphonic acid, 1 M hexylene glycol and 1 mM calcium chloride. The chromosome isolation buffer was modified during the course of these experiments and now consists of 25 mM Trizma base (adjusted to pH 7.5 with hydrochloric acid), 0.75 M hexylene glycol, 0.5 mM calcium chloride and 1.0 mM magnesium chloride. The 0.5 ml suspension of mitotic cells is sheared at 4°C in a Virtis homogeniser to disrupt the cell membrane and free the metaphase chromosomes. Staining is accomplished by adding to the sheared suspension an equal volume of 0.2 mM ethidium bromide (EtBr) in the same isolation buffer. The final suspension contains approximately 10^8 metaphase chromosomes in 1 ml of buffer at a final EtBr stain concentration of 0.1 mM.

Flow Cytometry and Sorting

The isolated and stained chromosomes are analysed on the Lawrence Livermore Laboratory flow cytometer (FCM) or sorted on the Becton-Dickinson (B–D) flow sorter. For measurement of the relative fluorescence intensity on the FCM, about 5×10^5 chromosomes are analysed at a rate of approximately 1000 per s. The fluorescence distributions are computer fitted by a least squares method using multiple Gaussian functions superimposed on an exponential function. Each Gaussian function represents a distinct group of chromosomes; the Gaussian mean representing the mean chromosome group fluorescence and the area representing the chromosome group frequency of occurrence. The exponential function is a measure of the background, i.e. fraction of the total area in the distribution underlying the chromosomal peaks (Moore 1975; Carrano, Van Dilla and Gray 1977). We ascribe this background to fluorescent debris and/or chromosomal fragments. The output of this fitting program includes the mean, area, and coefficient of variation of each chromosome peak, and the background area.

In order to preserve the chromosome morphology during flow sorting, the chromosome isolation buffer is used as the sorting fluid. Approximately

10 000 chromosomes per sort are collected directly on microscope slides and fixed by slowly dropping 3 : 1 absolute methanol : glacial acetic acid onto the chromosomes, followed by two further changes of the same fixative. The slides are then stained with a 7.5% Giemsa solution in distilled water.

Results

The Flow Karyotype

The karyotype of clone 650A of the Chinese hamster M3-1 cell line consists of 23 chromosomes. Chromosomal banding analysis distinguishes

Table 1. Comparison between the Giemsa-banded and DNA flow karyotype of the Chinese hamster M3-1 clone 650A.

Giemsa-banded karyotype		DNA flow karyotype		
Chromosome type	Chromosome frequency	Peak designation	Relative mode	Chromosome frequency
1	2	A	1.0	1.92
2	2	B	0.86	1.79
4,t(X;5)	3	C	0.54	3.03
5	1	D	0.46	1.02
6,7,Y	5	E,F[1]	0.36	5.29
8	2	G	0.29	2.02
9,M1	3	H	0.23	3.02
10,11,M2	5	I	0.15	4.90

[1] These peaks are considered together since it is often difficult to resolve their individual modes and areas.

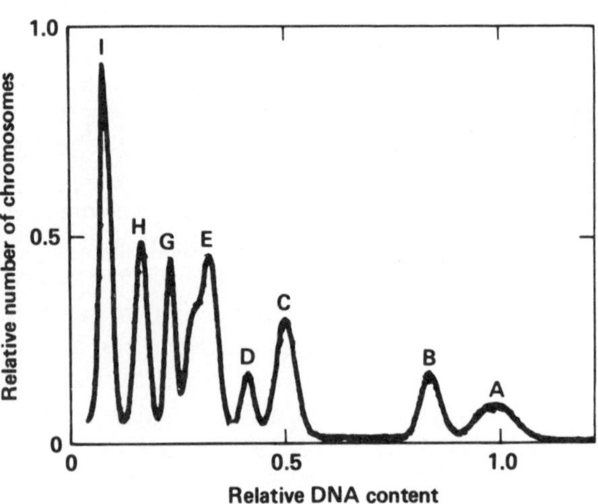

Figure 1. The flow karyotype of Chinese hamster M3-1 clone 650A. The isolated chromosomes were stained with ethidium bromide. Approximately 5×10^5 chromosomes were analysed to obtain this distribution. The line drawn through the data points was generated by computer analysis.

14 distinct types (Gray *et al.* 1975a), which are indicated in table 1. If each of the chromosome types possessed a unique amount of DNA then one would expect 14 distinct peaks in the FCM-generated chromosomal DNA distribution. Figure 1 shows the distribution actually obtained after staining the isolated chromosomes with EtBr. Nine distinct peaks are present, indicating that some chromosomes have very similar amounts of DNA, which are not resolved in this distribution. The chromosome types that constitute each peak have been determined by flow sorting and subsequent banding analysis as well as by independent scanning cytophotometric measurements of chromosomal DNA content (Gray *et al.* 1975a; Carrano, Van Dilla and Gray 1977). The chromosome types ascribed to the peaks in the fluorescence distribution, along with computer estimates of peak means (approximately proportional to DNA content) and peak areas (proportional to relative frequency of occurrence), are shown in table 1. There is excellent agreement between the actual chromosome group frequencies determined by banding analysis of metaphase cells and that determined from the FCM distribution. The relative DNA content and chromosomal frequency therefore constitute the quantitative DNA flow karyotype.

Analysis of Homogeneous Aberrations

If both the relative DNA content and chromosomal frequency can be established for each peak in the FCM distribution, it should be possible to karyotype a cytogenetically identical cell population. To test this hypothesis, we acutely irradiated clone 650A cells with either 300 or 1000 rad of 300 kVp X-rays and immediately cloned several sub-populations. These clones were allowed to grow for approximately two months in culture, at which time their chromosomes were isolated for FCM analysis. The flow karyotypes of two clones, 650AAA and 650AB, derived by this protocol, are shown in figure 2,

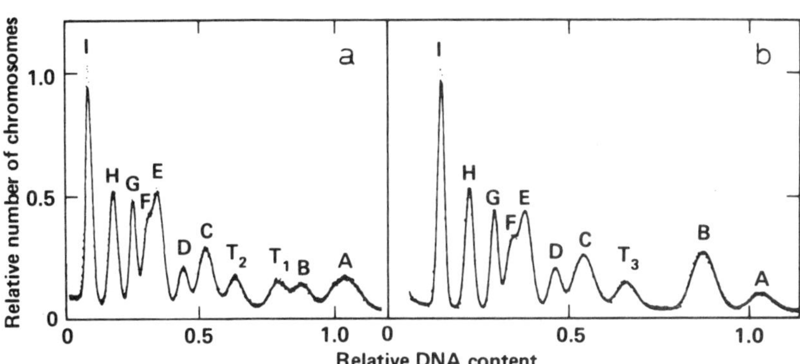

Figure 2. The flow karyotype of Chinese hamster M3-1 clone 650AAA (a) and clone 650AB (b). The chromosomes were stained with ethidium bromide. The two distributions were analysed at different amplifier gain settings and thus the length of the abscissae differ. The computer-generated fitted line is drawn through the data points.

Table 2. The DNA flow karyotypes of Chinese hamster clones 650AAA and 650AB.

650AAA				650AB			
Peak designation[1]	Chromosome type[2]	Relative mode	Chromosome frequency	Peak designation[1]	Chromosome type[2]	Relative mode	Chromosome frequency
A	1	1.0	2.12	A	1	1.0	0.93
B	2	0.86	0.93	B	2,der(1)	0.85	2.80
T_1	der(2)[3]	0.78	1.11	T_3	der(4)	0.65	1.19
T_2	der(4)	0.63	1.14	C	4,t(X;5)	0.53	2.30
C	4,t(X;5)	0.53	2.02	D	5	0.46	0.97
D	5	0.46	0.88	E,F	6,7,Y	0.36	4.81
E,F	6,7,Y	0.36	5.03	G	8	0.29	2.17
G	8	0.29	2.29	H	9,M1	0.23	3.16
H	9,M1	0.23	2.89	I	10,11,M2	0.15	4.67
I	10,11,M2	0.15	4.59				

[1] The peak designation, relative mode and chromosome frequency were determined from analysis of the FCM distributions in figure 2.
[2] The chromosome type was determined from Giemsa-banded metaphase cells.
[3] The term *der* is used to indicate a derivative chromosome, in this case, a structurally rearranged chromosome resulting from a translocation. The number in parenthesis indicates the chromosome from which the centromere was derived (Paris Conference, 1971).

a and b respectively. The results of the computer fit to these distributions are given in table 2.

From figure 2 and table 2 it is evident that, for both clones, peaks D through I are unaltered, either in relative mode or area, compared to clone 650A. The FCM distribution of clone 650AAA chromosomes (figure 2a) differs from that of clone 650A (figure 1) in that two new peaks, T_1 and T_2, are present and areas of peaks B and C are changed. Peaks B and C have each lost one chromosome while peaks T_1 and T_2 each contain one chromosome. If the amount of DNA in chromosome 1 is expressed on an arbitrary scale as 100 units, the relative means show that the chromosome in peak T_1 has 8 units less DNA than the chromosome in peak B, while the chromosome in peak T_2 has 10 units more than the chromosomes in peak C. Thus there appears to have been reciprocal exchange involving chromosome 2 in peak B and chromosome 4 or t(X;5) in peak C. The net effect is such that the chromosome from peak B lost 8–10 units of DNA and the chromosome from peak C gained a like amount of DNA. The new peaks T_1 and T_2 contain the derived chromosomes and peaks B and C, from which the derived chromosomes originated, are reduced in area. This prediction from the flow karyotype, of a reciprocal translocation between chromosomes in peaks B and C, is confirmed by the Giemsa-banded karyotype of metaphase chromosomes (table 2) that demonstrates a reciprocal translocation between chromosomes 2 and 4.

The flow karyotype of clone 650AB (figure 2b) has been previously reported (Gray *et al.* 1975b) and is summarised here. In this case compared to clone 650A, (figure 1) peaks A and C have each lost one chromosome, peak B has gained one chromosome and peak T_3 has been formed with one chromosome. On a scale of 100 units for chromosome 1, the chromosome in peak T_3 has 12 units more DNA than the chromosomes of peak C. The additional chromosome in peak B has 15 units less DNA than the chromosome in peak A. These results again suggest a reciprocal translocation of about 12 to 15 units of DNA involving chromosome 1 in peak A and either chromosome 4 or t(X;5) in peak C. Banding analysis confirms the translocation to be reciprocal between chromosomes 1 and 4 and the derivative chromosomes have been identified in peaks B and T_3.

Analysis of Heterogeneous Aberrations

Aberrations induced by clastogenic agents include asymmetrical and symmetrical exchanges, deletions, and inversions of chromosomal material. They normally occur at random so that a population of exposed cells contains many cells with different abnormal karyotypes in addition to the unaffected cells with a normal karyotype. Unless selection had intervened, one would not expect the flow karyotype of these cells to exhibit clearly defined peaks other than those exhibited by the prevalent normal karyotype. In order to determine empirically the shape of the FCM distribution of chromosomes isolated from cells with heterogeneous aberrations, we acutely irradiated clone 650A with 350 kVp X-rays. Chromosomes were isolated from these cells within 7 h after irradiation, a period of time in which only cells from G_2 and S would be arriving at metaphase. Hence the isolated chromosomes would contain chromatid aberrations.

The FCM distributions of clone 650A chromosomes exposed to 75 rad and 150 rad of X-rays are shown in figure 3, a and b respectively. Comparing these distributions to the unirradiated clone 650A distribution in figure 1, it is evident that the background, that is, the area underlying the peaks, is increased after irradiation and that this increase is dose-dependent. This result is expected for random chromosome breakage.

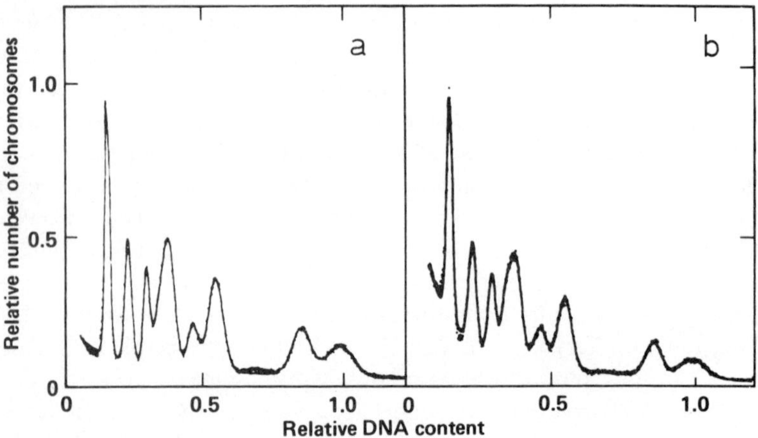

Figure 3. The flow karyotype of X-irradiated Chinese hamster M3-1 clone 650A chromosomes: (a) clone 650A cells received 75 rad of X-rays 7 h prior to chromosome isolation; (b) clone 650A cells received 150 rad of X-rays 7 h prior to chromosome isolation. The computer-generated fitted line is drawn through the data points.

If breakage is random and all breakpoints have an equal probability of exchange, then the larger chromosomes would show more exchanges. Thus, exchanges should predominate in large chromosomes (at high DNA amounts) and fragments at small DNA amounts. To test this, cells from clone 650A were given an acute dose of 300 rad of X-rays, and chromosomes were isolated from metaphase cells accumulated during a 5 h colcemid arrest from 15 to 20 h after irradiation. This metaphase population consists of cells that were in G_1 at the time of irradiation or were in G_2/M and are now in their second mitosis. Thus the aberrations are of the chromosome type. The FCM distribution generated by these EtBr-stained chromosomes is shown in figures 4 and 5. Chromosomes from each peak and valley of the distribution were sorted onto microscope slides, stained and scored visually for aberrations. The aberration frequency is expressed as dicentrics or deletions per scored chromosome. 2000 chromosomes were scored for each sort.

The frequency of dicentric chromosomes as a function of DNA content is shown in figure 4. These aberrations predominate at high DNA amounts and their frequency is increased in the valleys compared to the peaks. This is

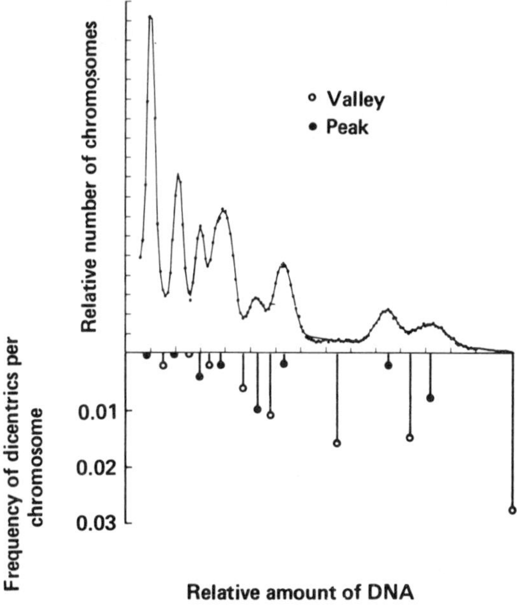

Figure 4. The frequency of dicentric chromosomes in each peak and valley of the flow karyotype of irradiated (300 rad) clone 650A cells. Chromosomes were isolated 20 h after irradiation.

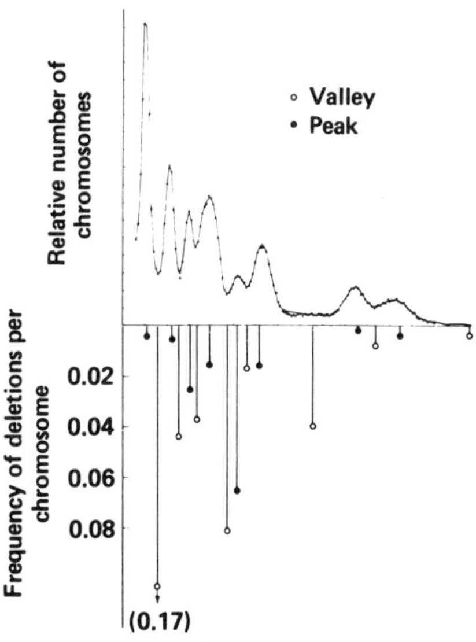

Figure 5. The frequency of chromosome deletions in each peak and valley of the flow karyotype of irradiated (300 rad) clone 650A cells. Chromosomes were isolated 20 h after irradiation.

simply a reflection of the fact that adjacent peak and valley regions have similar numbers of dicentrics but very different numbers of normal chromosomes, and the frequency is defined as the ratio of dicentrics to total chromosomes. Figure 5 shows the frequency of chromosome deletions (acentric fragments) as a function of DNA content. These aberrations predominate at low DNA amounts, and their frequency is increased in the valleys compared to the peaks for the same reasons as for dicentrics. These results from flow sorting strengthen the hypothesis that the increased background in the flow karyotype following irradiation is due to the presence of chromosomal aberrations. Other supporting evidence has been discussed previously (Van Dilla, Carrano and Gray 1976; Carrano, Van Dilla and Gray 1977).

These results suggest a quantitative relation between the background of the FCM distribution and the frequency of aberrations in a cell population. Figure 6 summarises the results of seven independent experiments to determine whether a quantitative correlation exists. For these experiments clone 650A cells were irradiated with doses of X-rays from 25 to 300 rad. Prior to chromosome isolation, a portion of the metaphase cells were prepared on

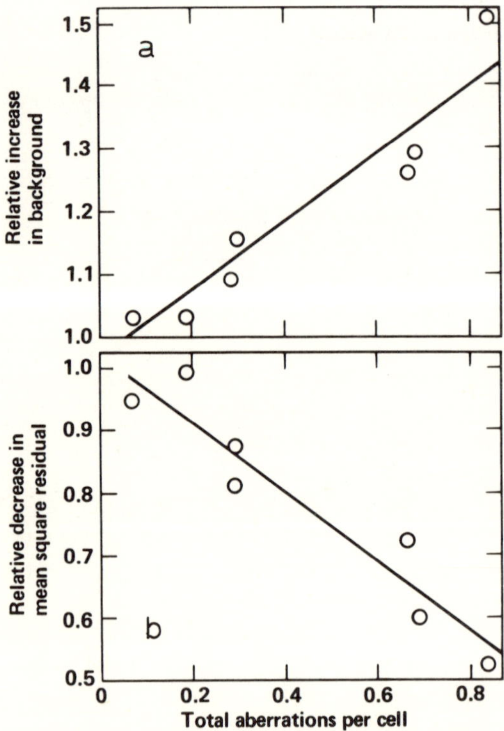

Figure 6. (a) The relation between the background of the flow karyotype and the frequency of aberrations in the metaphase cells from which the chromosomes were isolated. (b) The relation between the mean square residual and the aberration frequency. Each data point is pooled from seven independent experiments. A least squares linear regression line is drawn through each data set.

microscope slides for aberration analysis. Chromosomes were isolated from the remaining cells, stained with EtBr and analysed on the FCM. The background from each distribution was obtained from the computer fit. The mean squared residual, also obtained from the computer analysis, is the sum of the squared differences between the data points and the computer generated fit, divided by the number of channels over which the data was collected; it is an indication of the goodness-of-fit between the data points and the mathematical functions used to fit these data points.

Figure 6a demonstrates an increase in the background as a function of the aberration frequency. A 50% increase in the background occurs at an aberration frequency of about 0.8 per cell. We also observed that the mean squared residual (figure 6b) decreases with increasing aberration frequency for unexplained reasons. Unfortunately the experiment-to-experiment coefficient of variation for the background is approximately 13% and for the mean square residual is about 6%. Thus the quantitative estimation of the metaphase cell aberration frequency from the increased background is not yet possible. At present these parameters serve as a qualitative indication of clastogen damage for high levels of exposure; it is not likely that the sensitivity of this method will be increased to the extent necessary for monitoring populations exposed to low levels of clastogens. For this purpose, we have suggested (Van Dilla, Carrano and Gray 1976) a slit-scan approach to chromosomal flow cytometry (see below).

Discussion

Flow analysis of isolated chromosomes has been demonstrated in at least three laboratories (Gray *et al.* 1975a; Stubblefield, Cram, and Deaven 1975; Deaven, Stubblefield and Jett 1976; Otto and Oldiges 1977). The preparative technique is rapid. From mitotic cell collection to stained isolated chromosomes requires 1.5 h. The flow karyotype can be generated by the FCM in about 10 min and computer analysis of the distribution requires a further 10 min. It is possible therefore to establish a karyotype in approximately 2 h, considerably faster than the traditional 'cut and paste' method. Thus flow analysis offers the potential for rapidly screening a large number of samples. It would be of extreme utility in screening animal or human populations for heritable aberrations resulting from exposure to environmental pollutants. In the investigations reported here we show that DNA flow karyotyping can be applied to the analysis of homogeneous aberrations. In an earlier study it was found that flow chromosome analysis could be applied to detect mosaicism (Van Dilla, Carrano and Gray 1976; Carrano, Van Dilla and Gray 1977). It was possible, in that study, to detect the presence of the translocated chromosome der(4), in clone 650AB when it was present in a mixed population consisting of 5% clone 650AB cells and 95% clone 650A cells.

In principle, flow karyotype analysis should detect any homogeneous chromosomal alterations that involve a gain, loss or exchange of unequal amounts of DNA. Reciprocal exchanges of equal DNA amounts and inversions are not detectable. In addition, the detection of chromosomal exchanges of unequal DNA amounts, deletions or additions, is very much dependent upon the resolution of the chromosomes on the FCM. In clones 650AAA and 650AB, for example, we were unable to state definitively from the flow karyotype whether chromosome 4 or the t(X;5) chromosome was involved in the translocation, since the two were found in the same peak. In recent experiments, using the dye 33258 Hoechst, we have been able to resolve these chromosomes as well as the chromosomes with similar DNA contents in peaks E–F, H, and I of clone 650A. These results will greatly facilitate flow karyotype analysis with this species.

There are at least three approaches to the analysis of heterogeneous chromosomal aberrations. The first depends solely on a quantitative estimate of a dose dependent alteration in one of the parameters derived from the FCM flow karyotype, e.g. the background or mean square residual. Unfortunately this approach has, up to the present, served only as a qualitative indicator of chromosome damage. Further progress must be made in the preparative, cytochemical and instrumental procedures for handling isolated chromosomes to minimise the sample-to-sample and experiment-to-experiment variation in chromosome shear, in the cytochemistry and in the optimisation of the fluorescence signal-to-noise ratio.

The second approach to heterogeneous aberration detection is based upon flow-sorting enriched fractions of aberrant chromosomes onto microscope slides followed by conventional cytogenetic analysis. The feasibility of this method is dependent upon the degree of enrichment of the aberrant chromosomes. In order to relate the aberration frequency in the sorted chromosomes to that in metaphase cells precise calibration is necessary. For example, one may desire to know the quantitative relation between the frequency of dicentrics in the valleys to the metaphase cell frequency. If such a correlation existed it would be possible to score the aberrant chromosomes sorted from a single valley to attain the metaphase cell frequency. The scoring of the sorted chromosomes could either be performed manually by a trained cytogeneticist, or might even be facilitated by slide-based scanning instrumentation. The isolated chromosomes eliminate one of the major obstacles to the image-scanning systems; namely, the discrimination between overlapping chromosomes or chromatids and a true dicentric chromosome in a single metaphase cell. The isolated chromosome morphology is preserved during sorting and the chromosome density on the slide can be adjusted to eliminate overlap. We believe that, if a significant enrichment of aberrant chromosomes can be obtained by flow sorting, this method holds promise for a semi-automated or fully automated system for clastogen screening.

The third approach to the detection of aberrations focusses on the dicentric chromosome, and requires further instrument development. The general principle calls for a flow cytometer capable of detecting the centromere(s) in isolated chromosomes. In the proposed instrument each stained chromosome would pass lengthwise through a very thin laser beam. The resulting fluorescence signal would be a chromosome profile with a dip at the centromere region(s). The number of dips per chromosome profile would be determined, and two such dips per chromosome would indicate a dicentric. Alternatively, fluorescent stains specific for the centromere regions might also be developed. Such instrumentation is currently being designed in our laboratory.

Finally, purified chromosomes are important for biochemical analysis. Studies conducted with clone 650A of the Chinese hamster chromosomes show that DNA can be recovered from isolated sorted chromosomes, that such DNA can be transcribed and that the transcribed messenger can be hybridised to chromosomes *in situ* (Sawin, Scherberg and Carrano 1977). Such studies are potentially applicable to gene mapping. For example, unique sequence messenger RNA could be initially hybridised (in solution) to DNA from highly purified sorted chromosome fractions to map the gene to a specific chromosome. Once the chromosome is identified it is then feasible to obtain translocation or deletion variants of that chromosome that could be uniquely distinguished and sorted by the flow systems. This would further localise the gene to a specific chromosome region.

Flow chromosome analysis has potential both for karyotype analysis and biochemistry. We hope that this manuscript stimulates sufficient interest in other laboratories so that we might share the excitement of future development.

Summary

Using clonal derivatives of the Chinese hamster M3-1 cell line, we demonstrate the potential of flow systems to karyotype homogeneous aberrations (aberrations that are identical and present in every cell) and to detect heterogeneous aberrations (aberrations that occur randomly in a population and are not identical in every cell). Flow cytometry (FCM) of ethidium bromide-stained isolated chromosomes from clone 650A of the M3-1 cells distinguishes nine chromosome types from the fourteen present in the actual karyotype. X-irradiation of this parent 650A clone produced two sub-clones with an altered flow karyotype, that is, their FCM distributions were characterised by the addition of new peaks and alterations in area under existing peaks.

From the relative DNA content and area for each peak, as determined by computer analysis, we predicted that each clone had undergone a reciprocal translocation involving chromosomes from two peaks. This prediction was confirmed by Giemsa-banding the metaphase cells. Heterogeneous aberrations are reflected in the flow karyotype as an increase in background, that is an increase in area underlying the chromosome peaks. This increase is dose-dependent but, as yet, the sample variability has been too large for quantitative analysis. Flow sorting of the valleys between chromosome peaks produces enriched fractions of aberrant chromosomes for visual analysis. These approaches are potentially applicable to the analysis of chromosomal aberrations induced by environmental contaminants.

Acknowledgements

The authors are indebted to Tim Merrill and Jason Minkler for their technical assistance. This work was performed under the auspices of the U.S. Energy Research and Development Administration Contract No. W-7405-ENG-48 and supported in part by U.S.P.H.S. grants GM 20291 and GM 20901. Reference to a company or product name does not imply approval or recommendation of the product by the University of California or the U.S. Energy Research and Development Administration to the exclusion of others that may be suitable.

References

Carrano, A.V., J.W.Gray, D.H.Moore II, J.L.Minkler, B.H.Mayall, M.A. Van Dilla & M.L.Mendelsohn (1976) Purification of the chromosomes of the Indian muntjac by flow sorting. *J. Histochem. Cytochem. 24*, 348–54.

Carrano, A.V., M.A. Van Dilla & J.W.Gray (1977) Flow cytogenetics: a new approach to chromosome analysis, in *Flow Cytometry and Sorting* (eds M.Melamed, P.Mullaney & M.L.Mendelsohn). New York: Wiley.

Castelman, K.R., J.Melnyk, H.J.Frieden, G.W.Persinger & R.J.Wall (1976) Karyotype analysis by computer and its application to mutagenicity testing of chemicals. *Mutat. Res. 41*, 153–62.

Deaven, L.L., E.Stubblefield & J.H.Jett (1976) Karyotype analysis of Chinese hamster chromosomes by flow microfluorometry, in *Automation of Cytogenetics*, Asilomar Workshop (ed. M.L.Mendelsohn) pp.165–9. Springfield, Va.: NTIS.

Gray, J.W., A.V.Carrano, L.L.Steinmetz, M.A. Van Dilla, D.H.Moore II, B.H.Mayall & M.L.Mendelsohn (1975a) Chromosome measurement and sorting by flow systems. *Proc. Natl Acad. Sci. U.S.A. 72*, 1231–4.

Gray, J.W., A.V.Carrano, D.H.Moore II, L.L.Steinmetz, J.Minkler, B.H.Mayall, M.L. Mendelsohn & M.A. Van Dilla (1975b) High-speed quantitative karyotyping by flow microfluorometry. *Clin. Chem. 21*, 1258–62.

M.L.Mendelsohn, ed. (1976) *Automation of Cytogenetics*, Asilomar Workshop. Springfield, Va.: NTIS.

Moore II, D.H. (1975) Use of residuals in fitting normal (Gaussian) distributions. UCRL-76507, TID, Lawrence Livermore Laboratory, Livermore, Calif.

Otto, F. & H.Oldiges (1977) Preconditions and performance of chromosomal DNA measurements for rapid karyotype analysis in mammalian cells, in *Proc. Third International Symposium on Pulse Cytophotometry*, Vienna, 30 Mar–1 Apr, p.60.

Paris Conference (1971) *Standardization in Human Cytogenetics*. Birth Defects: Original Article Series VIII, 7. New York: The National Foundation.

Sawin, V.L., N.Scherberg & A.Carrano (1977) Hybridization in situ of ^{125}I-cRNA transcribed from sorted metaphase chromosomes, in *Proc. ICN-UCLA Symposium on Molecular Human Cytogenetics*, Keystone, Colorado, 6–11 Mar (in press).

Stubblefield, E., L.S.Cram & L.Deaven (1975) Flow microfluorometric analysis of isolated Chinese hamster chromosomes. *Exp. Cell Res. 94*, 464–8.

Van Dilla, M.A., A.V.Carrano & J.W.Gray (1976) Flow karyotyping: current status and potential development, in *Automation of Cytogenetics*, Asilomar Workshop (ed. M.L.Mendelsohn) pp.145–64. Springfield, Va.: NTIS.

Wray, W. & E.Stubblefield (1970) A new method for the rapid isolation of chromosomes, mitotic apparatus, or nuclei from mammalian fibroblasts at near neutral pH. *Expl. Cell Res. 59*, 469–78.

D. MASON and D. RUTOVITZ

The Economics of Automatic Aberration Scoring

In assessment of low-dose effects of radiations, or other agents causing chromosome aberrations, 500–1000 cells are often scored for rings, dicentrics and fragments. If translocations, etc., are also noted, equal confidence in the dose estimate could be obtained from a smaller sample, and if this is done with banded rather than homogeneous preparations a slightly greater event yield might enable a further reduction in the numbers of cells required. It is easily verified that the human operator can carry out the task of scoring dicentrics alone between two and three times faster than doing a full karyotype on homogeneous preparations, and perhaps ten times faster than a banding analysis. Therefore most laboratories with a considerable amount of scoring to do will opt for a dicentric/fragment/ring count only.

In an experiment carried out in our laboratory we have found that a trained technician scores on average 60 cells per hour, though this rate can probably not be sustained for an entire day. The per day throughput is typically between 200 and 300 cells. To preserve operator sanity it is probably essential to interleave scoring with other tasks, so that the technicians doing the scoring are usually employed in preparing the cultures.

Automated Cell Finding

What contributions can automation make in this field? Following in the footsteps of Dr Wald (Herron *et al.* 1972) we considered that the primary aid with which to supply the human operator must be in the location of dividing cells. At a meeting in Asilomar in December 1975, one of us described a reasonably inexpensive 'cytogeneticists' microscope' that we were building with this purpose in mind (Farrow, Green and Rutovitz 1975). The principal in this work is our colleague, D. K. Green (for other similar devices see Brenner *et al.* 1976, Le Go 1972, Lubs and Ledley 1973). The essentials of the microscope were completed soon after that meeting, and it has since been tested fairly extensively. It consists of a Reichardt microscope with Cambridge Instruments' motorised stage, to which we have added a photodiode-array scanner, logic to compensate for diode variations, a dynamic threshholding unit and an interline comparator. The logic is interfaced through a teletype line to a computer—any computer capable of accepting teletype signals at a fast rate is suitable. For stage control by the operator, a rolling ball and joystick are provided. The special-purpose logic detects places where an above-threshold segment in one scan line is not followed by a contiguous above-threshold segment in the next line. These are the places where connected dark 'objects' come to an end. Detection is done using a 1.5 μm scanning grid encoded at half resolution, i.e. in terms of a 3 μm grid. The encoded positions of the end points are sent to the computer, which looks for clusters of them, i.e. for regions of the slide where there are many small dark objects (the logic includes large segment rejection). Certain other characteristics of the end point distribution can be used to improve the extent to which the detected clusters correspond to usable metaphases.

The machine searches the slide at approximately 1 cm^2 min^{-1}, and as it does so a list of coordinates of possible metaphases are stored in the computer memory. When an operator-specified number of candidate cells have been found, or when the available slide area has been exhausted, control is returned to the operator, who can then step through the metaphases one by one.

Using this machine on preparations from this laboratory, we have found that on average about 50% of the machine-selected locations do indeed correspond to dividing cells. We have also found that technician throughput improves by up to 100%, i.e. operators are able to score up to 120 cells per hour, though usually somewhat less than this (Green, Bayley and Rutovitz 1977). That in itself might justify the use of such machinery, but it all depends on the cost. If we are enhancing an operator's performance by, say, 2/3, then the machine is worth 2/3 of an operator. In terms of current U.K. salaries and amortisation practices, this means that a machine could cost up to at most £20000, on the assumption that the running costs are fairly small. On a two-shift basis the machine could cost considerably more, but, although this is an accepted practice in computer installations, it is not very easy to set up shift work in a typical biological laboratory.

Our machine comes close to meeting this cost constraint, or did at last reckoning, although there are still a number of problems to be resolved. Our present implementation of the machine uses a rather noisy teletype and a computer of unnecessarily large physical size, but these are minor problems of re-engineering that can easily be overcome on a second trial version, or on a production instrument. More serious is that the number of metaphases missed is uncomfortably large on certain types of preparations (notably those that are lightly stained or contain much cellular debris) and that the yield, i.e. the number of cells found in relation to the number of usable cells on the slide, can be as low as 5%, though it is more usually 30%. This is quite impractical if one hopes to count a thousand cells and the slides are sparse, and it has been emphasised in a recent test of the metaphase finder carried out at the National Radiological Protection Board (Purrott and Stephenson 1978). The NRPB were particularly interested in testing the machine on sparse slides, where one might expect it to be most helpful, as the manual search time is longer. Unfortunately, they found that the yield obtained using their preparations was substantially less than that typically achieved in Edinburgh. The machine-aided scoring rate was no better than the unassisted scoring rate, principally because of the unacceptably high proportion of available cells missed. Obviously the machine still has major problems, but if the yield can be improved, the speed advantage of the machine should be pronounced for sparse slides. We are currently investigating ways of increasing yield and of making it less dependent on preparation type. On the positive side, Purrott and Stephenson demonstrated that the aberration counts obtained with the machine are comparable to those obtained by unassisted scoring, and also noted that scoring was less tiring for the operator because of the opportunity to rest during cell finding, so that the operator can score for longer periods using the machine. Despite the difficulties with the yield, we feel that an enhanced performance, obtained with metaphase-rich preparations, is encouraging.

Automatic Cell Scoring

Can automation be extended to the scoring task itself? It has been demonstrated by Wald *et al.* 1975, and we have confirmed, that dicentrics can readily be identified by programs with centromere recognition capabilities. Ring chromosomes and fragments can be recognised as well, but confining oneself to scoring dicentrics would not cause much loss of efficiency. Unfortunately many other objects can appear to display two centromeres. Whenever chromosomes are close enough together so that they are not separated by a clear sequence of low density values, a scanner and pattern processor will tend to pick them up as being a single entity. If each of the objects is a chromosome, that entity may have two genuine centromeres and another false one besides, namely the point at which the two parts touch. A human operator can usually recognise this configuration without trouble, but sometimes even the trained technician has considerable difficulty. Our machines, regrettably, are not nearly as clever as the human! It seems virtually impossible for them to take reliable decisions with such configurations without the assistance of the human operator. The overall timing of a dicentric scoring system will therefore depend in large part on the number of times the machine has to ask for assistance in distinguishing a pseudo-dicentric from a real one. In a study of the incidence of apparently multi-centric objects in 100 cells, in which there were no real dicentrics, it was found that there were slightly fewer than 1 per cell, mainly composite objects. This is not entirely a realistic figure, because in the experiment from which these data were obtained, the cost of ignoring a dicentric was not specially high. To obtain a very low false negative rate, which should operate as a computable bias in a system of this kind, the false positive rate may well have to increase, perhaps to double, so that the number of queries will be quite large: on average about 2 per cell.

Remembering that an unaided operator can score 60 cells per hour (on metaphase-rich slides, at least), and with the help of a metaphase-finding machine about 100 cells per hour, what sort of improvement in rate can we expect from an automatic system if, in every cell, the operator has to deal with two machine-presented questions as to whether a particular figure represents a dicentric or a pair of touching chromosomes? At first sight the conclusion is 'almost none'. Our experience in the use of the metaphase finder in seeking cells for karyotype analysis is that the process of cell culling, specification of the region to be scanned, focusing and digitising, even if carried out largely 'hands off' using a fast TV scanner system, takes about 20 s. With, say, 5 s per interaction, we have already spent 30 s per cell. That is, our fully automatic system achieves around 120 cells an hour, which is about as much as can be obtained with a cell-finding capability alone. Given that the scanning equipment and pattern processor required for dicentric detection is considerably more sophisticated than that required in finding metaphase cells, it is clear that the system is uneconomic.

On reviewing the situation, it seems that there are nevertheless avenues to a cost-effective system. 20 s per selected cell is the average amount of elapsed operator time for each cell accepted, including slide loading, searching, rejection of unsatisfactory fields, centring, placing of guard regions, and focusing. When material is selected for full karyotyping it is advantageous to spend a certain amount of time in rejecting broken cells, or those containing overlapping or entangled chromosomes, etc. This is because any perturbation in the karyotype that might result from such artefacts, or unanalysable objects, may necessitate the analysis of several additional cells to give sufficient confidence in the final result for an individual. The costs of embarking on the analysis of such cells can therefore be quite high. The position in aberration scoring is different, however. The incidence of dicentrics is low, and the probability of one being involved in one of the catastrophes mentioned is very much lower, so that we can relax the criteria for cell selection. In practice it is possible to do the final cell selection with the same low power objective (× 25) as is used during the cell search (Purrott and Stephenson 1978). A low power cull can be carried out more rapidly, as the operator has less information to deal with and there is no need for fine focusing, whether manual or automatic. If we also dispense with operator-controlled field limitation, often employed to eliminate neighbouring cells and artefacts from the scan, the cell selection need occupy a very much smaller amount of operator time. Indeed, one can go further and allow attempted analysis of all fields detected by the cell finder, although the confusion that this would cause due to the analysis of débris would have to be carefully assessed. With operator cell-selection at low power, technician time per cell so far will now be below 5 s. If the estimates of interaction time given above are still valid, we then require at least 15 s operator time per cell. This is a considerable improvement, and if actually achieved would mean a throughput of over 200 cells per hour, in which case the machine would prove economic.

It will be noticed that we have been emphasising operator time, not machine time. Despite the recent great reduction in the price of memory and fast logic elements, the machine time will not be a negligible factor in our costings without very careful systems engineering and construction of special-purpose analysis equipment, for example along the lines described at this symposium by Le Go (this volume) and by Farrow and Tucker (1976) and Duff *et al.* (1973).

An average of two interactions per cell can actually mean zero interactions on a particular cell, or ten. It is essential not to waste operators' time waiting for the machine to ask questions, and machine time must not be wasted waiting for operators to answer them (or to have a cup of tea). It is therefore essential that interaction and processing be completely asynchronous. This is an exercise in computer systems work, and, with our colleague James Piper, we have developed a satisfactory system for asynchronous interaction.

Although cell selection can be done at low power, scanning will have to be done at maximal optical resolution. Consequently there will be problems of focusing and centring, which can be done automatically, but even so take time. Again, if the operator is not to be sitting waiting for the machine to work through the selected cells before generating queries, and if the operator is not to be used once more to position the cells in the field, the machine will have to be capable of automatically recalling the fields selected at low power, and centring on them at high power. It will have to focus automatically, digitise and carry on to the next cell without asking the operator for help. Unfortunately this rather modest-sounding ambition—the automatic recall and scanning of previously located fields without operator intervention—while easily demonstrated in principle, is like many other tasks of automation in biology considerably more difficult to translate into a routine working system. Certain malfunctions occur when attempting an operation of this nature. Having searched with one objective one relocates and scans the cells with another at high power. On switching over, the operator can locate the first cell to ensure that all is well and overcome errors in parcentrality. When the machine begins its untended cruise over the oily surface of the slide, any dirt floating in the oil may attach itself to and follow the objective from cell to cell. Being out of focus, the dirt is not actually visible, but it alters the density levels and hopelessly confuses the auto-focus mechanism. Other problems may arise when, for example, the objective relaxes slightly on its seating and the field-coordinate relationship is disturbed, or when the oil runs dry. If the microscope's power supplies, earths, etc., are not as clean as they should be, electrical interference from neighbouring laboratories may cause the stage drive motors to take a few random steps, or the microscope lamp to go down, giving uneven and unusable field illumination.

None of these malfunctions causes much difficulty if an operator is in attendance. However, as there are so many potential snags in some of the automatic microscope's subsystems when working with high numerical aperture, and especially oil immersion, lenses, that open-loop working calls for an unduly costly engineering standard, and closed-loop working in this context requires a capability to detect and react sensibly to a very wide range of error conditions. When an error is detected an audible alarm can be set off and the machine can then wait for an operator, but too many spurious alarms will reduce the cost-benefit advantage of the machine.

If the search/scan time is reduced to a minimum, what can be done about the intervention time? The karyotyping system we have been developing (Hilditch 1970, Rutovitz *et al.* 1977) is designed to exploit the fact that while all shape-recognition algorithms fail on some of the configurations they are supposed to deal with, they usually succeed on those closest to some idealised shape for which the algorithm is designed, and of which any particular input object is a rough approximation. If closeness to the 'canonical configuration' can be inferred from measurements taken coincidentally or deliberately, it is often possible to estimate the likelihood of success of the algorithm; table 1 illustrates this point. The actual performance is rather bad, but the performance measure is quite accurate and can be used to divide specimens into an 'accepted' class and one requiring further investigation, in such a way that we will know fairly precisely the residual error in the accepted class. This technique can be applied immediately to dicentric *vs* composite discrimination. Thus, with careful statistical analysis, good engineering, and above all proper systems design, there is at least the possibility of attaining quite high throughput figures. The system will require extremely careful evaluation, however, as there are obvious possibilities of bias arising from analysis of cells selected at low power only, which may comprise badly confused material and artefacts; and also from the procedure used to reduce the number of interactions.

Table 1. Self-monitoring of shape analysis procedures.

Acceptance level (%)	15	35	60	85
Incidence of residual errors (%)	0.07	0.20	1.3	9.1

For finding centromeres by profile analysis the chromosome orientation must be reasonably well determined; within 20° has been found acceptable. Using a discriminant based on the area, aspect ratio and symmetry of the shape in relation to the axis chosen, the table indicates the likely percentage of cases in which the angle error will exceed 20° in chromosomes accepted as being within the tolerance limit, as a function of the acceptance level. For example, accepting 35% of chromosomes (there are alternative procedures), we should expect one residual error in approximately 11 cells.

Envoi

At a meeting on the Automation of Cytogenetics organised at Asilomar, California, in November 1975, sponsored by the U.S. Energy Research and Development Administration, one of us presented a paper describing a system for metaphase finding and aberration scoring (Farrow, Green and Rutovitz 1975). It was said, with some justice at the time, that attempts at automation have been in progress in this field for so long that people do not wish to hear any more about what we think we might do, but only about what we actually have done. The metaphase equipment, or as we like to term it, our 'cytogeneticists' microscope' was then almost complete, and although it may not yet meet all requirements for automatic scoring, fulfils its specifications reasonably well on metaphase-rich spreads. We did not, however, proceed with the implementation of the dicentric scorer as proposed at that time, mainly because we had not fully appreciated that the rather adverse timings given for cell selection and setup could be overcome by the operator protocol we have described. For the rest, one must determinedly face up to the engineering problems of 'hands-off' operation.

We have presented these reflections partly in the hope that the non-specialists, and perhaps even some of the specialists, may find this critique useful. Mainly, however, in the expectation that would-be users and our colleagues, or competitors, in automation will point out to us further difficulties, or flaws in our argument, before we become involved in this still substantial program of work. In any case we believe that cost-effective slide-scanning aberration-scoring machines can be built, and that they will function along the lines described in this paper.

References

Brenner, J.F., B.Dew, J.B.Horton, T.King, P.W.Neurath & W.D.Selles (1976) An automated microscope for cytological research: a preliminary evaluation. *J. Histochem. Cytochem. 24*, 100–11.

Duff, M.J.B., D.M.Watson, J.T.Fountain & G.K.Shaw (1973) A cellular logic array for image processing. *Pattern Recognition 5*, 229–47.

Farrow, A.S.J. & J.H.Tucker (1976) A real-time TV to computer image input system using run coding. *J. Histochem. Cytochem. 24*, 112–21.

Farrow, A.S.J., D.K.Green & D.Rutovitz (1975) A cytogeneticists' microscope and a proposed system for aberration scoring, in *Proc. Conf. Automation of Cytogenetics Asilomar*, pp.68–71.

Green, D.K., R.Bayley & D.Rutovitz (1977) A cytogeneticists' microscope. *Micro Acta 79*, 237–45.

Herron, J., R.Ranshaw, J.G.Castle & N.Wald (1972) Automatic microscopy for mitotic cell location. *Comput. Biol. Med. 2*, 129–35.

Hilditch, C.J. (1970) The principles of a software system for karyotype analysis, in *Human Population Cytogenetics* (eds P.A.Jacobs, W.H.Price and P.A.M.Law) pp.297–325. Edinburgh: University Press.

Le Go, R. (1972) Un systeme de selection automatique des metaphases. *C.R.Acad. Sci. 274*, 108.

Purrott, R.J. & B.D.Stephenson (1978) An assessment of the Edinburgh metaphase finder. NRPB Memorandum M-37.

Lubs, H.A. & R.S.Ledley (1973) Automated analysis of differentially stained human chromosomes. *Nobel 23*, 61–76.

Rutovitz, D., A.S.J.Farrow, D.K.Green & D.C.Mason (1977) Computer-assisted measurement in the cytogenetic laboratory, in *Pattern Recognition — Ideas in Practice* (ed. B. Bachelor), (in press).

Wald, N., C.C.Li, J.M.Herron, L.Davis & S.R.Fatora (1975) Automated analysis of chromosome damage, in *Proc. Conf. Automation of Cytogenetics Asilomar*, pp.39–45.

List of Participants

AWA, Dr A. A.
Cytogenetics Section, Dept of Clinical Laboratories, Radiation Effects Research Foundation, 5-2 Hijiyama Park, Hiroshima 730, Japan

BARTSCH-SANDHOFF, Dr Margret
Department of Human Genetics, Institut für Humangenetik, Universitätsklinikum Essen, Hufelandstrasse 55, Essen 1, Germany

BAUCHINGER, Dr M.
Institut für Biologie, Ges. f. Strahlen- und Umweltforschung mbH, D-8042 Neuherberg-München, Ingolstädter Landstrasse 1, Germany

BOCHKOV, Dr N. P.
Institute of Medical Genetics, Kashirskoye Shosse 6a, 115478 Moscow, U.S.S.R.

BREWEN, Dr J. G.
Biology Division, ORNL, P.O. Box Y, Oak Ridge, Tennessee 37830, U.S.A.

BRØGGER, Dr A.
Genetics Laboratory, Norsk Hydro's Institute for Cancer Research, Montebello, Oslo 3, Norway

BUCKTON, Miss Karin E.
MRC Clinical and Population Cytogenetics Unit, Western General Hospital, Crewe Road, Edinburgh EH4 2XU, U.K.

CARRANO, Dr A. V.
Biomedical Division L-523, Lawrence Livermore Laboratory, Livermore, California 94550, U.S.A.

CARTWRIGHT, Mr E. C.
British Nuclear Fuels Ltd, Windscale and Calder Works, Sellafield, Seascale, Cumberland, U.K.

COHEN, Dr M. M.
Department of Human Genetics, Hadassah-Hebrew University Medical Centre, Jerusalem, Israel

DE NETTANCOURT, Dr D.
C.E.C. – Directorate General for Research, Sciences and Education, D.G. XII – Biology Medical Research, 200 rue de la Loi, 1049 Brussels, Belgium

DIEZ, Dr J.
Fundacion de Genetica Humana, Salta 661, Buenos Aires, Argentina

DOLOY, Dr Marie Therese
Commissariat a l'Energie Atomique, Centre d'Etudes Nucléaires, Départment de Protection, Section de Radiopathologie, B.P. No. 6, 92260 Fontenay-aux-Roses, France

DOLPHIN, Dr G. W.
National Radiological Protection Board, Harwell, Didcot, Oxfordshire OX11 0RQ, U.K.

DRESP, Dr J.
Gesellschaft für Strahlen und Umweltforschung MHB, D-8042 Neuherberg, Post Oberscheissheim, Ingolstädter Landstrasse 1, Munchen, Germany

DUFFIELD, Dr D. P.
ICI Ltd, Mond Division, Castnerkellner Works, P.O.B. 9, Runcorn, Cheshire, U.K.

EVANS, Professor H. J.
MRC Clinical and Population Cytogenetics Unit, Western General Hospital, Crewe Road, Edinburgh EH4 2XU, U.K.

FAED, Dr M. J. W.
Cytogenetics Laboratory, Department of Pathology, Ninewells Hospital and Medical School, Dundee DD2 1UB, U.K.

FISCHER, Dr Patricia
Institute for Cancer Research, University of Vienna, Borschkegasse 8a, A1090 Vienna, Austria

HAINES, Dr D. O.
Employment Medical Advisory Service, Health and Safety Executive, Baynards House, 1 Chepstow Place, Westbourne Grove, London W2 4TF, U.K.

HANSMANN, Dr I.
Institut für Humangenetik, Universität Göttingen, Nikolausberger Weg 5a, 3400 Göttingen, Germany

HANSTEEN, Dr Inger-Lise
St. Josephs Hospital, 3900 Porsgrunn, Norway

HARNDEN, Professor D. G.
Department of Cancer Studies, University of Birmingham, Medical School, Birmingham B15 2TJ, U.K.

HEDDLE, Dr J. A.
Department of Biology, York University, Downsview (Toronto) Ontario M3J 1P3, Canada

HOLMBERG, Dr M.
Research Institute of National Defense, Dept. 4, S-10450 Stockholm 80, Sweden

JAGER, Miss Martine
Comeniusstr. 835, Amsterdam, The Netherlands

KEMMER, Dr W.
Institut für Biophysik der Universität des Saarlandes, Boris Rajewsky Institut, D-6650 Homburg (Saar) Germany

KIRKLAND, Dr D.
Department of Cytogenetics and Immunology, Royal Marsden Hospital, London SW3 6JJ, U.K.

KOMAROV, Dr E.
Control of Environmental Pollution and Hazards, Division of Environmental Health, World Health Organization, 1211 Geneva 27, Switzerland

KUČEROVÁ, Dr Maria
Genetics Laboratory, Institute of Hygiene and Epidemiology, 10042 Praha 10, Šrobárova 48, Czechoslovakia

LAWLER, Dr S. D.
Department of Cytogenetics and Immunology, Royal Marsden Hospital, Fulham Road, London SW3 6JJ, U.K.

LE GO, Dr R.
Commissariat a l'Energie Atomique, Centre d'Etudes Nucléaires, Département de Protection, Section de Radiopathologie, B.P. No. 6, 92260 Fontenay-aux-Roses, France

LINDSTEN, Professor J.
Department of Clinical Genetics, Karolinska Hospital, S-10401 Stockholm 60, Sweden

LINIECKI, Dr J.
Medical Research Center, Division of Nuclear Medicine and Radiobiology, 91425 Lodz, Sterlinga 5, Poland

LLOYD, Dr D. C.
National Radiological Protection Board, Harwell, Didcot, Oxfordshire OX11 0RQ, U.K.

MARSHALL, Dr R. R.
MRC Cell Mutation Unit, University of Sussex, Falmer, Brighton BN1 9QG, U.K.

MCLEAN, Dr A. S.
National Radiological Protection Board, Harwell, Didcot, Oxfordshire OX11 0RQ, U.K.

NATARAJAN, Dr A. T.
Department of Radiation Genetics and Chemical Mutagenesis, Sylvius Laboratory, University of Leiden, Post Bus 722, Leiden, The Netherlands

NACHEVA, Dr E.
Institute for Cancer Research, University of Vienna, Borschkegasse 8A, A1090 Vienna, Austria

OTT, Dr H.
Commission of the European Communities, D.G. XII-C-1, 200 rue de la Loi, 1049 Bruxelles, Belgium

PARRINGTON, Dr Jennifer M.
MRC Human Biochemical Genetics Unit, The Galton Laboratory, University College London, Wolfson House, 4 Stephenson Way, London NW1 2HE, U.K.

PERRY, Dr P.
MRC Clinical and Population Cytogenetics Unit, Western General Hospital, Crewe Road, Edinburgh EH4 2XU, U.K.

PRESTON, Dr R. J.
Biology Division, Oak Ridge National Laboratory, Oak Ridge, Tennessee 37830, U.S.A.

PROSSER, Dr J. S.
National Radiological Protection Board, Harwell, Didcot, Oxfordshire OX11 0RQ, U.K.

PURCHASE, Dr I. F. H.
Central Toxicology Laboratory, ICI Ltd, Alderley Park, Nr. Macclesfield, Cheshire, U.K.

PURROTT, Dr R. J.
National Radiological Protection Board, Harwell, Didcot, Oxfordshire OX11 0RQ, U.K.

RUTOVITZ, Dr D.
MRC Clinical and Population Cytogenetics Unit, Western General Hospital, Crewe Road, Edinburgh EH4 2XU, U.K.

SANKARANARAYANAN, Dr K.
Department of Radiation Genetics and Chemical Mutagenesis, State University of Leiden, Sylvius Laboratories, Wassenaarseweg 72, Leiden, The Netherlands

SASAKI, Dr M. S.
Department of Cytogenetics, Medical Research Institute, Tokyo Medical and Dental University, Yushima, Bunkyo-Ku, Tokyo 113, Japan

SAVAGE, Dr J. R. K.
MRC Radiobiology Unit, Harwell, Didcot, Oxfordshire OX11 0RD, U.K.

SCHMID, Professor Dr. W.
Abt. für Medizinische Genetik, Universitäts-Kinderklinik Zürich, Kinderspital, Steinwiesstr. 75, CH-8032 Zürich, Switzerland

SCOTT, Dr D.
Paterson Laboratories, Christie Hospital, Manchester M20 9BX, U.K.

SEABRIGHT, Dr Marina
Wessex Regional Cytogenetics Unit, General Hospital, Salisbury, Wiltshire, U.K.

STEFFEN, Dr J. A.
Department of Immunology, Maria Sklodowska-Curie Memorial Institute of Oncology, Ul. Wawelska 15, 02-034 Warszawa, Poland

STEVENSON, Dr A. C.
Dept. of Pathology, Royal Northern Infirmary, Inverness IV3 5SF, U.K.

SWAN, Dr A. A. B.
Central Toxicology Laboratory, ICI Ltd, Alderley Park, Nr. Macclesfield, Cheshire SK10 4TJ, U.K.

VIG, Dr B. K.
Nevada Mental Health Institute, P.O. Box 2460, Reno, Nevada 89505, U.S.A.

WALD, Dr N.
Department of Industrial Environmental Health Sciences, University of Pittsburgh, 130 DeSoto Street, Pittsburgh, Pa. 15261, U.S.A.

WOLFF, Dr S.
Laboratory of Radiobiology, University of California, San Francisco, California 94143, U.S.A.

Index